Flow Control of Congested Networks

NATO ASI Series

Advanced Science Institutes Series

A series presenting the results of activities sponsored by the NATO Science Committee, which aims at the dissemination of advanced scientific and technological knowledge, with a view to strengthening links between scientific communities.

The Series is published by an international board of publishers in conjunction with the NATO Scientific Affairs Division

A Life Sciences	Plenum Publishing Corporation
B Physics	London and New York
C Mathematical and Physical Sciences	D. Reidel Publishing Company Dordrecht, Boston, Lancaster and Tokyo
D Behavioural and Social Sciences	Martinus Nijhoff Publishers Boston, The Hague, Dordrecht and Lancaster
E Applied Sciences	
F Computer and Systems Sciences	Springer-Verlag Berlin Heidelberg New York
G Ecological Sciences	London Paris Tokyo
H Cel Biology	

Series F: Computer and Systems Sciences Vol. 38

Flow Control of Congested Networks

Edited by

Amedeo R. Odoni

Operations Research Center, Room 33–404
Massachusetts Institute of Technology
Cambridge, Massachusetts 02139, USA

Lucio Bianco

Istituto di Analisi dei Sistemi ed Informatica
Consiglio Nazionale delle Ricerche
00185, Rome, Italy

Giorgio Szegö

Universitá degli Studi di Roma "La Sapienza"
Facoltá Scienze Politiche
00185 Rome, Italy

Springer-Verlag Berlin Heidelberg GmbH

Proceedings of the NATO Advanced Research Workshop on Flow Control of Congested Networks, held in Capri, Italy, October 12–18, 1986.

ISBN 978-3-642-86728-6   ISBN 978-3-642-86726-2 (eBook)
DOI 10.1007/978-3-642-86726-2

Library of Congress Cataloging in Publication Data. NATO Advanced Research Workshop on Flow Control of Congested Networks (1986: Capri, Italy) Flow control of congested networks. (NATO ASI series. Series F, Computer and system sciences; vol. 38) Proceedings of the NATO Advanced Research Workshop on Flow Control of Congested Networks held in Capri, Italy, October 12–18, 1986. 1. Traffic engineering—Data processing—Congresses. 2. Telecommunication—Congresses. 3. Urban transportation–Congresses. 4. Aeroautics, Commercial—Congresses. I. Odoni, Amedeo R. II. Szegö, G. P. III. Bianco, L. (Lucio), 1941-. IV. Title. V. Series: NATO ASI series. Series F, Computer and systems sciences; vol. 38. HE336.AEN38 1986 629'.04 87-26430

© Springer-Verlag Berlin Heidelberg 1987

Originally published by Springer-Verlag Berlin Heidelberg 1987

Softcover reprint of the hardcover 1st edition 1987

2145/3140-543210

PREFACE

This volume is a compendium of papers presented during the NATO Workshop which took place in Capri, Italy, October 12-18, 1986 on the general subject of "Flow Control of Congested Networks: The Case of Data Processing and Transportation", and of which we acted as co-chairmen.

The focus of the workshop was on flow control methodologies, as applied to preventing or reducing congestion on:
(1) data communication networks;
(2) urban transportation networks; and
(3) air traffic control systems.

The goals of the workshop included:
-- review of the state-of-the-art of flow control methodologies, in general, and in each of the three application areas;

-- identification of similarities and differences in the objective functions, modeling approaches and mathematics used in the three areas;

-- examination of opportunities for "technology transfers" and for future interactions among researchers in the three areas.

These goals were pursued through individual presentations of papers on current research by workshop participants and, in the cases of the second and third goals, through a number of open-ended discussion-and-review sessions which were interspersed throughout the workshop's programme. The full texts or extended summaries of all but a few of the papers given at the workshop are included in this volume. Paralleling the structure of the workshop, the papers are organized into three groups:
(1) flow control of urban transportation networks (papers 1-9);
(2) flow control of data communication networks and some related or emerging methodological issues (papers 10-16); and
(3) flow control of air traffic (papers 17-21).

A very general summary of the principal findings and conclusions of the workshop might be as follows:

1. The field of flow control of urban transportation networks is, by far, the most "mature" and well established of the three. This is reflected in both the number and comprehensiveness of the papers on the subject in this volume. In addition to being widely used in such applications as network equilibration, planning of investments in urban highway infrastructures and setting of traffic signals, research in this general area has made important contributions to the theory of nonlinear programming and to the development of algorithms for large-scale, nonlinear optimization. Yet, major opportunities still exist for advances in methodology modeling and applications. Such promising areas identified during the workshop include: extensions of the theory of variational inequalities in the context of network equilibration; further progress in the development of efficient algorithms; applications of expert systems; further extensions to existing stochastic and dynamic traffic flow models; development of integrated, multi-modal models; closed-loop, dynamic control of access to highways and freeways.

2. The area of data communication networks is only now reaching the stage of a well-recognized field, but has made major progress in recent years. Its importance has grown immensely as a result of the explosive growth in the volume of "packets" of digital information transmitted throughout the world. This area brings together elements from such fields of inquiry as: network flow algorithms; queueing theory; parallel-processing and distributed algorithms; linear and nonlinear optimization; and theory of computational complexity. (Presentations touching all these aspects were given during the workshop.) It is clear that all theoretical and algorithmic work in this area must be closely wedded to familiarity with available technology and with system architectures. Keeping up with or simply staying abreast of rapid technological progress is one of the principal challenges that researchers in this area face.

3. Work on flow control of aircraft is only now beginning to address system-wide problems. To date, most research has been concerned with analysis, modeling and control problems associated with individual elements of the air traffic control (ATC) system, e.g. airports, terminal area airspace, en route sectors, etc. The problem of specifying mathematically the objective function and the constraints is very difficult, given the importance of distributive objectives (e.g. optimization on an individual aircraft basis) in addition to aggregative (e.g. minimization of system-wide user costs). Problems are of a strongly discretized nature and stochastic as well as dynamic considerations play an essential role. National and international ATC agencies are currently in the process of procuring large quantities of various types of expensive hardware but the state-of-the-art in flow control "software" and algorithms lags behind significantly.

4. Opportunities for "technology transfer" among the three areas do exist, but they are not of the simple "off-the-shelf" type. The workshop could not identify techniques or algorithms which could be immediately transferred from any one of these three fields to one of the other two. The reasons have to do with the fact that the three areas of application face models (e.g. with respect to the use of continuous vs. discrete variables), objectives (e.g. system-optimizing vs. user-optimizing objective functions) and constraints (e.g. types of control that can be exercised) which are fundamentally different from each other.

5. On the other hand, remarkable "synergies" might result from more extensive cooperation among researchers in these three areas. They, as well as researchers in a number of other related areas, have to deal with problems of: a fundamentally _stochastic_ and _dynamic_ nature; immense computational size; applicability to systems with enormous investment costs and wide social implications. It is through contributions to _generic_ methodologies dealing with the control of flows on large-scale, stochastic and dynamic systems that such synergies can best be achieved. The fact that the three fields attract some of the brightest individuals in science and technology suggests that there is excellent potential for major advances.

Overall, the workshop was judged by a consensus of the participants to have been extremely successful. In addition to the high quality of most individual presentations, it afforded ample opportunity for exchanges of ideas during the scheduled discussion sessions and on an informal one-to-one basis. There was general agreement that it would be most desirable to conduct a follow-up workshop on the same or a similar theme in a few years' time.

Cambridge and Rome, July 1987

AMEDEO R. ODONI
LUCIO BIANCO
GIORGIO SZEGÖ

ACKNOWLEDGEMENT

The Editors wish to express their very sincere thanks to Dr. Agostino Scognamiglio, general secretary of the Workshop, for his constant help and support, to the city of Capri which hosted the event, to ALITALIA for travel support to some participants from U.S. and to Dr. Enza Fratello head of the secretariat office, who had to overcome many organizational problems.

We also wish to thank Ms. Abigail Crear for assisting us in putting this volume together.

Amedeo Odoni
Lucio Bianco
Giorgio Szegö

TABLE OF CONTENTS

Congested Transportation Networks and Variational Inequalities*

Stella Dafermos
Lefschetz Center for Dynamical Systems
Division of Applied Mathematics
Brown University
Providence, Rhode Island 02912, U.S.A.

1. Introduction

In the standard setting of the traffic equilibrium problem, a travel demand is associated with each origin/destination (O/D) pair of nodes of a network, a travel cost is assigned to each link of the network and the traffic pattern is sought with the equilibrium property that, once established, no user may decrease his personal travel cost by altering his travel decisions. Link travel costs may depend on the load pattern (congestion effect) and O/D travel demands may depend upon travel costs associated with O/D pairs (elasticity of demand).

The realization that the equilibrium conditions characterizing this problem can be cast into the form of a variational inequality opened new vistas for the qualitative study of equilibrium patterns (existence, uniqueness, sensitivity) as well as for the development of theoretically sound and computationally efficient algorithms. So far, the development of the theory has concentrated on the case of monotone travel cost and demand functions and in this context variational inequalities have proven as effective a tool as mathematical programming is for dealing with system-optimizing networks with convex travel cost functions. Here we give a summary of results.

2. The Model

To fix ideas, we briefly describe the general traffic equilibrium model (see Dafermos (1982)). We consider a network G with K links, Q paths and W O/D pairs of nodes. The typical link, path and O/D pair will be denoted by a, p and w respectively.

The flow is described by a path flow vector F in R^Q, with components F_p which induces a travel demand vector in R^W with components

$$d_w = \sum_{p \text{ joining } w} F_p \qquad (2.1)$$

*Research was supported by the National Science Foundation Grant DMS-8601778 and a link flow vector f in R^K with components

$$f_a = \sum_p \delta_{ap} F_p \qquad (2.2)$$

where δ_{ap} is 1 if link a is contained in path p and 0 otherwise.

The users' travel costs on the links of the network are determined by a vector c in R^K with components c_a. A user traveling on path p incurs a travel cost

$$c_p = \sum_a \delta_{ap} c_a . \qquad (2.3)$$

A flow pattern (f,d) compatible with (2.1) and (2.2) is in equilibrium if, once established, no user has any incentive to alter his travel arrangements. This state is characterized by the following conditions (Wardrop (1952); see also Pigou (1920)):

$$c_p \begin{cases} = v_w , & \text{if } F_p > 0 \\ \geqslant v_w , & \text{if } F_p = 0 \end{cases} \qquad (2.4)$$

which hold for every O/D pair w and every path joining w, where v_w is the equilibrium travel cost associated with the O/D pair w. We group together the travel costs v_w into a vector v in R^W.

We assume that the users' travel cost may depend upon the entire link flow,

$$c = \hat{c}(f) \qquad (2.5)$$

and the O/D travel demands are determined by the O/D travel costs,

$$d = \hat{d}(v) \qquad (2.6)$$

where $\hat{c}$ and $\hat{d}$ are known continuous functions.

An alternative model is derived if, in the place of (2.6),

$$v = \hat{v}(d) \qquad (2.7)$$

is specified. These two models are of course equivalent only when $\hat{d}(v)$ is invertible.

An alternative way of formulating these models is to focus attention primarily on path flows and path costs rather than on link flows and link costs. For details see Aashtiani and Magnanti (1981).

Interestingly, it turns out that the traffic network equilibrium model described above is much more general than it would seem at first glance. In fact, seemingly more complicated models on multimodal networks, multiclass-user networks, networks in which the users are free to choose their origins and/or their destinations as well as their routes, etc. can be reduced to the above prototypical model by a judicious construction of an appropriate (abstract) network (see Dafermos (1972),(1976)). Even models in economics that seem completely unrelated to traffic equilibrium problems, such as the well-known Samuelson-Takayama-Judge model of spatially separated markets may be formulated as a traffic equilibrium problem of the type described above, on an appropriately constructed network (Dafermos (1986)).

3.   Basic Facts from the Theory of Variational Inequalities.

Recently, a general theory has been developed for the traffic equilibrium problem based on the observation (Dafermos (1980)) that the governing equilibrium conditions (2.4), as reformulated by Smith (1979), have the form of a *variational inequality*:

Find $\bar{x} \in K$ such that $\phi(\bar{x}) \cdot (x - \bar{x}) \geqslant 0$ for all $x \in K$,   (3.1)

where K is a closed convex subset of R^n and $\phi(x)$ is a known function from K to R^n.

In the sequel we outline some basic facts from the theory of variational inequalities (see e.g. Kinderlehrer and Stampacchia (1980)).

When the feasible set K is bounded and $\phi(x)$ is continuous there exists at least one solution of (3.1). When K is not necessarily bounded, a solution of (3.1) exists providing $\phi(x)$ is continuous and *coercive*, i.e.,

$$\frac{(\phi(x) - \phi(x^0)) \cdot (x - x^0)}{|x - x^0|} \to \infty \text{ as } |x| \to \infty, \quad x \in K \qquad (3.2)$$

for some fixed $x^0 \in K$. Here $|\;|$ denotes the Euclidean norm.

Furthermore, the variational inequality (3.1) has at *most one solution* providing the function $\phi(x)$ is *strictly monotone*, that is

$$(\phi(x) - \phi(x^0)) \cdot (x - x^0) > 0 \qquad (3.3)$$

for all $x, x^0 \in K$, $x \neq x^0$.

The function $\phi(x)$ is *monotone* if the left hand side of (3.3) is greater or equal to zero for all $x, x^0 \in K$ and is *strongly monotone* if the left hand side of (3.3) is greater or equal to $k|x - x^0|^2$ for every $x, x^0 \in K$, where k is a positive constant.

The variational inequality (3.1) encompasses, in particular, the mathematical programming problem. In fact, let $\psi(x)$ be a continuously differentiable scalar valued function defined on some open neighborhood of K and let $\nabla \psi(x)$ denote its gradient. Suppose that there exists an $\bar{x} \in K$ such that

$$\psi(\overline{x}) = \min_{x \in K} \psi(x) \ . \tag{3.4}$$

Then $\overline{x}$ is a solution of the variational inequality:

Find $\overline{x} \in K$ such that $\nabla\psi(\overline{x}) \cdot (x-\overline{x}) \geqslant 0, \ x \in K$. $\qquad$ (3.5)

It is also easy to verify that if $\psi(x)$ is convex, strictly convex, or uniformly convex then its gradient mapping $\nabla\psi(x)$ is, respectively, monotone, strictly monotone, or strongly monotone.

Conversely, if the function $\phi(x)$, defined on an open neighborhood of K, is the gradient of a convex, continuously differentiable, function $\psi(x)$ then the variational inequality (3.1) is equivalent to the minimization problem (3.4), in the sense that $\overline{x}$ solves the variational inequality (3.1) precisely when $\overline{x}$ minimizes the funciton $\phi(x)$ over K. It is well known that $\phi(x)$ is a gradient mapping if and only if its Jacobian matrix $[\partial\phi/\partial x]$ is symmetric.

4. Variational Inequality Formulation of the Traffic Equilibrium Conditions.

As mentioned in Section 3, the traffic equilibrium problem has been formulated as a variational inequality problem. Specifically, consider first the special case in which the travel demand d is fixed and known a priori (and thus the feasible set is bounded). Dafermos (1980) recognized that the equilibrium conditions (2.4), as reformulated by Smith (1979), defined a variational inequality problem. Specifically, we have the following

Theorem 4.1. (Fixed travel demand; bounded feasible set). A feasible $\overline{f}$ is an equilibrium traffic pattern if and only if it satisfies the variational inequality

$$\hat{c}(\overline{f}) \cdot (f-\overline{f}) \geqslant 0, \quad \text{for all feasible } f . \tag{4.1}$$

Next consider the situation where the travel demand is elastic and, in particular, the O/D travel costs are determined by the O/D travel demands, ie., $v=\hat{v}(d)$ is specified (and thus the feasible set is unbounded). Then, as shown in Dafermos (1982), the equilibrium conditions (2.4) can be cast into the form of a (different) variational inequality. Specifically, we have the following

Theorem 4.2. (Elastic demand; $v=\hat{v}(d)$ known; unbounded feasible set). A feasible $(\overline{f},\overline{d})$ is in equilibrium if and only if it satisfies the variational inequality

$$\hat{c}(\overline{f}) \cdot (f-\overline{f}) - \hat{v}(\overline{d}) \cdot (d-\overline{d}) \geqslant 0, \quad \text{for all feasible (f,d).} \qquad (4.2)$$

Finally, consider the most general case in which the travel demand is elastic and the O/D travel demands are determined by the O/D travel costs, i.e., $d=\hat{d}(v)$ is given. Then, as shown in Dafermos and Nagurney (1984), the equilibrium conditions (2.4) can be cast into the form of a (still different) variational inequality. Specifically, we have the following

Theorem 4.3. (Elastic demand; $d=\hat{d}(v)$ known; unbounded feasible set). A feasible $\overline{x}=(\overline{f},\overline{d},\overline{v})$ with $\overline{v} > 0$ is in equilibrium if and only if it satisfies the variational inequality

$$\phi(\overline{x}) \cdot (x-\overline{x}) \geqslant 0, \quad \text{for all feasible x} \qquad (4.3)$$

where $\phi(x) = (\hat{c}(f),-v,d-\hat{d}(v))$.

Remark

In the special but important *separable case* it is assumed that $c_a = \hat{c}_a(f_a)$, i.e., the travel cost on a link a depends solely on the link flow f_a, and $d_w = \hat{d}(v_w)$, i.e., the travel demand associated with the O/D pair w depends only on the O/D travel cost v_w. Then the Jacobian $[\partial\phi/\partial x]$ of the function $\phi(x)$ in (4.1) and (4.2) is symmetric. It follows that in the separable case (4.1) and (4.2) are associated with mathematical programming problems. This was already observed by Beckmann, McGuire and Winsten in their seminal work (1956). By contrast, note that, even in the separable case, the Jacobian of $\phi(x)$ in (4.3) is never symmetric and therefore this problem cannot necessarily be formulated as a mathematical programming problem.

5. **Existence, Uniqueness and Sensitivity of Traffic Equilibria.**

In this section we outline a number of results on the existence, uniqueness and sensitivity of traffic equilibrium patterns characterized by the variational inequalities (4.1),(4.2) and (4.3). Let us first consider the model characterized by (4.1) (travel demand d fixed; feasible set bounded). In this case a direct application of the theory, reviewed in Section 3, establishes the existence of solutions under the sole assumption that the travel cost function $\hat{c}(f)$ is continuous. Consider now the models characterized by (4.2) and (4.3) (elastic demand models with unbounded feasible set). For these models the standard theory in Section 3 would establish the existence of equilibria, provided that, among other things, the negative O/D travel cost function $-\hat{v}(d)$ (in the case of (4.2)), or the negative O/D travel demand function $-\hat{d}(v)$ (in the case of (4.3)) were coercive. These conditions, however, are grossly unrealistic since it is clear that travel demand functions must tend to zero as the travel cost grows to infinity. Nevertheless, by exploiting the special structure of the traffic equilibrium problem, the existence of solutions of variational inequalities (4.2) and (4.3) has been recently established under quite weak and realistic assumptions on the travel cost and travel demand functions (Dafermos (1985)).

We now turn to the problem of uniqueness of equilibrium. A direct application of the theory in Section 3 shows that the variational inequality (4.1) has at most one solution, providing the travel cost function $\hat{c}(f)$ is strictly monotone. Also, applying the same ideas together with standard properties of monotone functions, it follows that the variational inequality (4.2) has at most one solution, providing the travel cost function $\hat{c}(f)$ and the negative O/D travel cost function $-\hat{v}(d)$ are strictly monotone (Dafermos (1982)). In the more general case of variational inequality (4.3), it turns out that strict monotonicity of $\phi(x)$ is a rare occurrence. However, by exploiting the special structure of the problem it has been shown that (4.3) has at most one solution, providing the travel cost function $\hat{c}(f)$ and the negative O/D travel demand function $-\hat{d}(v)$ are strictly monotone (Dafermos and Nagurney (1984)).

The assumption of strict monotonicity of the functions $\hat{c}(f)$, $-\hat{v}(d)$, $-\hat{d}(v)$ is realistic for many single mode networks and even for multimodal networks so long as the interaction among different modes is relatively weak. The relevance of the monotonicity assumption in establishing uniqueness of solutions for equilibrium problems was recognized quite early by Kuhn (1959).

Finally, a number of results on sensitivity and stability of equilibria characterized by (4.1),(4.2) and (4.3) have been obtained under appropriate monotonicity conditions (see Dafermos and Nagurney (1983),(1984), Smith (1984), Tobin and Friesz (1986)).

6.   Algorithms

The variational inequality formulation of the traffic equilibrium problem induces effective and theoretically sound algorithms for computing equilibria. Numerous algorithms have been proposed recently (see e.g. the review article by Magnanti (1982) and the dissertation by Hammond (1984)). Here we will outline a general iterative scheme devised by Dafermos (1983) which not only contains as special cases projection methods (Dafermos (1980),(1982), Bertsekas and Gafni (1982)), linear approximation methods (Pang and Chan (1981)) and relaxation algorithms (Dafermos (1982b), Florian and Spiess (1982)) proposed earlier but also induces new algorithms.

Any function $g(x,y) : K \times K \rightarrow R^n$ such that i) $g(x,x) = \phi(x)$ and ii) the Jacobian matrix $[\partial g(x,y)/\partial x]$ is positive definite for fixed $x,y \in K$ generates the following

Algorithm

Step 0 : start with some $x_0 \in K$

Step k : (k=1,2,...): Compute x_k by solving the variational inequality

Find $x_k \in K$ such that $g(x_k,x_{k-1}) \cdot (x-x_{k-1}) \geqslant 0$, $x \in K$. (6.1)

The above scheme generates a sequence $\{x_m\}$ in K whose convergence under appropriate assumptions on $f(x)$ (most notably monotonicity) follows from a contraction estimate,

$$\|x_{m+1}-x_m\|_m \leqslant \lambda \|x_m-x_{m-1}\|_{m-1} , \quad 0 < \lambda < 1, \tag{6.2}$$

where $\{\|\cdot\|_m\}$ is a sequence of judiciously constructed norms in R^n.

Recalling the discussion in Section 3, we note that in the special case where the Jacobian matrix $[\partial g(x,y)/\partial x]$ is also symmetric, the variational inequality problem (6.1) reduces to the mathematical programming problem $\min_{x \in K} F(x,x_{k-1})$ where $g(x,y) = \nabla_x F(x,y)$.

As an illustration, we derive below from the above described general scheme two popular methods for computing the traffic equilibrium characterized by (4.1).

Projection Method. This corresponds to

$$g(f,\hat{f}) = \hat{c}(f) + \frac{1}{\rho} G(f-\hat{f}), \quad \rho > 0 \tag{6.3}$$

where G is a fixed symmetric and positive definite matrix.

Observe that in this case (6.1) reduces to a quadratic programming problem.

Jacobi Method. This corresponds to $g(f,\hat{f})$ defined by

$$g_a(f,\hat{f}) = c_a(\hat{f}_1,...,f_a,...,\hat{f}_K). \tag{6.4}$$

Observe that in this case (6.1) reduces to a separable nonlinear programming problem.

Finally, we mention that recently a simplicial decomposition algorithm for variational inequalities which has been applied successfully to solve the traffic equilibrium problem with fixed demands (Lawphongpanich and Hearn (1984), Hearn, Lawphongpanich and Ventura (1984)). Simplicial decomposition algorithms solve at each iteration a master problem and a subproblem. The subproblem generates points contained in the feasible region of the original variational inequality problem; the master problem solves a restricted version of the original variational inequality over the convex hull of these points. Any variational inequality algorithm can be used to solve the restricted problem. Smith (1983) and Pang and Yu (1984) have also proposed simplicial decomposition schemes for solving the traffic equilibrium problem with fixed demands.

In summary, there is now a plethora of algorithms which can or have been applied to solve traffic equilibrium problems. Most of these algorithms have been applied to the traffic equilibrium problem with fixed demands. It is high time for a systematic comparison of these algorithms. One expects, based on preliminary results (see e.g. Fisk and Nguyen (1982), Nagurney (1984),(1986)) that such a study will not single out an "all-optimal" algorithm. Instead, it will provide practitioners with guidelines as to which might be the best method under the prevailing conditions (network complexity and size, desired accuracy, form of travel cost and travel demand functions, type of available computers etc.)

7.   Concluding Remarks

The development of the theory of traffic equilibria has concentrated so far on the case of monotone travel cost and demand functions. As mentioned above, the assumption of monotonicity is realistic for many single mode networks, as well as multimodal networks, so long as the interaction among different modes is relatively weak. The immediate open problem concerns situations in which the monotonicity assumption is unrealistic. These arise, for instance, in multimodal networks with strong interaction among different modes as well as in some special single mode networks involving freight transportation.

In the absence of monotonicity, the situation is considerably more complicated since, as a rule, multiple equilibria arise. Questions of great theoretical and practical importance are to determine among the equilibria which one(s) is stable, and which one will be attained from specified initial conditions of the transportation system and, finally, to provide clues on how transportation planners may steer the evolving travel pattern towards the equilibrium with the most desirable features. Such a study will not only shed light on the stability of equilibrium patterns but, beyond that, will indicate whether the basic "equilibrium" concept used in the traffic equilibrium model is valid. As a byproduct it will also provide new algorithms for computing the equilibrium patterns by means of an iteration scheme that simulates the process by which actual travel patterns evolve towards equilibrium.

References

Aashtiani, H.Z. and T.L. Magnanti (1981). Equilibria on a congested transportation network. *SIAM Journal on Algebraic and Discrete Methods* 2, 213-226.

Beckmann, M.J., C.B. McGuire and C.B. Winsten (1956). *Studies in the Economics of Transportation.* Yale University Press, New Haven, Ct.

Bertsekas, D.P. and E. Gafni (1982). Projection methods for variational inequalities and application to the traffic assignment problem. *Mathematical Programming Study* 17, 139-159.

Dafermos, S. (1972). The traffic assignment problem for multiclass-user transportation networks. *Transportation Science* 6, 73-87.

Dafermos, S. (1976). Integrated models for transportation planning. In M. Florian (Ed.), Traffic Equilibrium Methods, *Lecture Notes in Economics and Mathematical Systems* 118, Springer-Verlag, New York, 106-118.

Dafermos, S. (1980). Traffic equilibrium and variational inequalities. *Transportation Science* 14, 43-54.

Dafermos, S. (1982). The general multimodal network equilibrium problem with elastic demand. *Networks* 12, 57-72.

Dafermos, S. (1983). An iterative scheme for variational inequalities. *Mathematical Programming* 26, 40-47.

Dafermos, S. (1985). Equilibria on nonlinear networks. Lefschetz Center for Dynamical Systems Report No. 86-1. Division of Applied Mathematics, Brown University, Providence, R.I.

Dafermos, S. (1986). Isomorphic multiclass spatial price and multimodal traffic network equilibrium models. *Regional Science and Urban Economics* 16, 197-209.

Dafermos, S. and A. Nagurney (1984a). Sensitivity analysis for the asymmetric network equilibrium problem. *Mathematical Programming* 28, 174-184.

Dafermos, S. and A. Nagurney (1984b). Stability and sensitivity analysis for a combined network equilibrium model. In J. Volmoller and R. Hamerslag (Eds.) *Proceedings of the Ninth International Symposium on Transportation and Traffic*, VNU Science Press, Utzecht, The Netherlands, 217-231.

Fisk, C.S. and S. Nguyen (1982). Solution algorithms for network equilibrium models with asymmetric user costs. *Transportation Science* 16, 361-381.

Florian, M. and H. Spiess (1982). The convergence of diagonalization algorithms for asymmetric network equilibrium problems. *Transportation Research* 16B, 477-483.

Hammond, J.H. (1984). Solving asymmetric variational inequality problems and systems of equations with generalized nonlinear programming algorithms. Ph.D. Dissertation, M.I.T., Cambridge, MA.

Hearn, D.W., S. Lawphongpanich and J.A. Ventura (1984). Restricted simplicial decomposition. Computation and extentions. Research Report No. 84-38, Department of Industrial and Systems Engineering, University of Florida, Gainesville, Florida.

Kinderlehrer, D. and Stampacchia, G. (1980). *An Introduction to Variational Inequalities*, Academic Press, New York, N.Y.

Kuhn, H.W. (1959). Mathematical appendix. In A.H. Land (Author) Endowments and factor prices. *Economica*, 137-144.

Lawphongpanich, S. and D.W. Hearn (1984). Simplicial decomposition of the asymmetric assignment problem. *Transportation Research* 18B, 123-133.

Magnanti, T.L. (1984). Models and algorithms for predicting urban traffic equilibrium. In Florian (Ed.) *Transportation Planning Models*, North-Holland.

Nagurney, A. (1984). Comparative tests of multimodal traffic equilibrium methods. *Transportation Research* 18B, 469-485.

Nagurney, A. (1986). Computational comparisons of algorithms for general asymmetric traffic equilibrium problems with fixed and elastic demands. *Transportation Research* 20B, 78-84.

Pang, J.S. and D. Chan (1981). Iterative methods for variational and complementarity problems. *Mathematical Programming* 24, 284-313.

Pang, J.S. and C.S. Yu (1984). Linearized simplicial decomposition methods for computing traffic equilibria. *Networks* 14, 427-438.

Pigou A.C. (1920). *The Economics of Welfare*. MacMillan, London.

Smith, M. (1983). An algorithm for solving asymmetric equilibrium problems with a continuous cost-flow function. *Transportation Research* 17B, 365-371.

Smith, M. (1984). The stability of a dynamic model of traffic assignment. An application of a method of Lyapunov. *Transportation Science* 18, 245-252.

Tobin, R.L. and Friesz, T.L. (1986). Sensitivity analysis for equilibrium network flow. Preprint.

Wardrop, J G. (1952). Some theoretical aspects of road traffic research. *Proceedings of the Institute of Civil Engineers*, Part II 1, 325-378.

Optimization Algorithms for Congested Network Models

D. W. Hearn
ISE Department
Univ. of Florida
303 Weil Hall
Gainesville, FL 32611

S. Lawphongpanich
Visiting Scholar
Sloan School
Mass. Inst. of Tech.
Cambridge, MA 02139

J. A. Ventura
IE Department
Univ. of Missouri
Columbia, MO 65211

Abstract

This paper describes recently developed nonlinear programming algorithms for certain large-scale congested network models. The techniques include Restricted Simplicial Decomposition (RSD) applied to the single commodity flow problem (RSDNET) and the standard traffic assignment problem (RSDTA), and the basic simplicial strategy applied to the network variational inequality problem (SDVI). Computational results are presented for each method, including tests conducted on large networks from real world models.

Introduction

This paper describes recently developed optimization algorithms for certain large-scale congested network flow problems. Algorithms and computational results are given for each of the following models:

- The single commodity minimum cost network flow problem with a pseudoconvex objective which need not be separable.

- The standard traffic assignment problem which seeks the equilibrium, or system optimal, flow of multicommodities on a network where travel time on each arc is an increasing convex function of the total arc flow, and there are no upper bounds on the arcs.

- The asymmetric traffic assignment problem which is formulated as a variational inequality problem with network constraints.

NATO ASI Series, Vol. F38
Flow Control of Congested Networks
Edited by A. R. Odoni et al.
© Springer-Verlag Berlin Heidelberg 1987

The three algorithms employ the strategy of simplicial decomposition (SD) and its restricted version (RSD) and are named RSDNET, RSDTA and SDVI, respectively.

The RSDNET Algorithm

The nonlinear single commodity minimum cost flow problem may be stated as

P1: $\min \{f(x) : Bx = b, u \geq x \geq 0\}$

where B is an mxn node-arc incidence matrix, b is a vector in R^m, u is a vector of upper bounds, x is a vector of arc flows in R^n, and $f(x)$ is a continuously differentiable pseudoconvex function. Letting S be the bounded feasible region, then any feasible point may be expressed as a convex combination of the extreme points of S, of which there are only a finite number. Thus the problem may be restated as

P2: $\min \{f(A\beta) : \Sigma \beta_i = 1; \beta_i \geq 0, i = 1,2,\ldots,N\}$

where N is the total number of extreme points, and column a_i of the matrix A corresponds to the ith extreme point of S.

The restricted simplicial decomposition (RSD) strategy is a column generating method which addresses the problem in the form P2. It alternates between a linear programming subproblem which generates extreme points of S and a master problem that optimizes the objective function over the convex hull of a restricted subset of the generated points. A parameter, r, controls the maximum number of points in the restricted set. This has the advantages of conserving computer memory requirements in the storage of extreme points and also controlling the maximum number of variables in the nonlinear master problem. Thus RSD combines the capabilities of linear programming methods for solving large-scale problems with the powerful methods of nonlinear programming for solving problems with a limited number of variables.

The following notation is needed to state the RSD algorithm formally:

(1) W_s = the set of retained extreme points.

(2) W_x = a set which may be empty or contain one of the iterates.

(3) $|W|$ = the cardinality of W, an arbitrary set of points.

(4) $H(W)$ = the convex hull of W = $\{x : x = \Sigma\beta_i z_i, \Sigma\beta_i = 1, \beta_i \geq 0,$ and $z_i \in W\}$

(5) $\delta f(x)$ = the gradient of $f(x)$.

(6) xy = the inner product of the vectors x and y.

    The RSD strategy may be employed with any linearly constrained problem. It is initiated with a parameter $r \geq 1$ and a feasible point x^0. If a feasible point is not known, one may be generated by solving the linear subproblem of Step 1 with any cost vector. For example, let x^0 = arg min $\{\delta f(x)y : y \in S\}$ for any x such that $\delta f(x)$ is defined.

Restricted Simplicial Decomposition (RSD)

<u>Step 0</u>: Let x^0 be a feasible point. Set $[W_s]^0 = \emptyset$, $[W_x]^0 = \{x^0\}$, and $k = 0$.

<u>Step 1</u>: Let y^k = arg min $\{\delta f(x^k)y : y \in S\}$

        If $\delta f(x^k)(y^k - x^k) \geq 0$, x^k is a solution and terminate.

        Otherwise,

            i) If $|[W_s]^k| < r$, set $[W_s]^{k+1} = [W_s]^k \cup \{y^k\}$, and
                $[W_x]^{k+1} = [W_x]^k$.

           ii) If $|[W_s]^k| = r$, replace the element of $[W_s]^k$ with the
                minimal weight in the expression of x^k as a convex
                combination of W^k with y^k to obtain $[W_s]^{k+1}$ and let
                $[W_x]^k = \{x^k\}$.

    Set $W^{k+1} = [W_s]^{k+1} \cup [W_x]^{k+1}$, and go to Step 2.

<u>Step 2</u>: Let x^{k+1} = arg min $\{f(x) : x \in H(W^{k+1})\}$ and x^{k+1} be written as

$$x^{k+1} = \sum_{i=1}^{r'} \beta_i z_i \quad \text{where } r' = |W^{k+1}| \text{ and } z_i \in W^{k+1}.$$

    Discard all elements z_i with weight $\beta_i = 0$ from $[W_s]^{k+1}$ or $[W_x]^{k+1}$. Set $k = k + 1$ and go to Step 1.

As constructed, the feasible region for the master problem, $H(w^{k+1})$, always contains the current iterate, x^k, and the incoming extreme point, y^k. When x^k is not a solution, $y^k - x^k$ is a descent direction, thereby ensuring a decrease in the objective function value at the new iterate, x^{k+1}. The global convergence proof (Hearn, Lawphongpanich and Ventura, 1984) then follows from the fact that x^{k+1} solves the master problem.

For problem P1 the linear program of Step 1 is a single commodity minimum cost flow problem for which there are many efficient algorithms. In our implementation, we have employed the primal simplex code NETFLO (Kennington and Helgason, 1980), modified to operate with real variables and to allow for restarting from the prior solution.

As mentioned earlier, the nonlinear master problem could also be solved by many algorithms such as successive quadratic programming methods which are available in computer packages such as NPSOL (Gill et al., 1983) and NLPQL (Schittkowski, 1984). Since the master must be solved repeatedly, it is important that a superlinearly convergent method be used, and this requires the storage of an adequate approximation of the Hessian matrix. However, this matrix is at most $(r+1) \times (r+1)$ in RSD. Further, the constraints are simple, making the projected (Quasi-) Newton method of Bertsekas(1982) a natural choice. This technique has proven to be very effective in reducing the relative error in the master problem to 10^{-6} for small problems and 10^{-4} for large problems.

An interesting question concerns the finiteness of the method, i.e., under what conditions will the algorithm terminate after a finite number of master problem iterations. This is answered by the following result:

Theorem: Let x^* be the unique solution of P1, and I^* be the optimal
      facet (of S) containing x^*. If

$$r \geq \text{dimension } (I^*) + 1$$

    then RSD converges to x^* after a finite number of master iterations.

Proof: (Hearn, Lawphongpanich, and Ventura, 1985)

Thus in applications the convergence rate of the overall algorithm will be that of the projected Quasi-Newton method (superlinear) if r is sufficiently large. While this would rarely occur on large scale problems, it does establish when rapid convergence can be expected on small problems. The example of Figure 1 illustrates this result. The optimal solution is in the facet defined by the extreme points (2,0,2,0,4) and (2,0,0,2,2), and thus the dimension of the facet is one. As shown in Table 1, when RSD is executed with r = 1, the convergence is slow, but when r = 2, convergence is achieved in two iterations. (Note that when r = 1, RSD reduces to the well-known algorithm of Frank and Wolfe (1956), which is known to converge arithmetically unless the solution is at an extreme point.)

The experimental computer code RSDNET has been developed for testing the RSD algorithm on the single commodity problem P1. This code utilizes a standardized input format and built-in function set developed by Dembo (1983), offers the automatic setting or user over-ride of all tolerances and parameters, and has restart capability. Computational results with two sets of test data are given in Table 2. In the Table the size of each network is displayed as (nodes/links/source+sink nodes). The final column shows the reason for termination in each case.

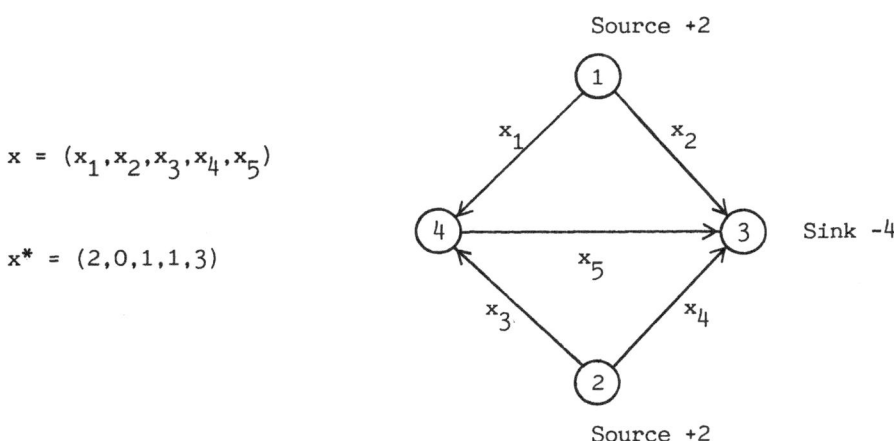

$$x = (x_1, x_2, x_3, x_4, x_5)$$

$$x^* = (2,0,1,1,3)$$

$$f(x) = 2x_1 + 15x_2 + 4x_3 + 16x_4 + 1/2(x_5)^2 + 1/3(x_5)^3$$

$$f(x^*) = 37.5$$

Figure 1. A 4 - Node Problem

Table 1.   RSDNET Solutions for 4-Node Problem

```
RSDNET (QUASI-NEWTON). PROBLEM: 4-Node Problem, MAXPJ=30, r=1


  MAX. NUM. EXTRE. POINTS =          1
  INITIAL OBJ. FUNCTION   =        41.3333340


                                                              CUM
   ITER        OBJ. FUN     LOWER BOUND     % RE   EX   PR    PRJ
    10          37.710         37.269       1.182   2   11    90
    20          37.630         37.362       0.716   2   13   220
    30          37.594         37.402       0.515   2   10   328
    40          37.574         37.423       0.403   2    9   423
    50          37.561         37.437       0.331   2   12   544
    60          37.552         37.447       0.280   2   12   664
    70          37.545         37.454       0.244   2   12   778
    80          37.540         37.459       0.215   2   12   885
    90          37.536         37.464       0.193   2   11   988
   100          37.533         37.467       0.175   2   12  1089


      FROM    TO       FLOW
       1       4    0.19675622E+01
       1       3    0.32437758E-01
       2       4    0.10277790E+01
       2       3    0.97222103E+00
       4       3    0.29953412E+01
```

```
RSDNET (QUASI-NEWTON). PROBLEM: 4-Node Problem, MAXPJ=30, r=2


  MAX. NUM. EXTRE. POINTS =          2
  INITIAL OBJ. FUNCTION   =        41.3333340


                                                              CUM
   ITER        OBJ. FUN     LOWER BOUND     % RE    EX   PR   PRJ
    1           37.982         11.333      235.138   2    5    5
    2           37.500         36.982        1.400   3   28   33
    3           37.500         37.500        0.000   2    0   33


   FROM    TO       FLOW
    1       4    0.20000000E+01
    1       3    0.00000000E+00
    2       4    0.10000001E+01
    2       3    0.99999990E+00
    4       3    0.30000001E+01
```

EX = No. of Extreme Points   MAXPJ = Max. Projections
PR = No. of Projections   RE = Relative Error
CUM PRJ = Cumulative Projections   Computer: PC

Table 2. RSDNET Results for Electrical and Water Distribution Problems

PROBLEM	r	CPU SEC	ITER	CUM PRJ	OBJ FUN	% RE	TERM
S-1	1	4.44	100	100	6.940507	0.0882	ITER=100
	2	4.70	100	100	6.939835	0.0785	ITER=100
(237/454/2)	4	5.43	100	100	6.937602	0.0463	ITER=100
(LB = 6.934391)							
S-4	1	12.69	100	100	1.598765	2.08	ITER=100
	2	15.10	100	100	1.598158	2.04	ITER=100
(852/2264/2)	4	19.01	100	100	1.597045	1.97	ITER=100
(LB = 1.566194)							
Dallas	1	34.28	50	92	-205071.9		
	1	56.14	88	131	-205676.7		ND
(667/1796/425)	2	31.85	50	50	-205563.1		
	2	54.88	90	90	-205888.4		ND
	4	31.48	50	50	-205509.3		
	4	61.68	103	103	-206179.8		ND

Computer: IBM 3081D

Problems S-1 and S-4 are large-scale resistor networks from physicists' studies of networks of conducting fibers in an insulating polymer (Balberg and Binnenbaum, 1983); the objective function measures the effective resistance of the network. The lower bounds(LB) on this measure were obtained from solving the dual (Ventura and Hearn, 1984) by a conjugate gradient method. The Dallas problem is a well-known water distribution problem from Collins et al. (1978). No lower bound is known for the objective, but the value -206179.8, obtained by RSDNET in approximately one minute of cpu time, is the lowest yet reported. Termination occurred when the algorithm could not decrease the objective in the projected direction by at least 10^{-4} (ND).

The RSDTA Algorithm

The standard traffic assignment problem is a nonlinear multicommodity problem of the form

P3:   min $\{\Sigma\ f_a(x_a)\ :\ x = \Sigma\ x^p,\ x_a = \Sigma\ x_a^p\ ,\ Bx^p = b_p\ ,\ x_a^p \geq 0\}$

B is again a node-arc incidence matrix for some network, b_p is a demand vector for commodity p, x_a^p is the flow of commodity p on arc a, x_a is the total flow on arc a, x^p is the vector of all arc flows for commodity p, and x is the aggregate vector of arc flows. A commodity is the flow emanating from a single origin. The convex function $f_a(x_a)$ represents congestion effects on arc a as a function of total arc flow. See Steenbrink (1974) for further formulation details and alternative forms of the objective function corresponding to different model criteria.

The application of the RSD strategy to problem P3 yields the algorithm RSDTA. It differs in one very important way from RSDNET: the linear program of Step 1 <u>decomposes</u> into a set of shortest path problems, one for each commodity, thus making that step more efficient. This decomposition is possible because there are no upper bounds on the variables of P3 and is the familiar shortest path decomposition of the Frank-Wolfe algorithm for the traffic assignment problem (see LeBlanc, et al., 1974).

Table 3 summarizes the results on three traffic assignment problems from the literature, two of which are based on street networks for the Canadian cities of Hull and Winnipeg. The quantity LB associated with each problem is the largest lower bound generated by RSD using the tangent plane inequality

Table 3. RSD Computational Results for Traffic Assignment Problems

PROBLEM	r	CPU SEC	ITER	CUM PRJ	OBJ FUN	% RE	TERM
ND1	1	1.19	50	50	85167.6	0.165	ITER = 50
(13/19/4)	2	1.21	50	85	85072.5	0.0528	ITER = 50
(LB = 35027.6)	3	0.87	32	60	85038.8	0.0132	ND
Nguyen & Dupuis(1984)	4	0.22	7	13	85027.6	0.0	RE ≤ 10E-6
Hull	1	33.23	100	109	34817.27	0.0331	ITER = 100
(501/798/23)	3	37.17	89	150	34808.72	0.00856	RE ≤ 10E-4
(LB = 34805.74)	5	20.50	41	95	34808.88	0.00902	RE ≤ 10E-4
Nguyen & Dupuis(1984)	7	15.97	31	74	34808.42	0.00769	RE ≤ 10E-4
	9	19.05	35	76	34808.25	0.00721	RE ≤ 10E-4
	11	17.36	32	66	34808.07	0.00669	RE ≤ 10E-4
	≥ 12	15.08	27	62	34808.18	0.00701	RE ≤ 10E-4
Winnipeg	1	485.37	73	808	893068	0.237	CPU = 500
(1052/2836/147)	4	490.33	69	136	891968	0.113	CPU = 500
(LB = 890958)	9	487.72	58	135	891726	0.0862	CPU = 500
Florian & Nguyen(1976)	14	485.37	60	105	891706	0.0840	CPU = 500

Computer: IBM 3081D

(equivalently, the duality gap; see Hearn, 1982). The percent error relative to this bound is shown in the column labeled RE. The reason for termination of each run is shown in the final column, whether maximum allowed iterations (ITER), maximum cpu seconds (CPU), relative error within the preset tolerance (RE), or three consecutive master iterations could not decrease f(x) with a step of at least 10^{-4} in the projected direction (ND).

In the traffic assignment problems, the majority of the effort (80%-90%) is spent on the shortest path calculations. Using the number of shortest path calculations as the sole criterion, RSD compares very favorably with other algorithms (see Table 4). In fact, there is only one instance in which another method, RESTRICTION on the Hull network, outperforms RSD. However, as described in Guelat (1983), RESTRICTION can be considered as a variant of RSD which allows for restarting at every sixth iteration.

Table 4. Number of Shortest Path Calculations to achieve the Indicated Error

Algorithm		% Relative Error					
		ND1		Hull		Winnipeg	
		.10%	.05%	.10%	.05%	1.0%	0.5%
RSD	r=1	87	NA	43	71	29	43
	r=2	22	55				
	r=3	11	11	15	19		
	r=4	6	6			18	26
	r=5			12	16		
	r=9			12	16	15	20
Guelat(1983)	FW	NA	NA	37	70	29	45
	Partan	11	28	15	21	18	27
	Restriction	9	9	10	13	17	25
	Quadratic Approx.			22	27	25	36
*Dembo and Tulowitzki(1983)	FW			65	171	46	80
	TQP-FW			32	56	35	52
	Partan			23	41	34	47
	TQP-PT			32	60	31	48

NA = The algorithm terminated before the indicated relative error was obtained.
Quadratic Approx. = Guelat's variation of Truncated Quadratic Programming.
Restriction = A strategy similar to RSD with restart every 6 iterations.
TQP-FW = Truncated Quadratic Programming using FW on the subproblem.
TQP-PT = Truncated Quadratic Programming Using Partan on the subproblem.
*The relative error as reported in Dembo and Tulowitzki (1983) is based on
    the lower bound calculated at the current iterate.

The most important conclusion from Tables 3 and 4 is that RSDTA uniformly outperforms the widely used Frank-Wolfe method ($r = 1$) with a relatively small value of r. This may be surprising, considering the size of the Hull and Winnipeg networks, but is partially explained by the fact that solutions to traffic assignment problems involve a small number of utilized paths. Since there is a one-to-one correspondence between paths and extreme points, a small number of utilized paths implies a small number of extreme points in the optimal facet.

The SDVI Algorithm

A nonlinear network model of increasing interest is the variational inequality problem

P4:   Find x^* such that $C(x^*)(x-x^*) \geq 0$ for all $x \in S$

where S is a closed convex set of feasible flow patterns and $C(x)$ is a vector function defined on S. The problem may be a single commodity congested flow problem with S defined as in P1 or multicommodity with S defined by the constraints of P3. The vector $C(x) = (c_1(x), \ldots c_a(x), \ldots c_n(x))$ contains a congestion function $c_a(x)$ for each arc which may depend on the entire flow vector x, in contrast to the functions $f_a(x_a)$ of the standard traffic assignment problem P3 which only depend on the flow on the individual arcs. Dafermos (1980) and Smith (1979) initiated the generalization of the traffic assignment problem to the form P4 and established the conditions for existence and uniqueness of the solution x^*.

To apply the simplicial decomposition strategy to P4 we assume that S is compact and restate the problem as

P5:   Find β^* such that $A^t C(A\beta^*)(\beta-\beta^*) \geq 0$ for all β satisfying

$\Sigma \beta_i = 1$, $\beta_i \geq 0$, $i = 1,2,\ldots N$

where N is the number of extreme points of S and A is a matrix with those points as columns. The enumeration of the extreme points is impractical, and simplicial decomposition generates them "as needed."

The SDVI algorithm (Lawphongpanich and Hearn, 1984) requires a small positive tolerance δ and a convergent sequence $\{\epsilon_k\}$ where $\epsilon_k > \epsilon_{k+1} > 0$ and $\epsilon_k \to 0$ as $k \to \infty$.

Simplicial Decomposition for Variational Inequalities (SDVI)

__Step 0__: Let x^0 be any feasible point, $[W_s]^0 = \emptyset$, $[D]^0 = \emptyset$, $U^0 = \infty$, and $k = 0$

__Step 1__: Let $y^k = \arg \min \{C(x^k)y : y \in S\}$ and $G(x^k) = C(x^k)(x^k - y^k)$.

If $G(x^k) = 0$, x^k is a solution and terminate

Otherwise,

i) If $G(x^k) \geq U^k - \delta$ set $[W_s]^{k+1} = [W_s]^k \cup y^k$

ii) If $G(x^k) < U^k - \delta$ set $[W_s]^{k+1} = [W_s]^k \backslash [D]^k \cup y^k$

Let $U^{k+1} = \min \{U^k, G(x^k)\}$ and go to Step 2.

__Step 2__: Find $x^{k+1} \in H([W_s]^{k+1})$ such that

$C(x^{k+1})(x - x^{k+1}) \geq -\epsilon_k$ for all $x \in H([W_s]^{k+1})$

Let D^{k+1} = columns of $[W_s]^{k+1}$ with zero weight in the expression

of x^{k+1} as a convex combination of columns of $[W_s]^k$.

Set $k = k + 1$ and go to Step 1.

Several features of SDVI should be noted. In RSD the progress of the algorithm is measured by decreases in the problem objective function, $f(x^k)$. Since P3 does not have an objective, we utilize the gap function, $G(x^k)$, to monitor SDVI (see Hearn, Lawphongpanich and Nguyen, 1984). This function is easily shown to be positive unless x^k solves [P4].

However, it does not decrease monotonically, and the number of retained extreme points in $[W_s]^k$ cannot be restricted as in RSD. Instead, the only dropping of extreme points occurs in Step 1 when the bound U^k decreases by δ. The parameters ϵ_k allow for approximate solving of the master variational inequality in Step 2 by _any_ convergent method. In computational testing we have used the projection method (Bertsekas and Gafni, 1982, and Dafermos, 1980). Finally, we note determination of y^k in Step 1 of SDVI is a linear program just as in RSD. Thus when SDVI is applied to the variational inequality formulation of the traffic assignment problem, i.e., S is defined by the constraints of P3, Step 1 decomposes into shortest path problems as in RSDTA.

Table 5 shows the results of using SDVI to solve traffic assignment problems ND1, ND2, ND3 and Hull from the paper of Nguyen and Dupuis (1984). Note that ND1 and Hull were also solved by RSDTA -- any minimization problem of the form P3 may be converted to the variational inequality form P4 by defining $c_a(x) = df_a(x_a)/dx_a$. Comparison of the two algorithms for ND1 and Hull shows that SDVI required more time to achieve objective values approximately the same as those achieved by RSDTA. However, the results on these examples suggest that simplicial decomposition remains a viable technique for large variational inequality problems because the growth in the number of retained columns is modest. The simplicial decomposition approach for variational problems is an important area of current research. See Pang and Yu (1934) and Marcotte (1985) for other uses of the approach.

Table 5. Results of SDVI on Nguyen and Dupuis Problems

PROBLEM	CPU SEC	OBJ FUN	ITER	EX	% RG	TERM
ND1 (13/19/4)	0.0910	94252.0	1	2	113.6	
	0.5178	85027.6	9	6	0.0067	RG ≤ 10E-4
ND2 (13/19/4)	0.0882		1	2	267.7	
	0.2329		5	4	0.0065	RG ≤ 10E-4
ND3 (13/19/4)	0.0856		1	2	194.1	
	0.2604		6	4	0.0064	RG ≤ 10E-4
Hull (501/798/23)	0.8600	39873.72	1	2	136.2	
	74.5900	34815.39	17	11	0.0652	

EX = Number of extreme points in master problem.
RG = Relative Gap at iteration k = $G(x^k)/C(x^k)y^k$.
All problems are from Nguyen and Dupuis (1984).

Computer: IBM 3033
Note: CPU times are approximately 10% slower than the IBM 3081D.

Summary

Simplicial decomposition provides an effective means of solving large-scale congested network models by combining linear network algorithms with second-order nonlinear programming techniques. The algorithms RSDNET, RSDTA and SDVI have successfully solved models with thousands of variables which arise in real applications of electrical networks, water distribution networks and traffic networks.

Acknowledgement

This research was supported in part by NSF Grants ECE-8420830 and ECS-8516365.

REFERENCES

Balberg I. and Binnenbaum N. (1983) Computer study of the percolation threshold in a two-dimensional anisotropic system of conducting sticks. Physical Review B 28, 3799-3812.

Bertsekas D. P. (1982) Projected Newton methods for optimization problems with simple constraints. SIAM Journal of Control and Optimization 20, 221-246.

Bertsekas D. P. and Gafni E. M. (1982) Projection methods for variational inequalities with application to the traffic assignment problem. Math. Prog. Study 17, 139-159.

Collins M., Cooper L., Helgason R., Kennington J. and LeBlanc L. (1978) Solving the pipe network analysis problem using optimization techniques. Management Sci. 24, 747-760.

Dafermos S. (1980) Traffic equilibrium and variational inequalities. Transpn Sci. 14, 42-54.

Dembo R. (1983) NLPNET: User's guide and system documentation. School of Organization and Management, Yale University, New Haven, CT, SOM Working Paper Series B #70.

Dembo R. and Tulowitzki U. (1983) Computing equilibria on large multicommodity networks: An application of truncated quadratic programming algorithms. School of Organization and Management, Yale University, New Haven, CT, SOM Working Paper Series B #65.

Florian M. and Nguyen S. (1976) An application and validation of equilibrium trip assignment methods. Transpn Sci. 10, 374-389.

Frank M. and Wolfe P. (1956) An algorithm for quadratic programming. NRLQ 3, 95-110.

Gill P. E., Murray W., Saunders M. A. and Wright M. H. (1983) User's guide for SOL/NPSOL: a Fortran package for nonlinear programming. Department of Operations Research, Stanford University, California, Report SOL 83-12.

Guelat J. (1983) Algorithms pour le probleme d'affectation du traffic d'equilibre avec demandes fixes: Comparasons. Center de Recherche sur les Transports, Universite de Montreal, Montreal, Publication 299.

Hearn D. W. (1982) The gap function of a convex program. Opr. Res. Lett. 1, 67-71.

Hearn D. W., Lawphongpanich S. and Nguyen S. (1984) Convex programming formulations of the asymmetric traffic assignment problem. Transpn Res. 18B, 357-365.

Hearn D. W., Lawphongpanich S. and Ventura J. A. (1984) Restricted simplicial decomposition: computation and extensions. Univ. of Florida ISE Department Research Report 84-38. (To appear in Math. Prog. Studies.)

Hearn D. W., Lawphongpanich S. and Ventura J. A. (1985) Finiteness in restricted simplicial decomposition. Opr. Res. Lett. 4, 125-130.

Kennington J. L. and Helgason R. V. (1980) Algorithms for Network Programming. John Wiley & Sons, New York.

Lawphongpanich S. and Hearn D. W. (1984) Simplicial decomposition of the asymmetric traffic assignment problem. Transpn Res. 18B, 123-133.

Lawphongpanich S. and Hearn D. W. (1986) Restricted simplicial decomposition with application to the traffic assignment problem. Ricera Operativa 38, 97-120.

LeBlanc L. J., Morlok E. K. and Pierskalla W. P. (1974) An accurate and efficient approach to equilibrium traffic assignment on congested networks. Transportation Research Record 491, Interactive Graphics and Transportation Systems Planning, 12-33.

Marcotte, P. (1985) A new algorithm for solving variational inequalities with application to the traffic assignment problem. Math. Prog. 33, 339-351.

Mulvey J. M., Zenios S.A. and Ahlfeld D. P. (1985) Simplicial decomposition for convex generalized networks. Engineering Management Systems Program, Princeton University, Princeton, NJ, Report EES-85-8.

Nguyen S. and Dupuis C (1984) An efficient method for computing traffic equilibria in a network with asymmetric transportation costs. Transpn Sci. 18, 185-202.

Pang J. S. and Yu C. S. (1984) Linearized simplicial decomposition for computing traffic equilibrium on networks. Networks 14, 427-428.

Schittkowski K. (1984) NLPQL: A Fortran subroutine for solving constrained nonlinear programming problems. Institut für Informatik, Universität Stuttgart, Germany, Report.

Smith M. (1979) Existence, uniqueness and stability of traffic equilibria. Transpn. Res. 13B, 295-304.

Steenbrink P. (1974) Optimization of Transport Networks. John Wiley & Sons, New York.

Ventura J. A. and Hearn D. W. (1984) Computing the effective resistance in a system of conducting sticks. Univ. of Florida ISE Department Research Report 84-43.

TRAFFIC ASSIGNMENT
FOR LARGE SCALE TRANSIT NETWORKS

S. Nguyen

Département d'informatique et de recherche opérationnelle

et Centre de Recherche sur les Transports

Université de Montréal

Montréal, Canada

S. Pallottino

Istituto per le Applicazioni del Calcolo "Mauro Picone"

Consiglio Nazionale delle Ricerche

Roma, Italia

ABSTRACT

This paper provides insight into a new traffic assignment framework for transit networks. This framework, based on an innovative graph structure—called hyperpath, permits the application to transit networks of the latest theoretical and algorithmic developments achieved in the urban car network field.

KEYWORDS: Traffic assignment, Transit network, Hyperpath.

NATO ASI Series, Vol. F38
Flow Control of Congested Networks
Edited by A. R. Odoni et al.
© Springer-Verlag Berlin Heidelberg 1987

INTRODUCTION

Consider an urban public transportation system consisting of a set of distinct transit lines, and stops where passengers board and alight transit carriers (buses). Consider a passenger waiting at a stop i served by several transit lines which can transport this passenger to his destination j. The decision problem faced by the passenger is whether to board an arriving carrier or to wait for a faster one. From the assignment point of view, the problem reduces to that of determining the subset of transit lines at stop i, such that the passenger always boards the first arriving carrier of this set, to get to destination j. This subset of transit lines is often referred to as the *passenger's attractive set*. The above *adaptive* component of choice behavior (i.e. to board the first arriving carrier of a subset of perceived equivalent lines) cannot be captured by the standard car network assignment framework, in which an alternative is represented by a single path connecting a pair of centroïds. Clearly, a new definition of the concept of alternative for a trip maker on a transit network must be made.

This paper describes a new traffic assignment framework for large scale transit networks based on a multipath concept—called a hyperpath. In broad terms, a hyperpath connecting two points is a collection a distinct itineraries with fixed utilization rates. This framework provides two major advantages over standard assignment models for transit networks. First, from a modeling point of view, all existing traffic assignment models for car networks can now be applied to transit networks simply by substituting the hyperpath flow space for the ordinary path flow space. Such modeling flexibility has not been obtained until now. Secondly, from a computational point of view, the intimate relationship between a path and a hyperpath enables the adaptation of the most advanced techniques for the computation of shortest hyperpaths and assignment flows in car networks to that of hyperpaths and flows in transit networks.

The next section formally introduces the notion of hyperpath and defines the hyperpath flows and costs. Section three describes various assignment models based on this multipath framework, and the last section discusses the computation of shortest hyperpaths. For a detailed discussion of the results stated here the reader is referred to Nguyen and Pallottino (1985, 1986, 1987).

2. A MULTIPATH FRAMEWORK

Consider an urban transit system, abstracted into a graph, in which each transit line is identified by a unique sequence of arcs, and each stop, either served by a single transit line or by multiple lines, is represented by a subgraph, in which each line is connected to an artificial node by a pair of boarding and alighting arcs. This artificial node is referred to in the sequel as a *stop node*. Centroïds are connected to stop nodes by walking arcs, and pedestrian paths are integrated into the transit system as transit lines with appropriate characteristics.

Let $(\mathcal{N}, \mathcal{L})$ denote the modeling graph, in which $\mathcal{N}$ is the set of nodes (centroïds, stops, line nodes) and $\mathcal{L}$ is the set of arcs (line segments, boarding and alighting arcs, centroïd connectors). In this graph, the subset of stop nodes $\mathcal{N}^B \subset \mathcal{N}$ and that of boarding arcs $\mathcal{L}^B \subset \mathcal{L}$ will assume a predominant role.

It is assumed that each transit line has a fixed line frequency, and this frequency φ_{ij} will be associated with all the boarding arcs $(i,j) \in \mathcal{L}^B$ to this transit line. Also, an average in-vehicle travel time or a generalized cost c_{ij} is associated with every link $(i,j) \in \mathcal{L}$.

In summary, the following notation will be used :

- node set $\quad\quad\quad \mathcal{N} = $ (centroïds, stops, line nodes)
- arc set $\quad\quad\quad\;\; \mathcal{L} = $ (line segments, boarding arcs,
  $\quad\quad\quad\quad\quad\quad\quad\quad$ alighting arcs, centroïd connectors)
- stop-nodes $\quad\;\; \mathcal{N}^B \subset \mathcal{N}$
- boarding-arcs $\quad \mathcal{L}^B \subset \mathcal{L}$
- in-vehicle time $\quad C = (c_{ij}) \quad or \quad (c_l)$
- line frequency $\quad \Phi = (\varphi_{ij}), \quad (i,j) \in \mathcal{L}^B$
- waiting time $\quad\;\; \tau = (\tau_i), \quad i \in \mathcal{N}^B$
- trip demand $\quad\;\; T = (t_{rs})$
- hyperpath flow $\quad Y = (y_{krs})$
- link flow $\quad\quad\; V = (v_{ij})$
- hyperpath cost $\quad U = (u_k)$
- trip proportion $\quad P = [p_{krs}]$

For the present transit network $(\mathcal{N}, \mathcal{L}, C, \Phi)$ a routing alternative is defined as an acyclic subgraph with given arc traversal probabilities, called a hyperpath. More precisely, consider a single pair of centroïds, say pair (r, s).

Definition. A subgraph $H = (X, E, \pi)$, where $X \subset \mathcal{N}$, $E \subset \mathcal{L}$, and $\pi = (\pi_{ij})$ a real value vector of dimension $|\mathcal{L}|$, is a hyperpath connecting r and s if:

- H is acyclic with at least one arc,

- node r has no predecessor and s no successor,
- for every node $i \in X - \{r, s\}$, there is a path from r to s traversing i,
- the characteristic vector π satisfies:

$$\sum_j \pi_{ij} = 1, \qquad \forall i \in X$$

$$\pi_{ij} \geq 0, \qquad \forall(i,j) \in \mathcal{L}$$

$$\pi_{ij} = 0, \qquad \forall(i,j) \notin E.$$

For every stop node i of hyperpath H, the set of boarding arcs $E_i^+ \subset E$ identifies the attractive set of transit lines for the subgroup of passengers who travel on this hyperpath. Each component π_{ij} represents the probability that arc (i,j) is traversed by a passenger who arrived at node i.

Now let

$$\lambda_q = \text{Prob(choosing path } q \mid H),$$

$$\beta_i = \text{Prob(traversing } i \in X \mid H),$$

$$d_{ij} = \text{Prob(traversing } (i,j) \in E \mid H),$$

then clearly

$$\lambda_q = \prod_{(i,j) \in E} \pi_{ij}^{\delta_{qij}},$$

where δ_{qij} is equal to 1 if (i,j) belongs to path q, and 0 otherwise.

One can also show that the node and arc probabilities, β_j and d_{ij}, satisfy the following equations (Nguyen and Pallottino, 1985):

$$\beta_r = \beta_s = 1,$$

$$\beta_j = \sum_i \beta_i \pi_{ij}, \quad \forall j \in X - \{r\}$$

$$d_{ij} = \beta_i \pi_{ij}, \quad \forall(i,j) \in E,$$

which show that the arc probabilities d_{ij} can be computed in a natural topological order which obviates the enumeration of all elementary paths defining the hyperpath. Without this property the hyperpath framework would not be operational.

Let y denote the passenger flow on hyperpath H. This flow is subdivided into elementary path flows, which induce in turn the aggregated arc flows :

$$v_{ij} = d_{ij} y, \quad \forall(i,j) \in E.$$

Let us now consider the cost structure of a hyperpath $H = (X, E, \pi)$. In addition to the ordinary arc cost c_{ij}, for every arc $(i, j) \in E$, there is also a node traversing cost τ_i, for every stop node $i \in X^B$ ($X^B \subseteq \mathcal{N}^B$), which represents a generalized waiting time. Note that the node traversing cost τ_i varies among hyperpaths connecting the same pair of centroïds. In general, it is assumed that τ_i is a function $f(\pi)$ of the characteristic vector π with the property that any two hyperpaths H_k and $H_{k'}$ having identical arc-set $E_{ik}^+ = E_{ik'}^+$ produce equal traversing costs at node i:

$$E_{ik}^+ = E_{ik'}^+ \quad \Rightarrow \quad \tau_{ik} = \tau_{ik'}.$$

The travelling cost u_k of any hyperpath H_k will be defined as:

$$u_k = \sum_{ij \in E} c_{ij} d_{ijk} + \sum_{i \in X} \tau_{ik} \beta_{ik}.$$

Let i be any node of H_k, a hyperpath connecting nodes i and s is a sub-hyperpath of H_k if it is a maximal sub-graph of H_k which has this property. Furthermore, let $u_k(i, s)$ denote the cost of the unique sub-hyperpath of H_k connecting $i \in X_k$ and s, then it is shown in Nguyen and Pallottino (1985) that:

$$\begin{cases} u_k(s, s) = 0 \\ u_k(i, s) = \tau_{ik} + \displaystyle\sum_{(i, j) \in E_i^+} \pi_{kij}(c_{ij} + u_k(j, s)), \quad \forall i \in X_k - \{s\}. \end{cases}$$

It may be interesting to note that since H_k is acyclic, the above expression provides an easy way to compute the cost of hyperpath H_k. This can be achieved by scanning the nodes of H_k in a topological order, starting from end node s.

We now have all the ingredients to apply to transit networks the standard car network assignment models, such as the 'all-or-nothing', the deterministic equilibrium model and stochastic models.

3. TRANSIT NETWORK ASSIGNMENT MODELS

Existing assignment models are built upon the premise that every traveller, being a rational decision maker, minimizes his perceived journey cost in the choice of his routing alternative.

Let $\tilde{u}_k$ denote the perceived cost of alternative (path or hyperpath) k:

$$\tilde{u}_k = u_k + \varepsilon_k, \quad E(\varepsilon_k) = 0.$$

Then in deterministic assignment models ("all-or-nothing" or Wardrop equilibrium) it is assumed that the perceived cost is exactly equal to the average cost $(\mathrm{Var}(\varepsilon_k) = 0)$. In this case, a single alternative is utilized when there is no congestion. In contrary, in stochastic assignment model $(\mathrm{Var}(\varepsilon_k) \neq 0)$ multiple alternatives are always utilized, and each is perceived as the minimum cost alternative by a subgroup of trip makers.

Let p_{krs} denote the proportion of the demand t_{rs} choosing alternative k, then independently from the definition of the trip proportion matrix $P = [p_{krs}]$ all models must satisfy the following fundamental equations:

$$\sum_{k \in I_{rs}} p_{krs} = 1, \qquad\qquad \forall rs$$
$$p_{krs} \geq 0, \qquad\qquad \forall k \in I_{rs},$$
$$y_{krs} = p_{krs}\, t_{rs}, \qquad\qquad \forall k \in I_{rs},$$
$$v_l = \sum_{rs} \sum_{k \in I_{rs}} d_{lk} y_{krs}, \qquad\qquad \forall l \in \mathcal{L},$$
$$u_k = \sum_{l \in E_k} c_l d_{lk} + \sum_{i \in X_k} \tau_{ik} \beta_{ik}, \qquad \forall k \in I_{rs},$$

where d_{lk} and β_{ik} are respectively the probability of using arc $l \in E_k$ and that of traversing node $i \in X_k$ defined earlier; and I_{rs} is the set of hyperpaths connecting centroïds r and s.

It may be interesting to contrast the above equations with similar equations for car networks. Let $\Delta = [\delta_{lk}]$ denote the usual multi-origin-destination pairs arc-path incidence matrix. The generic car network assignment map may then be expressed as:

$$V = \Delta Y$$

or

$$V = \Delta PT. \qquad\qquad (3.1)$$

Similarly, if we define an "arc-hyperpath incidence" matrix for transit networks as $D = [d_{lk}]$ then the above equation (3.1) yields the assignment map:

$$V = DPT.$$

Thus from a mathematical point of view the difference between the generic assignment map for car networks and that for transit networks is completely absorbed in the definition of the corresponding incidence matrices. This feature is definitely a major asset of the proposed assignment framework.

Although all assignment models for transit networks must satisfy the same basic set of flow conservation equations, they differ from each other by the assumptions underlining the specification of the trip proportion matrix $P = [p_{krs}]$. For instance, in the "all-or-nothing" model:

$$p_{krs} = \begin{cases} 1, & \text{if } k \text{ is the shortest hyperpath;} \\ 0, & \text{otherwise.} \end{cases}$$

We will return to the computation of shortest hyperpaths in the next section.

In stochastic assignment models the perceived cost is a random variable and the average trip proportion p_{krs} is equal to the probability that hyperpath k is a shortest hyperpath connecting r and s:

$$p_{krs} = \text{Prob}\,(\,\tilde{u}_k < \tilde{u}_h,\ \forall h \neq k\,),$$

where

$$\tilde{u}_k = u_k + \varepsilon_k, \quad E(\varepsilon_k) = 0,$$

and thus values of p_{krs} depend solely on the probability distribution assumed for the errors ε_k (see for example Dial, 1971; Daganzo and Sheffi, 1977).

Finally, in deterministic equilibrium models, the trip proportion matrix P is implicitly (and not necessary uniquely) defined by the Wardrop conditions (Wardrop, 1952):

$$p_{krs}^* > 0 \quad \text{only if} \quad u_k(V^*) = \min_{h \in I_{rs}} \{u_h(V^*)\},$$

$$V^* = DP^*T.$$

In a more compact form, the equilibrium model may also be expressed as the following variational inequality problem:

$$C(V^*)^T(V - V^*) + W^T(Y - Y^*) \geq 0 \quad \forall Y \text{ (feasible)},$$

where W is the vector of constant hyperpath node traversing costs:

$$W = [w_k] \quad \text{and} \quad w_k = \sum_{i \in X_k} \tau_{ik}\beta_{ik}.$$

A general algorithm for a subclass of deterministic equilibrium assignment models, which obviates the storing of hyperpath flows and thus has a space complexity $O(|\mathcal{L}|)$, is described in Nguyen and Pallottino (1985).

Note that the above framework is independent of specific values given to the conditional probabilities π_{kij} and the functional form of the node traversing cost τ_{ik}. These variables depend on the particular stochastic model adopted for the passenger and the transit carrier arrivals at a stop node and other assumptions. For instance, the following simplifying assumptions are frequently used in practice (see, for example, Chriqui and Robillard, 1975; Spiess, 1983):

- At a stop served by several lines, a passenger always boards the first arriving carrier of his choice set;

- Passengers arrive randomly at every stop node;

- All transit lines are statistically independent with given exponential distribution of headway, with mean equal to the inverse of line frequency.

In this case, the conditional probabilities π_{kij} and the node traversing cost (mean wait time) τ_{ik} at stop node i, take on specific values:

$$\pi_{kij} = \varphi_{ij}/\Phi_i , \quad \forall (i,j) \in E_K^B \tag{3.2}$$

$$\tau_{ik} \propto \Phi_i^{-1} \tag{3.3}$$

where φ_{ij} is the mean frequency of the transit line associated with boarding arc (i, j), and

$$\Phi_i = \sum_{(i,j) \in E_{ik}^+} \varphi_{ij}$$

is the combined frequency of the attractive set at stop node i, identified by the set E_{ik}^+ of boarding arcs.

The simplicity of these expressions allows the development of efficient shortest hyperpath techniques. This topic will be reviewed in the next section.

4. COMPUTING SHORTEST HYPERPATHS

If the characteristic vector π_k is exogenously defined for every hyperpath H_k connecting r and s, then finding a shortest hyperpath amounts simply to compute explicitly the cost of each hyperpath and determine the one with the smallest cost. We will review here the more interesting case where π and the waiting cost τ are defined by equations (3.2) and (3.3).

For clarity purposes, assume that $\tau_{ik} = \Phi_{ik}^{-1}$. Consider a given destination s, and let S_i be the length of a shortest hyperpath from node i to s. The following results have been obtained (Nguyen and Pallottino, 1985, 1986):

Theorem 1. (Generalized Bellman's equations)
$\{S_i\}$ *is a unique solution to:*

$$
S_i = \begin{cases}
0, & \text{if } i = s; \\
\min_{j \in N_i^+}\{S_j + c_{ij}\}, & \text{if } i \notin \mathcal{N}^B; \\
\min_{A_i \subseteq N_i^+}\{(\sum_{j \in A_i} \varphi_{ij}S_j + 1)/\sum_{j \in A_i} \varphi_{ij}\}, & \text{if } i \in \mathcal{N}^B.
\end{cases}
$$

The above generalized Bellman's equations may be solved with the following iterative procedure:

Basic shortest hyperpath algorithm

(1) (Initialization)

$$S_s := 0, \qquad S_i := +\infty, \quad \forall i \neq s.$$

(2) (Updating node labels)

> **if** $S_i > S_j + c_{ij}$ **for** $i \notin \mathcal{N}^B$
> > **then** $S_i := S_j + c_{ij}$;
> **if** $S_i > S_i^* = \min_{A_i \subseteq N_i^+}\{(\sum_{j \in A_i} \varphi_{ij}S_j + 1)/\sum_{j \in A_i} \varphi_{ij}\}$ **for** $i \in \mathcal{N}^B$
> > **then** $S_i := S_i^*$;

Repeat step (2) until no label can be further decreased.

From a practical point of view, this algorithm is operational only if the combinatorial problem embedded in step (2) (determination of S_i^*) can be solve efficiently. The following theorem:

Theorem 2.

$$
S_i = \left(\sum_{j \in A_i^*} \varphi_{ij}S_j + 1 \right)/ \sum_{j \in A_i^*} \varphi_{ij}
$$

if and only if

$$
\begin{aligned}
S_i \leq S_j, \quad \forall j \notin A_i^* \\
S_i \geq S_j, \quad \forall j \in A_i^*.
\end{aligned}
$$

provide a greedy type algorithm for computing the optimal set A_i^* of step (2).

Several labeling algorithms implementing the above basic shortest hyperpath procedure are described in Nguyen and Pallottino (1986, 1987).

CONCLUSION

This paper presents a multipath framework for traffic assignment modeling on large scale transit networks. It is shown that mathematical formulations of assignment models for transit networks based on this framework are identical to that developed for car networks. This feature permits the application to transit networks of the latest theoretical and algorithmic developments achieved in the urban car network field. In addition, under reasonable assumptions for large scale networks, the most advanced shortest paths techniques can be adapted successfully to the computation of shortest hyperpaths.

In conclusion, the multipath concept seems to provide a practical methodology for modeling and solving large scale traffic assignment problems, which strike a balance between computational tractability and ideal behavioral modeling.

REFERENCES

Chriqui, C., and Robillard, P. (1975). Common bus line. *Transportation Science* 9, 115–121.

Daganzo, C. F., and Sheffi, Y. (1977). On stochastic models of traffic assignment. *Transportation Science* 11, 253–274.

Dial, R. B. (1971). A probabilistic multipath traffic assignment model which obviates path enumeration. *Transportation Science* 5, 83–111.

Nguyen, S., and Pallottino, S. (1985). *Equilibrium Traffic Assignment for Large Scale Transit Networks*. Quarderno 14, IAC, C. N. R., Roma.

Nguyen, S., and Pallottino, S. (1986). Assegnamento dei Passeggeri ad un Sistema di Linee Urbane: Determinazione degli Ipercamini Minimi. *Ricerca Operativa* 38, 29–74.

Nguyen, S., and Pallottino, S. (1987). *Hyperpaths and shortest hyperpaths*. Publication 601, dept. informatique et de recherche opérationnelle, université de Monréal.

Spiess, H. (1983). *On Optimal Choice Strategies in Transit Networks*. Publication 286, Centre de Recherche sur les Transports, Université de Montréal.

Wardrop, J. G. (1952). Some theoretical aspects of road traffic research. *Proc. Inst. Civil Engr., Part II* 1, 325- -378.

Mathematical Programming Methods for Urban Network Control

G. Improta
Dipartimento di Informatica e Sistemistica
Cattedra di Ricerca Operativa
Universita' di Napoli
Via Claudio, 21
80125 Napoli
Italy

1. Introduction

The history of traffic light control (Webster and Cobbe 1966, Gazis 1974, Strobel 1982) began on December 1868 when the first traffic signal was installed in Westminster, London, in front of the Houses of Parliament.

The coordination of some traffic signals was first realized in Salt Lake City (USA) in 1918. This system was called "progression" because the green lights appear, to the drivers, to "progress" along the road.

The control system of Salt Lake City possessed three different synchronization schemes : the first and second ones for morning and afternoon peak-periods respectively, the third for average conditions. The transition from one to another scheme was carried out at fixed times during the day.

In the early 1930s a first traffic actuated control system was attempted in USA placing microphones at the side of the roads and using the sound of the horns as a traffic level measure.

Already in 1932 a first vehicle actuated control for a single junction was installed in London and three years later the first linked systems, consisting entirely of vehicle actuated controllers, were realized in London and Glasgow.

At the beginning of the 1950s one of the first systems based on a traffic-responsive transition, using traffic detector measurements, was installed in Baltimore (USA).

Between the end of 1950s and the first years of 1960s, a new age in the field of the traffic control was begun when in Toronto (Canada) a digital computer system for controlling traffic lights in an urban network was first used.

Work partially supported by
Progetto Finalizzato Trasporti,
Consiglio Nazionale delle Ricerche, Italy,
Grant no. 85.00157.93

NATO ASI Series, Vol. F38
Flow Control of Congested Networks
Edited by A. R. Odoni et al.
© Springer-Verlag Berlin Heidelberg 1987

Using these elements, a classification of the traffic control systems can be stated.

With reference to the system to be controlled, we can distinguish among : the single junction, when a junction is operated independently of the others; the arterial system when some junctions along a one-way or a two-way path are coordinated; the network, when the road system to be controlled has one or more loops.

Depending on the type of control used we can distinguish : fixed-time control, when the traffic control, calculated for average conditions, is independent of the time and of the actual characteristics of the traffic flows; pre-timed control when the traffic control changes automatically in some times of the day; actuated control when the control changes according to the characteristics of the traffic.

In any case, the classical comprehensive objectives of traffic control are (Gazis 1974) the improvement of the safety and the diminution of the discomfort suffered by drivers.

In other words the signal control is used as a tool for increasing the performance of traffic networks without changing the road structure and, like other tools of the Traffic Management (see Improta et al. 1986) is a non capital intensive technique to promote safe, efficient and convenient movement of persons and goods, making a better use of existing roads (Gartner and Garshwin, 1983).

Many performance criteria to evaluate traffic control systems (see tab.1) have been proposed , however the average delay and the number of stops are generally used since they also express in some way other criteria.

----

* Average delay per vehicle
* Maximum individual delay
* Fuel consumption
* Percentage of cars that are stopped as they go through the system
* Average number of stops before a vehicle goes through the system
* Throughput of the system
* Travel time through the system (average of maximum)
* Level of pollution

Table 1 - Some performance criteria in traffic control.

Bearing in mind the historical reasons for the introduction of signal control, it is obvius that the first objective associated to it was to reduce the confusion deriving from the interactions among increasing flows of cars and pedestrians.

The better arrangement of traffic streams, through the introduction of signal control, can however, especially for low levels of congestion, increase delay and stops. For these reasons, the efforts of researchers and technicians were and are directed at increasing the positive results of the traffic control, reducing at the same time the negative ones.

Traffic control is often thought of as a neutral factor respect to the users behaviour, and the current traffic regulation models assume path choice, and so link flows, to be independent of the control adopted. This can happen when origin-destination paths are strongly constrained and also when a same generalized level of congestion occurs, but it is not true in general.

Several researchers have shown that the traffic pattern can be a function of the control adopted, and this concept is also intuitive. It is obvious in fact that the signal control can change the values of the users perceived travel costs, inducing new flow equilibria for traffic networks.

This active role that can be carried out by traffic control suggests a development of traffic management and control strategies able (Robertson 1982) to influence user's choice of route for increasing the benefit of the community.

In the following the fundamental elements related to the optimization, by mathematical programming methodologies, of traffic signal, in the case of a given pattern of traffic and in the other of simultaneous traffic assignment and signal optimization, will be outlined and discussed and some suggestions for further development and research will be given.

2. Optimization of coordinated signal timing for a given pattern of traffic.

The optimization of coordinated traffic signal timing involves the two aspects outlined in the previous section, i.e. the coordination of arterials and networks.

This optimization is generally carried out using a three-step sequential decision process (Gartner 1976).

In the first step an appropriate master cycle is found in dependence of the requirements of the most loaded junctions.

In the second step, using the master cycle fixed before, the green splits for the junctions are calculated.

Finally a set of optimal offsets among signals is determined.

It is obviuos that this procedure cannot ensure a globally optimal solution since only a simultaneous decisional process could furnish it. However experience has shown that the signal timings deriving from sequential procedures can be good solutions for arterial and network signal coordination problems in practice.

2.1 Single junction control

A junction is defined "isolated" if there are no "interactions" with it and the other surrounding junctions. In practical terms this means that the mutual distances must be sufficient to eliminate the platooning effect generated by the traffic signals.

Interacting junctions can be treated as isolated (single junctions) only if the control objective can be assumed as independent of the characteristics of the arrivals (maximum capacity factor, minimum cycle time).

A junction can be described, schematically, through a set of approaches and a common crossing area. An approach is a part of a road leading to the junction such as the traffic in it has right of way simultaneously and a vehicle joining the back of the queue can expect to pass the signal at roughly the same time whichever lane it chooses.

The traffic at a junction is usually divided into streams. A stream is formed by all the users who cross the junction from the same approach, it is the smallest portion of the traffic which needs to be distinguished in the analysis.

We assume, in general, that the vehicle departures occur at a constant time-interval. An effective green-time g during which the vehicles can leave the stop-line, and an effective red-time r during which no departure occurs are hence considered.

Let s(t) be the flow leaving the stop line during the green and amber periods. A saturation flow s is defined for each stream, as the average flow that can cross the junction in the unitary time when a queue is present at the approach.

Let G, R, A be respectively the displayed green, red and amber for a stream. According to Webster (1958) and Webster and Cobbe (1966), we define:

$$g = (1/s) \int_0^{G+A} s(t)dt$$

$$r = G + R + A - g$$

we define also:

$$l = G + A - g \qquad = \qquad \underline{\text{lost time}}$$

The <u>cycle-time</u>, c, is the least time during which a complete succession of signals occurs. The cycle-time is, obviously, equal to the sum of the effective green and the effective red for each group.

Let q(t) be the flow reaching the stop line. We define <u>undersaturated approach</u> an approach for which:

$$\int_0^c q(t)dt \qquad \leq \qquad \int_0^{G+A} s(t)dt$$

If the arrival rate is assumed uniform with a value, q, and the discharge of the queue is supposed to be at an average flow equal to the saturation flow, the undersaturation condition becomes:

$$q.c \quad \leq \quad g.s$$

<u>Undersaturated junction</u> is a junction for which all the approaches are undersaturated.

When considering the control system, users are divided into <u>groups</u>, which are sets of streams that receive identical signals from the controller. The streams in a group may therefore use different approaches roads. The group represents the smallest unit which is considered in the control of the traffic on a junction.

The cycle can be subdivided into <u>stages</u> during which all the signals are constant.

The order in which n groups receive right of way in m stages is generally specified by a boolean matrix S(m.n), called <u>stage matrix</u>, such that: $s_{ij}=1$, if group j has rigth of way in stage i; $s_{ij}=0$, otherwise.

Two groups that cannot safely cross the junction at the same time are called <u>incompatible</u>. The <u>clearance-time</u> for two incompatible groups is the minimum time between the end of the amber of one and the beginning of the green of the other.

The compatibilities among n groups can be expressed by a boolean matrix A(n,n), called <u>compatibility matrix</u>, in which $a_{ij}=1$ if groups i,j are compatible ; $a_{ij}=0$ otherwise. A <u>compatibility graph</u> can be defined from matrix A. It is an unoriented graph with n nodes (one node for each stream) and one edge ij for each pair of compatible groups i and j.

A set of groups that are mutually compatible is called a <u>clique</u>. Usually the set of groups having the right to cross the junction during a stage forms a clique. An ordered succession of cliques is called a <u>sequence</u>.

Two dependent variables are of great importance in defining the performances of a single junction:

the capacity factor μ, i.e., for an assigned signal setting, the maximum common multiplier for all the rates of flow arriving at junction, for which the subsaturation still holds;

the delay, i.e. the difference between the travel time of a vehicle crossing the signalized junction and the time it would have taken if no other traffic were present and the stream in which it travels had constant right of way.

Corresponding to the introduced statements, the usual objectives of optimization are: maximization of the capacity factor of the junction and minimization of the total rato of delay.

A third objective corresponds to the evaluation of the critical cycle time, i.e. the minimum cycle time for which μ=1, i.e. the last cycle time which gives a useful operating condition for the junction.

Maximization of capacity factor is generally choosen objective for heavily congested junctions. A delay minimization control scheme can be adopted if the capacity factor is satisfactory. The critical cycle time is useful for the calculation of the master cycle in signal coordination.

The Highway Capacity Manual (1985) indicates as a measure of level-of-service for single junctions the stopped delay per vehicle (see table 2).

LEVEL OF SERVICE	STOPPED DELAY PER VEHICLE (SEC)		
A		≤	5.0
B	5.1	TO	15.0
C	15.1	TO	25.0
D	25.1	TO	40.0
E	40.1	TO	60.0
F		>	60.0

Table 2 - Level of service criteria for signalized junctions (H.C.M. 1985).

In recent years the evaluation by mathematical programming techniques of optimal values for control variables (stage composition, stage sequence green split and cycle time) has been carried out by various methods that can be grouped into two classes of different complexity.

In the first class, the flows and saturation flows of the streams being known, the stage matrix is preliminarily fixed and the optimal green times for the stages calculated (Allsop, 1971, 1972; Yagar, 1974; Improta, 1978).

In the second class (single junction control system design), the flows and saturation flows are still assumed known and the optimal timing is evaluated from the knowledge of only the crossing compatibilities of the different streams.

It is possible to divide the methods belonging to the second class (see Allsop, 1983) into two subclasses.

The first subclass is based on the analysis of possible sequences of clique of the compatibility graph. Any sequence of cliques respecting some imposed constraints is a possible stage succession. The optimization problem consists of calculating the optimal sequence, the green times and the cycle time. This is called a stage-based approach (Stoffers, 1968; Zuzarte Tully, 1977; Zuzarte Tully and Murchland, 1978; Cantarella and Improta, 1981).

Conversely, an endpoint and the green time for each stream are used in the second subclass to determine optimal cycle time, stage sequence and stage composition. This is called a stream-based approach (Improta and Cantarella, 1982; Gallivan and Heydecker, 1983; Improta and Cantarella, 1984; Cantarella and Improta, 1987; Heydecker and Dudgeon, 1987; Moller, 1987).

Control variables (stage composition and sequence, green times and cycle time) are generally calculated in sequence. Improta and Cantarella (1982,1984), Cantarella and Improta (1987), proposed the only model in which the optimization of all control variables is performed simultaneously.

According to the general statements previous introduced, mathematical programming methodologies for single junction control assume as possible objectives of optimization: the capacity factor maximization (linear); the total rate of delay minimization (non linear but convex or linear if using piecewise linearization of delay functions); the cyle time minimization (linear).

The methods for calculating the optimal green times for the stages use sets of constraints in which are:

- subsaturation constraints (for each approach);
- minimum green constraints (for each stage);
- maximum red constraints (for each stage);
- congruence constraint between cycle time and green times of stage;
- constraints on minimum and maximum value for the cycle time.

All constraints are linear and variables of optimization are real.

The methods for single junction control system design use

sets of constraints in which are:

- subsaturation constraints (for each approach);
- minimum green constraints (for each group);
- maximum red constraints (for each group);
- constraints on minimum and maximum value
  for the cycle time;
- incompatibility crossing constraints (for each
  pair of incompatibility groups).

Some models include real and binary variables with linear constraints. Other models use only real variables but can include some non linear constraints.

2.2 Arterial systems control

Figure 1 indicates an elementary traffic system in which two adjacent signalized junctions are connected by a one-way link. We suppose the master cycle and the green splits to be already calculated and the speed on the link to be a constant.

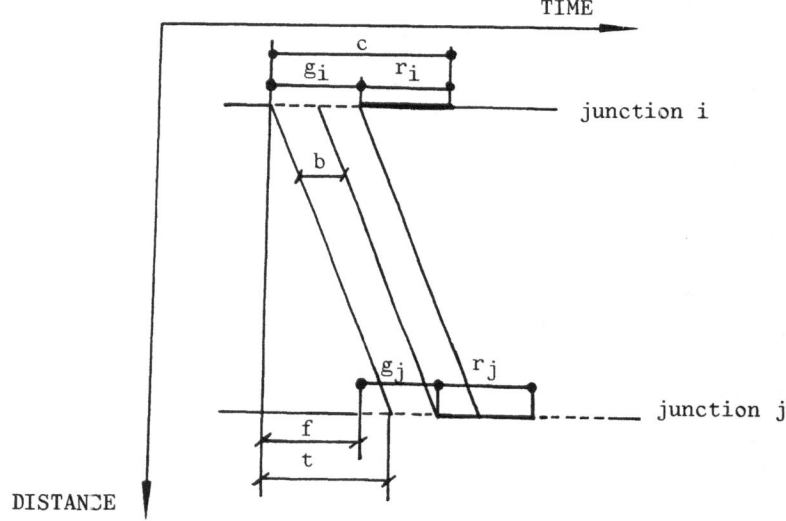

Figure 1 - Time-Distance relationship.

At the onset of each green period for the signal at i, a car platoon (having a time length g_i, on the hypothesis of fully saturated greens) will leave the stopline at i. The head of the platoon will reach the stopline at j a time t after.

According to the offset f between the signals at i and
j, a different part of the platoon will cross the stopline at j
in the long run.

The length of platoon for which is no stop due to the
traffic light at j (segment b indicated in fig. 1) is
generally indicated as <u>bandwidth</u>.

The bandwidth concept can be easily extended to one-way or
two-way paths with a greater number of junctions.

The Highway Capacity Manual (1985) defines urban and
suburban arterials as "signalized streets that primarily serve
through traffic and provide access to abutting properties as a
secondary function".

"Traffic signals force vehicles to stop and to remain
stopped for a certain time, and then release vehicles in
platoons. The delays and speed changes caused by traffic signal
operation considerably reduce the capacity of an urban arterial
and lower the quality of traffic flow".

H.C.M. indicates the average travel speed as a measure of
level of service for the arterials (see table 3).

ARTERIAL CLASS	I	II	III
Range of Free Flow Speeds (mph)	45 to 35	35 to 30	35 to 25
Typical Free Flow Speeds (mph)	40 mph	33 mph	27 mph
LEVEL OF SERVICE	AVERAGE TRAVEL SPEED (MPH)		
A	≥ 35	≥ 30	≥ 25
B	≥ 28	≥ 24	≥ 19
C	≥ 22	≥ 18	≥ 13
D	≥ 17	≥ 14	≥ 9
E	≥ 13	≥ 10	≥ 7
F	< 13	< 10	< 7

Table 3 - Arterials levels of service (H.C.M. 1985).

Little (1966) proposed a general formulation of the
arterials coordination as a mixed-integer linear program. This
proposal is the first use of mathematical programming
techniques in the field of the traffic control. Little et al.
(1966) indicated a mathematical formulation and a resolutive
algorithm for two-way arterial progressions.

The objective function is a weighted sum of the bandwidths of all arterials. Each bandwidth is linearly dependent on the offsets which are continous variables.

The offsets must satisfy the condition that adding them along each closed path through the network an integer multiple of the cycle must be obtained.

The presence of integer variables and the introduction of several redundant variables and constraints, apart from the low efficiency of the mixed-integer linear programming computer codes, generated many complications in the practical use of the method.

In recent years (Little et al., 1981) Little's methodology has been implemented in an off-line computer program, entitled MAXBAND, containing many special features.

2.3 The network coordination problem

The maximal bandwidth approach for signal light coordination have some practical advantages. Bandwidth methods use relatively little input and the space-time diagrams that can be obtained provide a simple graphic tool to evaluate, and eventually to modify, the obtained results.

The bandwidth-based regulation, often satisfactory for open road-system, can be inadequate when applied to networks. On the other hand optimization using bandwidth, mainly a geometric parameter, does not take into account real traffic flows and users delay which, in the following formulations, through the disutility oriented methods, becomes the basis of regulation.

The common element of Little's arterial technique and of many following mathematical programming based methodologies for network coordination, consists in the use of some consistency constraints on offsets, known as loop constraints.

We can refer to the simple road system indicated in figure 2 : a loop formed by three one-way links connecting three junctions, signalized with a common cycle time.

Let:

$g_i,(g_j),(g_k)$ effective green at signal i,(j),(k) on link ij,(jk),(ki);

$r_i,(r_j),(r_k)$ effective red at signal i,(j),(k) on link ij,(jk),(ki);

$f_{ij},(f_{jk}),(f_{ki})$ offset between signals at i and j, (j and k),(k and i) measured as the time from the starting point of a green at i, (j),(k) to the next starting point of green at j,(k),(i).

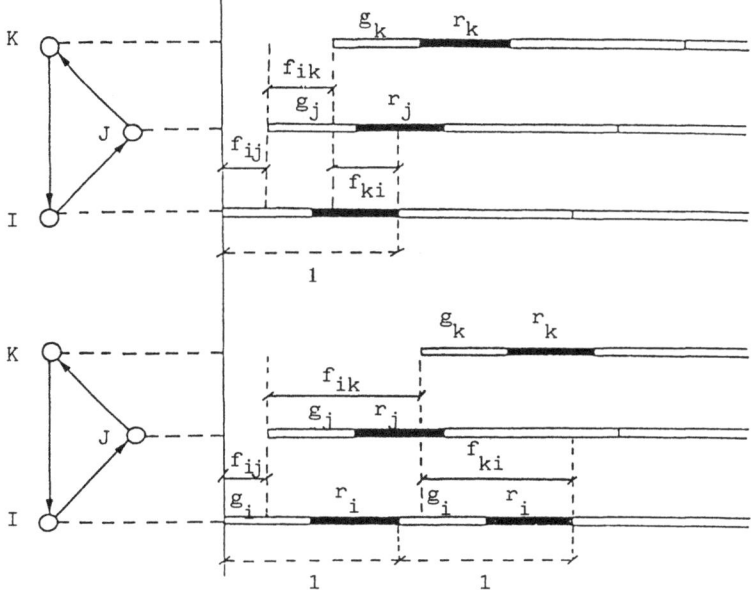

Figure 2 - An example of loop-constraint.

If the green and red times are expressed as a proportion of the cycle time it results :

$$g_i + r_i = g_j + r_j = g_k + r_k = 1$$

From figure 2a it follows :

$$f_{ij} + f_{jk} + f_{ki} = 1$$

From figure 2b it follows :

$$f_{ij} + f_{jk} + f_{ki} = 2$$

These last results can be generalized stating that : the algebraic sum of relative offsets associated with the links of any loop has an integer value.

Gartner (1972a) also demonstrated that, for a connected network, with v vertices and l links only l-v+1 loop constraints are linearly independent.

Let us denote by G the graph representing the network; T a tree of G consisting of (v-1) links; T* the cotree of T [(l-v+1) links]. A set of (l-v+1) independent loops (a fundamental set of loops) can be obtained from T adding to it the links of T* one by one.

2.3.1 Delay-offset relationships

It is obvious that a change of the offset between the signals at two adjacent junctions generally determines a change in delay and stops. For each assigned profile of the platoon reaching the stopline at j a delay-offset relationship for link ij can be obtained.

Wagner et al. (1969) indicate a numerical procedure to calculate the delay-offset relationship for a link, on the elementary hypothesis of rectangular platoons (i.e. in the case in which the departure rate in the green period at i, is assumed constant), without dispersion along the link.

The delay-offset relationship is avaluated using a step by step procedure in which, for each value of the offset between the signals at upstream and downstream junction, the delay is derived as the integral of the corresponding queue.

This approach has been generalized by Allsop (1969) and Gartner (1973).

Figure 3 indicates an example of delay-offset relationship calculated using the Wagner's procedure.

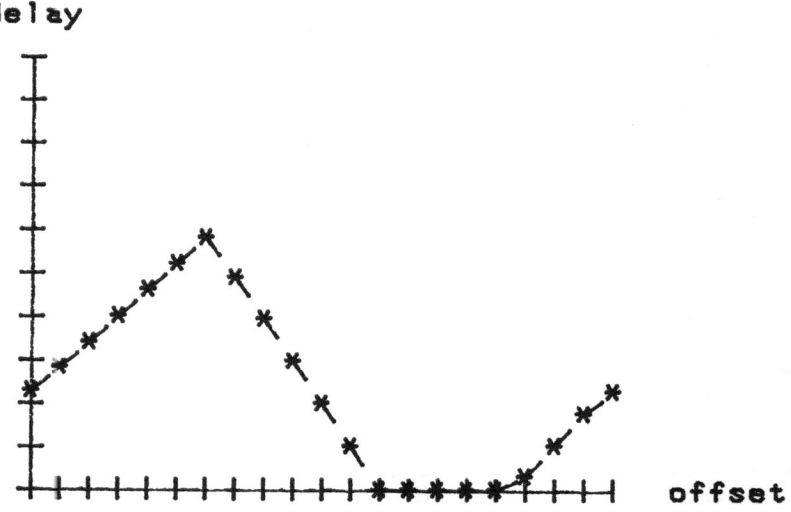

Figure 3 - Delay-Offset relationship (Wagner et al., 1969).

2.3.2 The combination and other derived methods

Knowledge of a delay-offset relationship for each link of the network is the basis of the combination method (Hillier, 1966) and of some other subsequent coordination procedures.

Combination requires the calculation in advance of : a master cycle; the green split at each junction; a delay-offset relationship, expressed in discrete form, for each link of the network.

The method supposes that :

a) the network can be represented by a series-parallel graph, i.e. that it can be reduced, using the simple condensation rules indicated in figure 4, to a single link (see fig. 5a) expressing the delay-offset relationship for the whole network;

b) the traffic delay along a link can be considered as a function only of the offset between the traffic signals at upstream and downstream junctions and does not depend on other offsets in the network;

c) the characteristics of the adopted traffic signal control do not affect the origin-destination patterns and the paths choosen by the users.

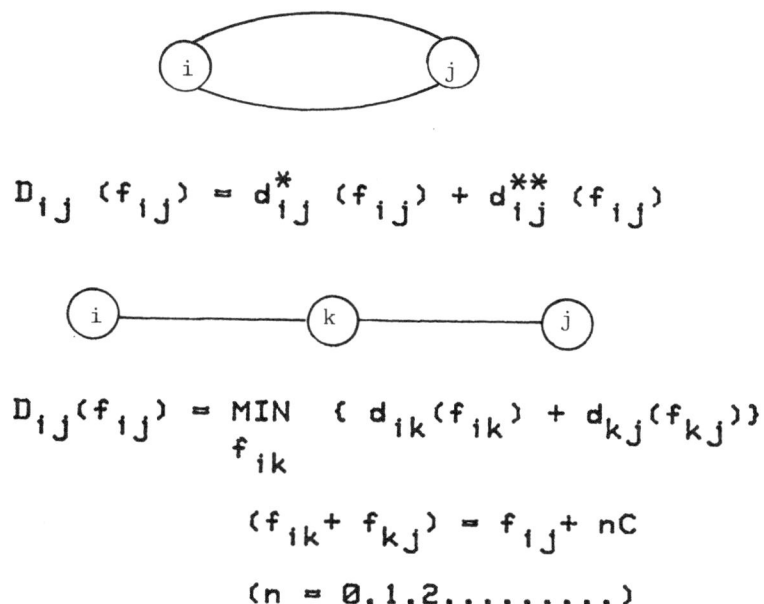

$$D_{ij}(f_{ij}) = d^{*}_{ij}(f_{ij}) + d^{**}_{ij}(f_{ij})$$

$$D_{ij}(f_{ij}) = \underset{f_{ik}}{\text{MIN}} \ \{ d_{ik}(f_{ik}) + d_{kj}(f_{kj}) \}$$

$$(f_{ik} + f_{kj}) = f_{ij} + nC$$

$$(n = 0,1,2,\ldots\ldots)$$

Figure 4 - Condensation rules used in combination method.

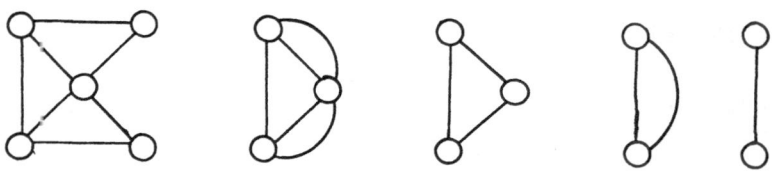

fig. 5.a Example of condensable network

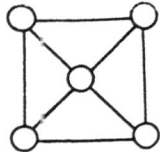

fig. 5.b Example of uncondensable network

Figure 5 - Examples of condensable and uncondensable networks.

We will discuss the second and the third hypothesis in the sequel. Now we would like to note that many networks, in practice, are uncondensable (see fig. 5b) and in these cases combination is unapplicable.

Allsop (1968) suggested an iterative procedure, by which uncondensable networks could be solved. Gartner (1972b), Gartner and Little (1975) proposed a dynamic programming approach. Improta and Sforza (1982) formulated the coordination problem through a bynary-integer programming model. The three methods are subject to problems of dimensionality but the last one is less dependent on the topology of the network. Moreover the branch and backtrack procedure used to solve the model has the advantage of finding good suboptimal solutions in a very short running time.

2.3.3 The network synchronization problem

An interesting observation concerns the optimization of the cycle time. The performance function for each link i j can be expressed as a sum of a <u>deterministic delay</u> $d^d_{ij}(f_{ij}, r_{ij}, C)$ corresponding to the average delay at j, and of a <u>stochastic delay</u> $d^s_{ij}(r_{ij}, C)$ expressed by an overflow queue, Q_{ij}, at downstream junction j .

Therefore the total delay in the network can be expressed by:

$$D = D_d + D_s$$

$$D_d = \Sigma_{ij} \ q_{ij} d^d_{ij}(f_{ij}, r_{ij}, c)$$
$$D_s = \Sigma_{ij} \ d^s_{ij}(r_{ij}, c)$$

Following Gartner (1976), we can consider a typical delay versus cycle time correspondence shown in figure 6.

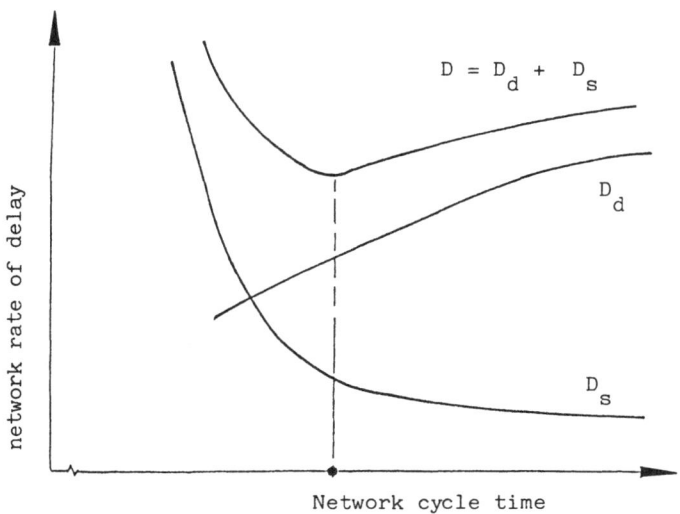

Figure 6 - Rate of the total delay vs. cycle time in a synchronized network (Gartner, 1976).

The deterministic component of delay generally increases as cycle time increases, the stochastic component decreases with it because of the increase in capacity at higher cycle times. The optimal cycle time for the network constitutes a least-cost equilibrium point between deterministic and stochastic delays.

A network for which the cycle time has been optimized is generally indicated as synchronized network. A three-step procedure for network coordination can be used for network synchronization by varying recursevely the cycle time in a fixed reasonable range of values.

The MITROP model (Gartner et al., 1975, 1976) represents at now the only proposal of simultaneous optimization of all control variables (cycle time, green splits and offsets).

MITROP is based on the resolution of a mixed-integer linear program. The aim of MITROP is the minimization of global delay in the network using convex piecevice linear delay functions. This convexity assumption, cannot be said to cover all the cases that can occur. Moreover the use of MITROP requires large computer memory and high running time.

2.3.4 Mathematical models for signalized networks synchronization and coordination

According to the statements introduced in the previous sections a mathematical model for network synchronization can be given:

$$D = \Sigma_{i,j} \, d_{ij} \, (f_{ij} \, , \, r_{ij} \, , \, c) \quad \text{MIN !}$$

subject to :

$$r_{ij} - l_{ij} = g_{kj} + l_{kj} \quad \forall \text{ pair of conflicting streams}$$

$$\Sigma_{i,j \in 1} \, f_{ij} = n_1 \, c \quad \forall \text{ loop 1 belonging to a fundamental set of loops}$$

$$g_{ij} + r_{ij} = c$$

$$q_{ij} \, c \leq s_{ij} \, g_{ij}$$

$$g_{ij} \geq g \, min_{ij} \quad \forall \text{ stream ij}$$

$$g_{ij} \, , \, r_{ij} >= 0$$

$$c_{min} \leq c \leq c_{max}$$

$$n_1 \text{ integer}$$

$$\text{where : } d_{ij} = \text{delay on link ij}$$
$$D = \text{delay on the network}$$

If the cycle time and the green splits are calculated in advance the model of network synchronization can be simplified into the network coordination model :

$$D = \Sigma_{i,j} \, d_{ij} \, (f_{ij}) \quad \text{MIN!}$$

subject to :

$$\Sigma_{i,j \in 1} f_{ij} = n_1 \, c \quad \forall \text{ loop 1 belonging to a fundamental set of loops}$$

$$n_1 \text{ integer}$$

2.3.5 Cyclic flow profiles

A simple example of cyclic flow profile can be introduced referring again to a one-way link joining two signalized junctions (see Gallivan, 1984).

We suppose that : the upstream junction controls only two conflicting (N-S,W-E) streams without turning traffic; the traffic queues are sufficient to ensure a constant discharge, with saturation flow rate, through the green times; there is no dispersion of platoons leading the junction.

Given these assumptions, an observer would see, in any point along the downstream roads, regular sequences of rectangular platoons separated from each other by the time of one cycle.

In practice, considering that there are turning flows, that the discharge of the queue does not occur with a constant rate, that the dispersion changes the shape of the platoon progressing along the road and that cycles can be no fully saturated, the flow profiles assume a variable shape like that shown in figure 7.

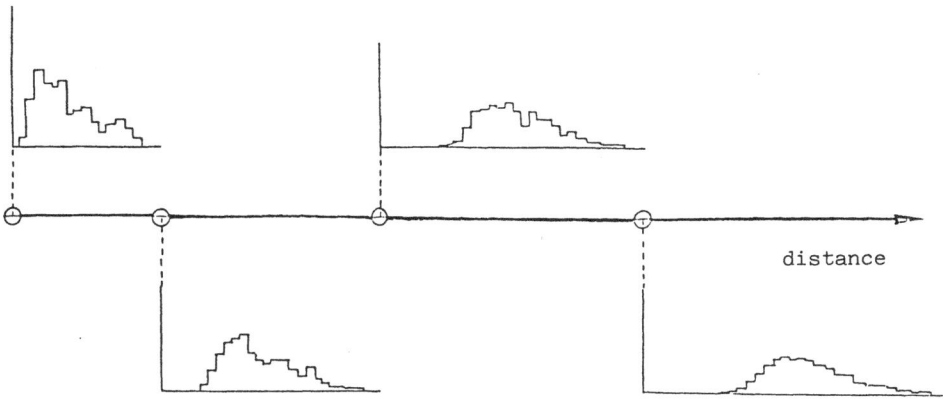

Figure 7 - Example of flow profile along a road.

The platoon shape, hence, depends on three different phenomena:

- distorsion and merging, at upstream junction;
- dispersion, along the link;
- division, at downstream junction;

Distorsion and merging. The distorsion is a change of the shape of the platoon entering a link, due to signal lights at upstream junction. The phenomenon is explained in figure 8.

Figure 8a indicates the shape of the platoon arriving at an approach of upstream junction. The platoon area expresses the total number of vehicles arriving during one cycle time.

During the red period a queue increases at the stop line (figure 8b). At the onset of the green period a car platoon leaves the stop line, at an average flow equal to the saturation flow s of the approach.

The queue decreases and vanishes a time t_0 after. If the arrival rate in each unit time after t_0 is not greather than the saturation flow of the approach, no queue can occur until the end of the green time.

The platoon leaving the approach (figure 8c) is then formed by a first part with average height equal to s until the time at which the queue vanishes. After this time the platoon shape coincides with the arrival pattern.

The shape of the platoon entering the link is defined by merging the platoons leaving the different approaches, feeding the link at upstream junction.

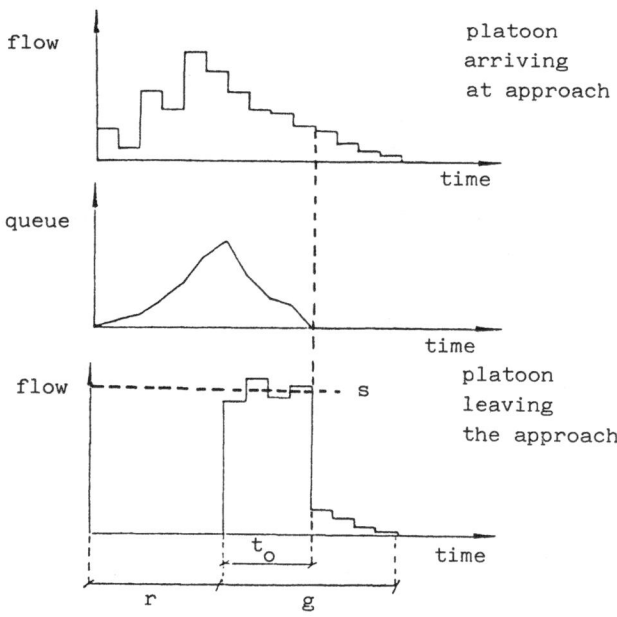

Figure 8 - Distorsion at upstream junction.

In some simplified models uniform arrivals at upstream junction are assumed. So doing the shape of the platoon entering the link does not depend on the offsets on links leading to upstream junction. In the most simplified model (see Gartner et al., 1975) a rectangular shape is assumed.

In a more realistic model, named "tadpole" (Huddart and Turner, 1969) for each approach i a rectangular platoon, with height equal to s, is assumed until the time at which the queue vanishes and then a rectangular shape with height equal to the mean arrival flow.

Roberthson (1969) assumes a shape of platoons rectangular with height equal to s, until t_0; equal to the arrival pattern after t_0.

<u>Dispersion</u>. The dispersion of a platoon moving along a link depends on the differences of vehicles performance and drivers' behaviour.

Pacey (1956) assumed that each vehicle in the platoon had a constant speed, and that such speeds were normally distributed. This model can give reasonable prediction of platoon dispersion but it use comparatively more complex formulae.

Roberthson (1969) developed a prediction method based on a recurrence relationship. It furnishes the dispersed shape of a platoon at any given distance from the entering line of a link. The method is computationally simple and is able to give realistic prediction of vehicle behaviour (Seddon, 1972).

Let:
$q_1[i]$ = departure flow in interval i at upstreem junction;
$q_2[i+t]$ = predicted arrival flow in interval (i+t) at the downstreams stop line;
t = 0.8 times the average journey time over the distance (measured in the same time intervals used for q_1 and q_2;
π = smoothing factor.

The recurrence relationship indicated by Robertson can be written:

$$q_2[i+t] = \pi \cdot q[i] + (1 - \pi) \cdot q_2[i+t-1]$$

An interesting property of Roberthson's method is that, in any point of the link, the shape of a platoon resulting from some platoons, merging at upstream junction, can be obtained by summing the dispersed shapes of these platoons.

Gartner et al. (1975) used an empirical factor, depending on distance, which increases the lenght of the platoon and decreases its height.

Figure 9 shows the effect of distorsion and dispersion on a delay-offset relationship calculated by Wagner's method.

<u>Division</u>. The division is a complex phenomenon that occurs when the platoon moves from the upstream to the downstream junction in dependence of the different paths choosen at downstream junction by users.

In current practice, constant turning percentages are used to divide the platoon arriving at downstream junction in some derived platoons with the same shape.

It is obvious that division can generate platoons with different shapes, but a more realistic model should require the knowledge of the actual paths used between each o/d pair.

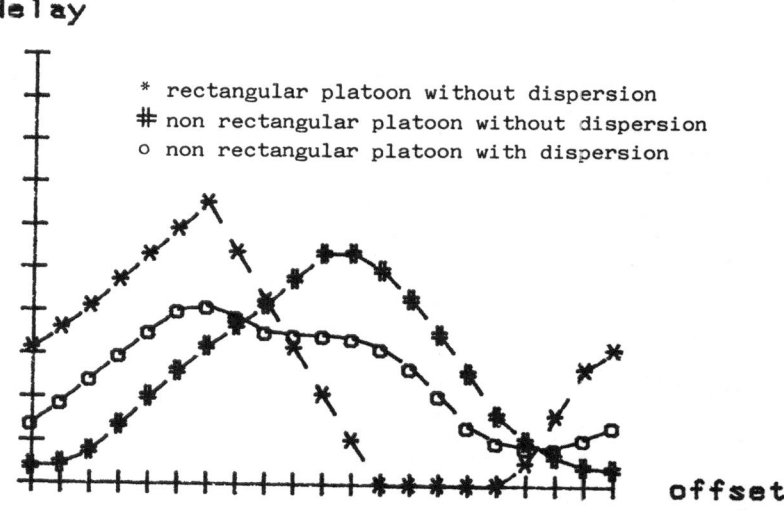

Figure 9 - Comparison among delay-offset relationships.

2.3.6 <u>The heuristic approach to the coordination problem</u>

As indicated in 2.3.2, several mathematical programming based methods for the removal of the signalised network condensability hypothesis can be profitably used.

Some criticism of the mathematical approach in network coordination can however be expressed. The principal one concerns the possibility of removing the second fundamental hypothesis of combination, also used in disutility oriented mathematical programming methods, i.e. the independence of delay to traffic along a link, of traffic signals on other links.

It is clear, from previous subsection, that the platoon shape along each link is a function of the regulation adopted for the upstream junction, but also of the regulations adopted on the links feeding the junction itself and so on going back along a path.

The delay on a link is, on the other hand, a function of the characteristics of the signals on the link and of the platoon shapes. Therefore the delay on a link depends also on the signal characteristics fixed on other links.

At present there are no mathematical approaches to the coordination problem able to model this situation.

The difficulty of expressing in a mathematical form the complex dependence existing between delay (or a more general link performance function) on a link and network control variables, suggested heuristic methods in wich the motion of cars in the network is somehow simulated.

Among these, the TRANSYT method (Robertson, 1969) is surely the most widely used in practical applications.

The concept of cyclic flow profile is fundamental to the development of TRANSYT (see also Vincent et al., 1980; Morton, 1985), in which all calculations are carried out by manipulations of flow profiles.

TRANSYT operates on the following assumptions : all major junctions in the network are controlled by signals or priority rules ; all signals have a common cycle time or a multiple of it ; the pattern of traffic is known and constant.

TRANSYT realized a "hill-climbing" procedure for minimizing a weighted sum of delay and stops on the network.

The method ensures the achievement of a good solution, but not necessarily the global optimum.

3. Combined traffic assignment and signal optimization

Heuristic methodologies, like TRANSYT, are useful to remove the second hypothesis on the base of combination and of mathematical programming based methods, but they are not able to remove the hypothesis of fixed pattern of traffic.

As already indicated in section 1, it is not generally true to assume traffic pattern and, particularly, flow distribution, as being independent of the adopted control.

Control signal system can change travel costs and delay at junctions, and then can influence traffic equilibrium on the network.

Let us consider a traffic network for which the origin-destination demands are assumed known. We assume true Wardrop's 1st principle (Wardrop, 1952) for which each driver chooses his path having in mind the minimization of his own generalized travel cost.

According to this "user optimum" principle, each link of the network will be charged by a resulting flow. The set of flows will be consistent with an equilibrium condition, so that no driver will be able to change his path improving his own travel cost.

Using this set of traffic flow values we could calculate a "system optimal" traffic control strategy i.e. a control strategy having the minimization of the total cost as objective. The transition from the previous to the new traffic control scheme could change the set of costs of users on links (road links and manoeuvres at junctions).

Reusing the user-optimum choice principle, we can obtain a new set of flows corresponding to the new equilibrium point consistent with the new set of costs. If the new set of flows is different from the previous one, a new system optimal traffic control strategy could be calculated and so on.

This recursive process will terminate only if a mutually consistent set of flows and control strategy are found, i.e. a set of traffic flows for which a regulation is calculated which in turn generates the same flow distribution.

Using this example we can underline some interesting points.

The regulation calculated using a fixed pattern of traffic is not necessarily optimal and not necessarily consistent with traffic flows. In the second case two different possibilities can occur: the first regulation calculated is not changed, hence a collective benefit lower than expected will be obtained; an iterative process of updating traffic control to the successive configurations of traffic flows is performed. In this case two possibilities hold. A mutually consistent point is not achieved and, in theory, the process of updating will continue indefinitely. A mutually consistent point is achieved but the collective benefit is not necessarily better than the initial one.

3.2 Mathematical formulation of network equilibrium

A transportation network can be represented by an oriented graph $G(N,A)$ where N is the set of nodes and A the set of links.

It is assumed, in general, that the unitary cost c_{ij} on a link ij can be expressed by a nondecreasing function of the flow f_{ij}. The travel demand D_{od}, for an o/d pair can be a decreasing function of the total cost C_{od} (elastic demand), or independent of cost (fixed demand).

Following Wardrop's equilibrium principles, Beckman et al. (1956) formulated the mathematical model of the network equilibrium. The set of constraints includes flows conservation, additivity and non-negativity conditions.

Assuming, for simplicity, that signal control does not affect the origin destination demand, we will consider only the case of fixed demand, for which Wardrop's, Ist principle, equilibrium corresponds to the integral cost function minimization:

$$z' = \Sigma_{ij} \int_o^{q_{ij}} c_{ij}(q) \, dq \qquad MIN!$$

and Wardrop's 2nd principle is equivalent to the total cost function minimization :

$$z'' = \Sigma_{ij} \, q_{ij} \, c_{ij}(q_{ij}) \qquad MIN!$$

Beckman et al. (1956) also formulated the existence, uniqueness and stability conditions of equilibrium.

In the model link cost c_{ij} is assumed as function only of flow f_{ij} on the link ij. In this way the Jacobian matrix $(\delta c_{ij}/\delta f_{hk})$ has non zero elements only on the leading diagonal and the model is indicated as a separable equilibrium model.

This assumption is not general. In urban networks, e. g., the cost on a link also depends on the delay at downstream junction and hence on the flows on the other links approaching the same junction. On the other hand, as indicated in subsection 2.3.6, for a signalized network the delay on a link is also a function of the signal characteristics fixed on other links.

Dafermos (1971) introduced a more general formulation in which the link cost is a function of the flow pattern on the network and the Jacobian is not diagonal but still symmetric (symmetric equilibrium model).

Smith (1979a,b) and Dafermos (1980), proposed a general equilibrium model for which the Jacobian is asymmetric and the model is called an <u>asymmetric equilibrium model</u>.

Lafermos (1980), Smith (1979a,b) Smith (1981a,b; 1982,1983,1985), Heydecker (1983) investigated about the conditions of existence, uniqueness and stability of the equilibrium of the asymmetric models.

3.2.1 Cost-flow relationships

An urban signalized traffic network can be expressed by an oriented graph, more detailed than in the case of a transportation network.

Each junction is represented by a subgraph having a number of nodes equal to the number of input-output points of the junction and a number of links equal to the number of possible manoeuvres.

Hence the set of links of such a graph (see figure 10) includes the subset of link roads and the subset of manoeuvres at junctions.

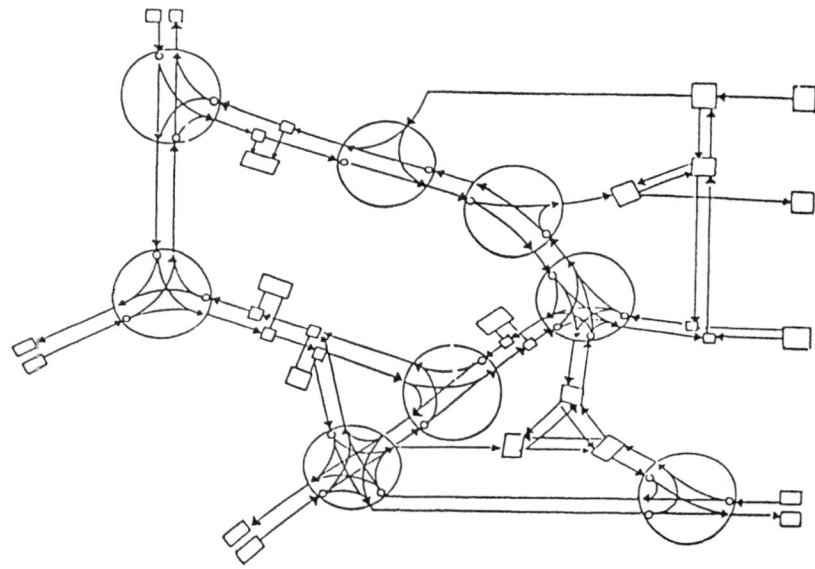

Figure 10 - Example of graph associated to an urban signalized traffic network.

For each link belonging to the first subset a speed-flow (or [travel time]-flow) function must be defined. For each link belonging to the second subset a delay-[control parameters]-flow function must be defined.

The functions most widely used in practical applications derive from both theoretical and empirical results. Therefore, they usually contain some parameters to be estimated from data collections in the study area (*).

Travel time on a road link

Let:

q   = the flow on a link;
Q   = the maximal flow on a link;
α = calibration parameter to be estimated from data collections.

The travel time t(q) on a road link can be obtained using the Davidson's formula (Taylor, 1984):

$$t(q) = t_o[1 + \alpha q/(Q - q)]$$

where t_o is the free flow travel time.

Delay at junction approach

If the mutual effect of adjacent junctions is negligible two term formulae can be used to evaluate delay d(q) at a single junction.

The first term expresses the component of delay due to uniform arrivals, the second accounts for the random fluctuations.

Let :

Γ = g/c the effective green expressed as a proportion of the cycle time;
s   = the saturation flow of the approach;
Q   = s.Γ the maximum number of vehicles that can cross the stopline in one cycle;
α,β = calibration parameters to be estimated from data collections.

The unitary delay can be expressed by:

$$d(q, c, \Gamma) = \alpha.[c(1-\Gamma)^2]/2 + \beta.q/[Q(Q-q)] \quad (**) \ (***)$$

In this way, if the cycle time and the green times are

(*)   For a general review on cost-flow functions see: Branston (1976), Boyce et al. (1981) for link roads; Wormleighton (1965), Allsop (1969), Hutchinson (1972), Catling (1977), Vincent et al. (1980) for junction approaches.
(**)   The well-known Webster's formula (Webster, 1958) can be obtained setting : $\alpha = 9/10 (1 - q/s)$, $\beta = 9/20$.
(***) In Doherty's expression (Catling, 1977) $\alpha = 1$ and β must

be estimated from data collections.
fixed, a delay-flow function is defined and can be used in
traffic equilibrium models. On the other hand, if the arrival
flow is fixed a unitary delay function is defined for control
variables optimization.

If the junctions are so close that the interactions cannot
be assumed negligible, the unitary delay for an approach can be
evaluated as the sum of two terms expressing respectively
deterministic and stochastic delay.

The first term, $d_d(q)$, depends on the arrival pattern,
that is on the shape of the platoons. It can be evaluated from
the queue trend and the departure law (see for example Allsop,
(1969)).

The stochastic delay $d_s(q)$ can be computed using formulae
derived from queue theory (Wormleighton, 1965) or simplified
expressions, like that implemrented in TRANSYT (Vincent et al.,
1980):

$$d_s(q,\Gamma) = T/4 \left\{ [(q-Q)^2 + 4q\epsilon/T]^{\frac{1}{2}} + (q-Q) \right\}$$

with:

$\epsilon = 1$ (vehicles)
$T = $ period considered in the computation

It is important to observe that many proposed cost – flow
functions (for link road and junction approaches) tend
asimptotically to infinity as q approaches Q. Can be useful to
assume a suitable modification to avoid computational
difficulties.

Let $c'(q)$ be the first derivative of $c(q)$. Following
Taylor (1984) a value $\delta < 1$ can be choosen and a new function
$c^*(q)$ can be adopted:

$$c^*(q) = \begin{cases} c(q) & q \leq \delta Q \\ c(\delta Q) + c'(\delta Q)(q-\delta Q) & q > \delta Q \end{cases}$$

It should be noted that δ becomes a calibration parameter
of the function.

3.3 Combined assignment and signal setting as a two level optimization problem

According to the statement introduced by Gartner et al.
(1980) for a general urban traffic management problem, the
combined assignment and signal optimization can be defined as a
two level optimization problem. The first level concerns the
decision of a public manager who estabilishes the control
system to be adopted, the second one concerns the behaviour of

the users in the choice of their path.

The manager's criteria for evaluating his decision are based on the public interest, the individual trip-maker is assumed to minimize only his own travel costs.

Thus the model consists of the optimization of a function of the flows and control parameters, expressing an evaluation of user total cost (system optimum). Flow distribution must guarantee the minimization of user travel cost for each assigned regulation (user optimum).

Signalized network equilibrium is therefore defined by control variables that minimize the collective cost (Wardrop's second principle - normative system) together with a flow pattern which satisfies Wardrop's first principle (descriptive system).

The solution of the two level optimization problem finds the minimum total cost solution among all those which respect Wardrop's first principle.

Assuming c_{ij} to be a function of flow q_{ij} on link ij and vector $\underline{p}$ of network control system parameteres, the total travel cost can be expressed by:

$$T_1(\underline{p},\underline{q}) = \Sigma_{ij} \; q_{ij}c_{ij}(q_{ij},\underline{p}) \qquad \text{where } \underline{q} = \{q_{ij}\}$$

The integral travel cost, whose minimization gives an user optimal flow pattern, is expressed by:

$$T_2(\underline{p},\underline{q}) = \Sigma_{ij} \int_0^{q_{ij}} c_{ij}(t,\underline{p}) \; dt$$

The two-level objective is therefore :

$$\begin{array}{c} \text{MIN} \\ \underline{p} \end{array} \{T_1(\underline{p},\underline{q}^*) : [T_2(\underline{p},\underline{q}^*) = \begin{array}{c} \text{MIN} \; T_2 \\ \underline{q} \end{array} (\underline{p},\underline{q})]\}$$

subject to conservation, additivity and non negativity constraints on flows and other constraints on regulation parameters.

3.4 <u>Solution of two-level optimization model</u> (*)

It is obvious that only a simultaneous optimization procedure can ensure the optimum of two-level, signal setting - assignement, problem.

This approach is used by Tan et al. (1979) and Marcotte (1983). The proposed models are based on the limitative

----

(*) For a more detailed information see Cantarella and Sforza

(1986).
hypothesis that the cycle time, the stage matrices at junctions
and the offsets are preliminary fixed. In this way the actual
decisional variables are only the green splits.

The problem can be alternatively approached to obtain
mutually consistent sets of flows and signal setting, by
recursive, [signal setting] - assignment, procedures somehow
simulating the iterative decisional process outlined in
section 3.

Allsop (1974), first indicated the possibility of using an
iterative procedure to obtain an improvement of the signalized
network performance expressed by a decrease in user's delay.

Successively Allsop and Charlesworth (1977) and
Charlesworth (1975,1977) introduced the computation of a
mutually consistent flow pattern and control policy for a set
of coordinated junction by a procedure consisting of: i)
initial assignment; ii) signal setting to minimize delay at
each junction; iii) estimation of cost-flow relationships
(using the module of simulation of TRANSYT); iv) equilibrium
assignment by TRAFFIC method (Nguyen, 1976); v) new signal
setting by TRANSYT; vi) repetition of steps iii),iv),v) until
no change in signal timing or negligible change in traffic
assignment occur.

Gartner et al. (1980) proposed two iterative procedures,
the first using the Webster's method and an equilibrium
assignment, the second based on MITROP method to calculate all
control variables and on an equilibrium assignment.

Cantarella et al. (1983) used an iterative procedure to
solve a problem in practice (see also Improta, 1985). An
interesting improvement of this procedure can be found in
Cantarella et al. (1986).

As indicated by Dickson (1981), the solution obtained
by iterative methods can be not optimal and the network total
cost, assumed as performance index, can increase during the
procedure.

4. Concluding remarks

Mathematical programming methods, for urban network
control, can be grouped in four classes, corresponding to
problems having increasing complexity:

- single junction control,
- coordination and synchronization of arterials,
- coordination and synchronization of network,
- simultaneous traffic assignment and signal setting.

Referring to Improta and Cantarella (1984), Cantarella and

Improta (1987) for an extended state of the art and opportunities of research for mathematical programming methods in single junction control, some research suggestions in the field of urban network control are given here.

The evolution of traffic signal coordination models can be thought as derivying from combination method by removing its limitative assumptions.

Mathematical programming based methodologies are able to calculate also uncondensable networks, removing in this way the first limitative hypothesis of combination.

On the contrary, at present, only, heuristic procedures are able to remove the second limitative hypothesis of combination, concerning the independence of delay on a link on characteristics of signal control on other links.

The link performance functions used in mathematical programming methodologies are based on the assumption that the delay on a link depends only on the offset between signals at upstream and downstream junctions. This hypothesis is not true in general. It is then necessary to define new formulations expressing the delay as a more general function of network regulation parameters.

Mathematical programming based and heuristic methodologies for network control are developed on the hypothesis of flow subsaturation. It is very frequent in urban networks, mainly in rush periods, that temporary oversaturations occur. This consideration indicates the necessity of further research: i) for defining adeguate models of flow distribution in transient regimes; ii) for formulating mathematical programming based control strategies able to limit the oversaturation effects.

Mathematical programming and heuristic methodologies are used for off-line calculations. Some resolutive algorithms proposed, in recent years, for mathematical programming methods have encouraging characteristics of computational efficiency. Experiments in the use of these methods for real time control should be very interesting.

Both, mathematical programming based and heuristic methodologies, are at present not able to remove the third hypothesis of combination method, and suppose a fixed pattern of traffic.

Simultaneous optimization methods, for combined traffic assignment and signal timing, are based on very limitative hypothesis and furthermore cannot be applied to real size networks. On the other hand iterative procedures can be applied to real, medium size, networks but not ensure the achievement of the global optimum.

In any case, the cost on a link of urban signalized networks is certainly function of the regulations adopted on other links and of the flows using other links. It is necessary

therefore to use methodologies for asymmetric traffic
equilitrium able to treat practical networks in reasonable
running times.

Ccst flow functions at present used, derive from rural
experiences and, hence, refer to uninterrupted flow conditions.
It is important to verify their validity for urban networks an
eventually to propose new better formulations.

REFERENCES

Allsop, R.E. (1968). Choice of Offsets in Linking Traffic
    Signals. Traff. Engng & Con.,9, 73-75.
Allsop, R.E. (1969). An Analysis of Delays to Vehicle Platoons
    at Traffic Signals. Fourth Int. Symp. on Theory of Traffic
    Flow, Karlsruhe, June 1968. Strassenbau und
    Strassenverkehrstechnik, 86, 98-104.
Allsop, R.E. (1971). Delay Minimizing Setting for Fixed Time
    Traffic Signals at a Single Road Junction. J. Inst. Maths
    Applics,8, 164-185.
Allsop, R.E. (1972). Estimating the Traffic Capacity of a
    Signalized Road Junction.Transpn Res.,6, 345-355.
Allsop, R.E. (1974). Some Possibilities for Using Traffic
    Control to Influence Trip Destination and Route Choice.
    Transportation and Traffic Theory. Proc. Sixth Int. Symp. on
    Transportation and Traffic Theory, Elsevier, Amsterdam, 345-
    374.
Allsop, R.E. (1983). Optimization of Timings of Traffic
    Signals. Proc. 1983 AIRO Conf. Guida Editori, Napoli, 103-
    120.
Allsop, R.E. and Charlesworth, J.A. (1977). Traffic in a Signal
    Controlled Road Network : An Example of Different Signal
    Timings Inducing Different Routing. Traff. Engng & Con.,18,
    262-264.
Beckman, M.J., Mcguire, C.B. and Winsten, C.B. (1956). Studies
    in the Economics of Trasportation. Yale University Press,
    New Haven, CT.
Boyce, D.E., Janson B.N. and Eash R.W. (1981). The Effect on
    Equilibrium Trip Assignment of Different Link Congestion
    Functions. Transpn Res.,15A, 223-232.
Branston, D. (1976). Link Capacity Functions: A Review. Transpn
    Res.,10, 223-236.
Cantarella, G.E. and Improta, G. (1981). Una Metodologia per il
    Progetto Globale di Intersezioni Semaforizzate. Atti delle
    Giornate di lavoro AIRO 1981, Torino, 33-48.
Cantarella, G.E. and Sforza, A. (1986). Methods for Equilibrium
    Network Traffic Signal Setting. To appear in Proc. of
    Advanced Workshop on : Flow Control of Congested Networks
    (A. Odoni, G. Szego eds.). Capri, october 12-18, 1986.
Cantarella, G.E. and Improta, G. (1987). Capacity Factor and
    Cycle Time Optimization: A Graph Theory Approach, Transpn
    Res.,B, (to appear).
Cantarella, G.E., Cosentino, M., Improta, G., Sforza, A.

(1986). Equilibrium Network Control System Design. Preprints 5th IFAC - IFIP - IFORS Int. Conf. on Control in Transportation Systems,Wien, 8-11 July, 1986, 395-403.

Cantarella, G.E.,Improta, G., Sforza, A., Sorrentini, M.C. (1983). Flow Equilibrium for a Signalized Urban Network. An Application. Proc. 1983 AIRO Conf. Guida Editori Napoli, 821-845.

Catling, I. (1977). Development of Junction Delay Formulae. LTR 1. Working Paper 2.

Charlesworth, J.A. (1975). Relations between Travel-Time and Traffic Flow for the Links of Road Networks Controlled by Fixed-Time Signals. Univ. of Newcastle upon Tyne. Transport Oper. Res. Group, Research Report n. 13.

Charlesworth, J.A. (1977). Mutually Consistent Traffic Assignment and Signal Timings for a Signal-Controlled Road Network. Univ. of Newcastle upon Tyne. Transport Oper. Res. Group, Working Paper n. 24.

Dafermos, S.C. (1971). An Extended Traffic Assignment Model with Applications to Two-Way Traffic. Transpn Sci., 5, 366-389.

Dafermos, S.C. (1980). Traffic Equilibrium and Variational Inequalities Transpn Sci.,14, 42-54.

Dickson, T.J. (1981). A Note on Traffic Assignment and Signal Timings in a Signal Controlled road network. Transpn Res., 15B, 267-271.

Gallivan, S. (1984). Flow Profiles in Coordinated Networks. Advanced Course on Optimization for Traffic Engineering and Control. AIRO, P.F.T.- C.N.R.,Capri october 18-25, 1984.

Gallivan, S. and Heydecker, B.G. (1983). Optimising the Control Performance of Traffic Signals at a Single Junction. 15th Ann. Conf. Universities Transport Study Group, Imperial Coll. London (unpublished).

Gartner, N.H. (1972a). Constraining Relations among Offsets in Synchronized Networks. Transpn Sci., 6, 88-93.

Gartner, N.H. (1972b). Optimal Synchronization of Traffic Signal Networks by Dynamic Programming. Proc. 5th. Int. Symp. on Theory of Traffic Flow and Transportation. (Ed. G.F. Newell). American Elsevier, New York, 281-295.

Gartner, N.H. (1973). Microscopic Analysis of Traffic Flow Patterns for Minimizing Delay on Signal-Controlled Links. HRB Record, 445, 13-15.

Gartner, N.H. (1976). Area Traffic Control and Network Equilibrium. Traffic Equilibrium Methods (Ed. M.Florian). Springer-Verlag, Berlin, 274-297.

Gartner, N.H. and Gershwin, S.B. (1983). Analitycal Models for Transportation System Management. Proc. 1983 AIRO Conf. Guida Editori Napoli, 793-804.

Gartner, N.H. and Little, J.D.C. (1975). Generalized Combination Method for Area Traffic Control. Transpn Res. Rec., 531, 58-59.

Gartner, N.H., Little, J.D.C., and Gabbay, H. (1975). Optimization of Traffic Signal Settings by Mixed-Integer Linear Programming. Part I: The Network Coordination Problem; Part II: The Network Synchronization Problem, Transpn Sci.,9, 321-363.

Gartner, N.H., Little, J.D.C. and Gabbay, H. (1976). Simultaneous Optimization of Offsets, Splits and Cycle Time.

Transpn Res. Rec., 596, 6-15.

Gartner, N.H., Gershwin, S.B., Little, J.D.C. and Ross, P. (1980). Pilot Study of Computer-Based Urban Traffic Management. Transpn Res., 14B, 203-217.

Gazis, D.C. (1974). Traffic Science. Wiley., New York, USA.

Heydecker, B.G. (1983). Some Consequence of Detailed Junction Modeling in Road Traffic Assignment. Transpn Sci., 17, 263-281.

Heydecker, B.G. and Dudgeon, I.W. (1987). Calculation of Signal Settings to Minimize Delay at a Junction. To be published on Proc. Thenth Int. Symp. on Transportation and Traffic Theory. Cambridge Mass. july 8-10th 1987.

HCM (1985). Highway Capacity Manual. Transportation Research Board., National Research Council Special Report 209.

Hillier, J.A. (1966). Appendix to Glasgow's Experiment in Area Traffic Control. Traff. Engng & Con., 7, 569-571.

Hutchinson, T.P. (1972). Delay at a Fixed Time Traffic Signal - II: Numerical Comparisons of Some Theoretical Expressions. Transpn Sci., 6, 286-305.

Huddart, K.W. and Turner, E.D. (1969). Traffic Signal Progerssion G.L.C. Combination Method. Traff. Engng & Con., 11, 320-327

Improta, G. (1978). Un Contributo sul Problema della Capacita' per le Intersezioni a Raso Semaforizzate con Sistema Fixed-Time. Atti del XVIII Cong.Naz. Stradale, Taormina, Italy, 349-356.

Improta, G. (1985). Traffic Engineering and Perspectives of Transportation Planning in the Region of Naples. The Practice of Trasportation Planning (Ed. M. Florian). Elsevier Science Publishers, Amsterdam, 161-209.

Improta, G., Allsop, R.E. and Heydecker, B.G. (1986). Network Models for Traffic Management. To be published on Proc. of "The Management and Planning of Urban Transport Systems from Theory to Practice" on International Seminar Montreal, august 25-29, 1986, (Ed. M. Florian).

Improta, G. and Cantarella, G.E. (1982). Signalized Junction Control System Design. EURO - TIMS XXV, Lausanne, july 1982 (unpublished).

Improta, G. and Cantarella, G.E. (1984). Control System Design for an Individual Signalized Junction. Transpn Res., 18B, 147-167.

Improta, G. and Sforza, A. (1982). Optimal Offsets for Traffic Signal Systems in Urban Networks. Transpn Res., 16B, 143-161.

Little, J.D.C. (1966). The Synchronization of Traffic Signals by Mixed-Integer Linear Programming. Opns Res., 14, 568-594.

Little, J.D.C., Martin, B.V., Morgan, J.T. (1966). Synchronizing Traffic Signals for Maximal Bandwidth. Highway Res. Rec., 118, 21-45.

Little, J.D.C., Kelson, M.D., Gartner, N.H.(1981). MAXBAND: A Program for Setting Signals on Arteries and Triangular Networks. Transpn Res. Rec., 795, 40-46.

Marcotte, P. (1983). Network Optimization with Continuous Control Parameters. Transpn Sci., 17, 181-197.

Moller, K. (1987). Calculation of Optimum Fixed-Time Signal Programs. To be published on Proc. Thenth Int. Symp. on Transportation and Traffic Theory. Cambridge Mass. july 8-

10th 1987.

Morton, G. (1985). The TRANSYT and SCOOT Methods of Traffic Control. Based on presentations given by D.I. Robertson. Advanced Course on Optimization for Traffic Engineering and Control. Capri, 18-25 october 1984.

Nguyen S. (1976). A Unified Approach to Equilibrium Methods for Traffic Assignment. Traffic Equilibrium Methods (Ed. M. Florian), Springer-Verlag, Berlin, 148-182.

Pacey, G.M. (1956). The Progress of a Bunch of Vehicles Released from a Traffic Signal. Road Res. Note No. RN/2665/GMP.

Robertson, D.I. (1969). TRANSYT Method for Area Traffic Control. Traff. Engng & Con., 11, 276-281.

Robertson, D.I. (1982). Research on Strategies for Traffic Control and Management. ARRB Proc., vol. 11, part.1, 83-89.

Seddon, P.A. (1972). The Prediction of Platoon Dispersion in the Combination Methods of Linking Traffic Signals. Transpn Res., 6, 125-130

Smith, M.J. (1979a). Traffic Control and Route-Choice: a simple example. Transpn Res., 13B, 289-294.

Smith, M.J. (1979b). The Existence, Uniqueness and Stability of Traffic Equilibria. Transpn Res., 13B, 295-304.

Smith, M.J.(1981a). The Existence of an Equilibrium Solution of the Traffic Assignment Problem when there are Junction Interactions. Transpn Res., 15B, 443-451.

Smith, M.J.(1981b). Properties of a Traffic Control Policy which Ensure the Existence of a Traffic Equilibrium Consistent with the Policy. Transpn Res., 15B, 453-462.

Smith, M.J. (1982). Junction Interaction and Monotonicity in Traffic Assignment. Transpn Res., 18B, 1-3.

Smith, M.J. (1983). The Existence and Calculation of Traffic Equilibria. Transpn Res., 17B, 291-303.

Smith, M.J. (1985). Traffic Signals in Assignment. Transpn Res., B, 19B, 155-160.

Stoffers, K.E. (1968). Scheduling of Traffic Lights - A New Approach. Transpn Res., 2, 199-234.

Strobel, H. (1982). Computer Controlled Urban Transportation - A Survey of Concepts, Methods and International Experiences. IIASA.

Tan, H., Gershwin, S.B. and Athans, M. (1979). Hybrid Optimization in Urban Traffic Networks. Report No. DOT-TSC-RSP- 79-7, MIT Cambridge, MA.

Taylor, M.A.P. (1984). A Note on Using Davidson's Function in Equilibrium Assignment. Traspn Res., 18B, 181-199.

Vincent, R.A., Mitchell, A.I. and Robertson, D.I. (1980). TRANSYT Version 8. Road Res. Lab. Report , LR 80, Crothorne, Berkshire.

Wagner, F.A., Gerlough, D.L. and Barnes, F.C. (1969). Improved Criteria for Traffic Signal System on Urban Arterials. N.C.H.R.P. Report 73, Highway Research Board.

Wardrop, J.G. (1952). Some Theoretical Aspects of Road Traffic Research. Proc. Instn Civil Engnrs, Part. II, 1, 325-378.

Webster, F.V. (1958). Traffic Signal Setting. Road Research Technical Paper, 39, HMSO, London.

Webster, F.V. and Cobbe, B.M. (1966). Traffic Signals. Road Research Technical Paper, 56, HMSO,London.

Wormleighton, R. (1965). Queues at a Fixed Time Traffic Signal with Periodic Random Input. C.O.R.S. Journal, 3, 129-141.

Yagar, S. (1974). Capacity of a Signalized Road Junction. Transpn Res., B, 6, 137-147.

Zuzarte Tully, I.M. (1977). Syntesis of Sequences for Traffic Signal Controllers Using Techniques of the Theory of Graphs, Ph.D. Thesis, University of Oxford Engineering Laboratory Report OUEL, 1189/77.

Zuzarte Tully, I.M. and Murchland, J.D. (1978). Calculation and Use of the Critical Cycle Time for a Single Traffic Controller. PTRC Summer Annual Meet, Proc., PTRC-P152, 96-112.

METHODS FOR EQUILIBRIUM NETWORK TRAFFIC SIGNAL SETTING

G.E. Cantarella, A. Sforza

Dipartimento di Informatica e Sistemistica
Universita' degli Studi di Napoli
via Claudio 21, 80125 Napoli, Italy
tel. 081/7683377

Abstract

Traffic signal setting as a tool for traffic management of urban networks is introduced. A brief state of art about Traffic Signal Setting and Network Equilibrium Assignment is reported. Then Equilibrium Network Traffic Signal Setting is analyzed together with the principal methods for its solution. Global optimization and iterative procedure approaches are described in detail. Finally, after some theoretical remarks, research perspectives are outlined.

Keywords.

Traffic signal setting, flow equilibrium, network design, urban networks

Work partially supported by
Progetto Finalizzato Trasporti,
Consiglio Nazionale delle Ricerche, Italy,
Grant no. 85.00157.93.

INTRODUCTION

Traffic signal setting is an important tool for traffic management of urban networks. It is well placed between the classical problems of traffic assignment and network design. Indeed, delays at junctions, and therefore cost-flow functions, depend on signal setting. So, a change in the control policy has a redistributional effect on flows until a new stationary flow pattern is reached. On the other hand, the design of the control system is a way to set capacities of network links, and so it can be classified as a network design problem.

These considerations account for the increasing interest of many researchers in interfacing traffic signal setting and assignment, evaluating theoretical and practical aspects of models and related algorithms.

The equilibrium network traffic signal setting requires a great number of input parameters:

i)   travel demand,
ii)   network topology,
iii) cost-flow function for each link,
iv)   crossing incompatibilities among movements, for each junction,
v)   saturation flow, lost time, minimum green time, maximum red time and maximum queue length, for each approach.

The decisional variables, describing the flow pattern and the signal setting, are the following:

i)   flow on each link,
ii)   cycle time for each junction,
iii) green timing and scheduling (or stage matrix) for each junction
iv)   offset for each pair of adjacent junctions.

For the time being, solution algorithms for the

simultaneous computation of all the above mentioned variables are not available. However, models and procedures optimizing only a subset of these decisional variables, while assuming the other ones as fixed, have been proposed. In the following sections models and algorithms for traffic signal setting and network equilibrium are briefly reviewed. Then equilibrium network traffic signal setting is analysed, together with the principal methods proposed in literature for its solution. Global optimization and iterative procedure approaches are described in detail. Finally, after some theoretical remarks, research perspectives are outlined.

1. TRAFFIC SIGNAL SETTING

Single Junction Signal Setting requires that flows on links leading to the junction are given and no interaction with surrounding junctions occurs. In this way the control parameters (cycle time, green times and schedule) are assumed as decisional variables. After Webster's method (Webster, 1958), some mathematical programming models for green timing, with a fixed stage matrix, were proposed, assuming as objective function total delay minimization, capacity factor maximization or cycle time minimization (Allsop, 1971, 1972; Improta, 1978).

Later, these methods were generalized to calculate the stage matrix (that is stage number and streams having the right to way in each stage) (Zuzarte Tully and Murchland, 1978; Cantarella and Improta, 1981; Gallivan and Heydecker, 1983). More recently, Improta and Cantarella (1984), Cantarella and Improta (1985, 1987) proposed a periodic scheduling model for simultaneous optimization of cycle time and green times and schedule.

The Network Signal Coordination requires the knowledge of flow pattern on the network and signal setting for each junction. In this way the optimal offsets for a network with interacting signalized junctions are assumed as decisional variables. Many authors approached this problem, among others, Hillier (1966), Allsop (1968), Gartner (1972 a,b), Gartner and Little (1975), Improta and Sforza (1982).

Network Signal Syncronization requires as known stage matrix at each junction and network flow pattern and assumes cycle time, green split and offsets as decisional variables.

With the following definitions:

p   vector of control parameters,
P   set of feasible vectors p,
q'   assigned flow pattern,
$d_i(p)$ unitary delay at approach i,
$D(p,q') = \Sigma_i\, d_i(p)q_i'$ total user delay,

a general model can be formulated as:

MIN $D(p)$; $p \in P$

The only mathematical programming method available for its solution, named MITROP, was proposed by Gartner et al. (1975, 1976). It is based on the solution of a mixed integer linear program requiring a great computational effort. Moreover, for each link, a simplified relationship between user delay and offset is adopted. Particularly, only rectangular platoons are permitted and their lenghts must be exogenously defined.

An heuristic approach to the same problem is provided by TRANSYT method, widely used in practical applications for its interesting features in traffic modelling (Robertson, 1969; Vincent et al. 1980).

A state of art about traffic signal setting is reported in Allsop (1983 a) and more recently in Improta (1986).

2. NETWORK EQUILIBRIUM ASSIGNMENT

The seminal works in the field of network equilibrium models are by Beckman et al. (1956) and Dafermos and Sparrow (1968). Travel demand D_{od} for each origin-destination pair is assumed fixed or elastic with the cost C_{od} between o and d. The cost-flow functions on links are supposed known.

In the case of fixed demand, defining:

$I(q)=\Sigma_i \int_o^{q_i} c_i(t)dt$ the integral travel cost

Q the set of the feasible flow patterns

the model can be expressed in the following way:

MIN $I(q)$; $q \in Q$

The solution of this convex programming model is an equilibrium flow pattern consistent with the user behaviour criterion about path choice, as expressed by Wardrop's well-known first principle (Wardrop, 1952).

In the model cost, c_i on link i is assumed as a function only of flow q_i. Therefore the objective function is separable and the Jacobian matrix $||\delta c_i/\delta q_j||$ is diagonal ($\delta c_i/\delta q_j=0$ for $i \neq j$).

The proposed solution algorithms refer to this convex and separable programming formulation of the model with a non linear objective function and linear constraints (Leventhal et al., 1973; Le Blanc et al., 1975; Nguyen, 1976).

This formulation and related solution algorithms can easily be extended to the more general case of symmetrical, but not necessarily diagonal, Jacobian, so defining the symmetric equilibrium model ($\delta c_i/\delta q_j=\delta c_j/\delta q_i \geq 0$). This can happen, for example, for a two-way street (Dafermos, 1971).

Neverthless this formulation is not general enough to cover all the possible cases. Indeed, in urban networks, cost on a link can depend on flows on other links in such a way that the Jacobian is asymmetric ($\delta c_i/\delta q_j \neq \delta c_j/\delta q_i$). Particularly, this can happen for minor streams at a priority junction, for opposed streams at a signalized junction, for a link jointing two adjacent and interacting junctions.

Smith (1979 a,b) developed a general framework for the asymmetric equilibrium model, which can handle link costs defined as a function of the whole flow pattern on the network. It can be formulated as a variational inequality model (Dafermos, 1980), or a non linear complementarity model (Aashtiani and Magnanti, 1981), as well as a convex programming model (Hearn et al., 1984).

Several solution approaches have been developed in the last few years for these asymmetric equilibrium models (Fisk and Nguyen, 1982; Florian and Spiess, 1982; Smith, 1983). Friesz (1985) reports an extended state of art about model formulation and solution algorithms. On the same argument Sheffi (1985) constitutes a very useful text book.

3. EQUILIBRIUM NETWORK SIGNAL SETTING

Traffic signal setting models assume path choice as well as link flows to be independent of the adopted regulation. This can occur when origin-destination paths are strongly constrained and independent of the congestion level , but it is not generally true.

Therefore a set of control parameter values must be chosen taking into account that users, travelling from an origin to a destination, react to the adopted control policy by choosing a new set of paths such that their individual cost is minimized

(Allsop, 1974; Gartner, 1976; Allsop, 1983 b; Fisk, 1984; Smith, 1985).

There are two main approaches to the solution of this problem. The first one refers to the solution of a global optimization model. The second one is based on iterative procedures which successively performs the stages of traffic signal setting and flow equilibrium assignment.

3.1 Global optimization models

This approach leads to continuous network design models (Magnanti and Wong, 1983), in which control parameters play the role of design variables. It should be noted that these models assumes as known parameters cycle time, stage matrix and link offsets. Therefore green times are the only decisional variables. These models require the optimization of an objective function of flows and green times, expressing an evaluation of total travel cost or an other performance index (e.g. fuel consumption or air pollution). Flow pattern must satisfy a user behaviour criterion like the widely used Wardrop's first principle. This requirement can be met in two different ways, reported in the following.

<u>Two level optimization model</u>

In this model green times are the explicit decisional variables at the outer level and link flows are the implicit decisional variables at the inner level.

Defining p,q,P,Q,I as above, and assuming travel cost c_i on link i as a function of flow q_i and control parameter vector p, total travel cost T is expressed by:

$$T(p,q) = \Sigma_i \ q_i c_i (q_i,p)$$

With the above assumptions, for a fixed p, the symmetric

equilibrium conditions are obtained. The two-level model is therefore expressed by:

$$\text{MIN}_p \ (T \ (p, \ q^*); \quad p \in P,$$

where q^* is such that

$$I \ (p, \ q^*) = \text{MIN}_q \ I \ (p, \ q); \quad q \in Q$$

Hence for each objective function evaluation at the outer level for a fixed p, the associated flow pattern is to be computed by performing an equilibrium assignment at the inner level. In this way the user behaviour criterion is expressed by the optimization of a duly defined objective function. In other words an equilibrium network traffic signal setting is performed minimizing a total cost function (Wardrop's second principle) with a flow pattern satisfying a user behaviour criterion (Wardrop's first principle).

This approach appeared in Tan et al. (1979) where the proposed model, defined by authors as a "hybrid optimization" model, is solved via a non linear programming algorithm. It assumes as variables only flows and green split, considering the other control parameters fixed or irrilevant.

One level optimization model

In this model both regulation parameters and link flows are explicit decisional variables. The user behaviour is expressed by a set of constraints for which several formulations were proposed. Among others it is worthwhile to mention Aashtiani (1979), Tan et al. (1979), Marcotte (1983).

The first model (Aashtiani, 1979) uses a non linear complementarity formulation:

$$\text{MIN} \ T(q,p); \quad p \in P, \quad q \in Q$$

$$h = A \cdot q, \quad C = A \cdot c$$

$$\Sigma_i \Sigma_{j \in B_i} (C_j - C_i^*) h_j + \Sigma_i (\Sigma_{j \in B_i} h_j - H_i) u_i = 0$$

$$C_j - C_i^* \geq 0 \qquad \forall \ j \in B_i, \ \forall \ i$$

$$\Sigma_{j \in B_i} h_j - H_i \geq 0 \qquad \forall \ i$$

$$h_j \geq 0 \qquad \forall \ j$$

$$C_i^* \geq 0 \qquad \forall \ i$$

where the following definitions are used:

A   path-link incidence matrix,
h_j flow on path j,
C_j total cost on path j,
B_i set of path for the i-th o-d pair,
H_i flow demand for the i-th o-d pair,
c_i^* cost of the i-th o-d pair at the equilibrium,
u_i Lagrange multiplier associated to i-th o-d pair,

The second model (Tan et al., 1979) can be formulated from the previous one with some manipulations (Fisk, 1984), and using the same definitions of the first model, as

$$\text{MIN } T(p,q); \qquad p \in P, \ q \in Q$$

$$h = A \cdot q, \qquad C = A \cdot c$$

$$C_j \geq (1/H_i) \ \Sigma_{k \in B_i} C_k h_k \qquad \forall \ j, \ \forall \ i$$

This model is solved via an iterative augmented lagrangian method, for avoiding the explicit enumeration of all paths.

The third model was proposed by Marcotte (1983), who adopted a variational inequality formulation for the user behaviour constraints. Denoting by Q' the set of all flow extremal solutions ("all or nothing" flow patterns), and by c(p,q) the link cost vector, the model can be written in the following way:

MIN T(q,p);   q ∈ Q, p ∈ P

c(p,q)·(q'-q)≥0   ∀ q' ∈ Q'

Marcotte (1983) suggested a solution approach which avoid the explicit enumeration of all extremal solutions.

3.2 Iterative Procedures

An iterative procedure finds a feasible solution of the problem solving sequentially an equilibrium assignment and a traffic signal setting until two successive flow patterns or signal settings are equal within a specified tolerance. If this convergence criterion is met in a finite number of iterations the final solution, which is a fixed point solution, can be said mutually consistent. This means that signal setting generates a set of link costs which determines a flow pattern such that signal setting is optimal for it.

Charlesworth (1975, 1977) and Allsop and Charlesworth (1977) introduced the computation of mutually consistent flow pattern and signal setting for a network of interacting junctions. In their procedure traffic signal setting is performed using TRANSYT procedure. The cost-flow functions are estimated evaluating delay costs for different values of flow, with the TRANSYT traffic simulation module, and then ·fitting these points with a polynomial function. The equilibrium assignment is performed using TRAFFIC procedure (Nguyen and James-Lefebre, 1975).

Gartner et al. (1980) proposed two iterative procedures. The first one assumes non interacting junctions. In this way offsets are not decisional variables and signal setting is performed by Webster method (Webster, 1958). In the second one signal setting stage is performed by using MITROP method, so evaluating all the control parameters. For this reason a great computational effort, due to the solution of a mixed integer linear program, is required. In both cases traffic assignment is performed using a technique based on Cantor-Gerla's method (Cantor and Gerla, 1974).

Recently a new iterative procedure has been proposed for equilibrium network traffic signal setting (ENETS, Cantarella, Improta and Sforza, 1985; Cantarella, Cosentino et al.,1986). It is based on methods for traffic signal setting previously developed by the same authors. Particularly, traffic signal setting is performed in two successive steps:
- Green timing and scheduling at each junction,
- Signal coordination on the network.

The green timing and scheduling at a single junction is based on a mixed binary linear program (Improta and Cantarella, 1984). This model is solved by a problem-oriented algorithm (SICCO, Single Intersection Capacity factor or Cycle time Optimization) described in Cantarella and Improta (1987). It should be noted that this method allows also the green scheduling (i.e. the stage matrix) be considered as a set of decisional variables.

Signal coordination for the whole network is performed by solving a discrete programming model with total delay minimization objective function (NETCO, Network Coordination). This model is solved by a Branch and Backtrack algorithm which behaves efficiently since it is well tailored to the model formulation. The details of the model and a first version of the algorithm are described in Improta and Sforza (1982). The iterative procedure ENETS uses a later development of this method not yet published.

The traffic signal setting so performed allows to estimate delay functions for movement links at the junctions. A traffic model, obtained with some simplifications from TRANSYT traffic model, is used to evaluate delay costs for different values of flow. The resulting points are fitted by a generalized Webster two term formula (Webster, 1958).

The flow assignment stage (NETEQ, Network Equilibrium) refers to the classical symmetric equilibrium model with fixed demand, and uses the Frank-Wolfe algorithm (Frank and Wolfe, 1956) or the Simplex Convex algorithm (Zangwill, 1967), both adapted to network flow problem.

Tests on simple networks suggested that ENETS would satisfactory behave even on real size networks. However the procedure needs to be further experienced with a set of randomly generated networks in order to achieve a detailed knowledge of its practical convergence. These experiments and the developments of models and algorithms for each component of the procedure constitute the base of future work.

Some well-known characteristics about the convergence of an iterative procedure should be reported. Some authors (Charlesworth, 1977; Tan et al., 1979) noted that the final solution strongly depends on the initial assignment. So the procedure does not necessarily converge to the optimal solution, as noted by Tan et al. (1979) with three simple examples. Moreover Dickson (1981) showed that the network total cost assumed as a performance index, can increase during the procedure, so leading to a non optimal solution.

On the other side, it is worthwhile to underline the modular structure of an iterative procedure. This feature allows to use methods already developed for solving the different problems composing the general one, updating a single component if a new method is proposed for its solution, or even changing a single model if a new approach is proposed for it.

From this viewpoint an interesting line of research can be devoted to find more sophisticated delay-offset functions in which delay on a link generally depends on control parameters and flow pattern of the whole network. In this case the need for an asymmetric equilibrium model in the assignment stage arises.

It should be noted also that the number of decisional variables is a larger set than permitted by one of the above described methods for solving a global optimization model, comprising green timing and scheduling and offsets between adjacent junctions. Furthemore it is useful to remark that an iterative procedure can be applied even for solving real size problems.

3.3 Some theoretical remarks

The solution of a global optimization model is defined by a set of control parameters minimizing the system total cost, with a flow pattern satisfying a user behaviour criterion.

A lower bound for the value of this optimal solution can be found solving a model in which constraints expressing the assumed user behaviour are eliminated. In other words this means that users would adopt paths indicated by the traffic management agency for the system optimization, instead of choosing paths minimizing their individual cost.

This model can easily be solved by using an iterative procedure in which the assignment stage is performed minimizing the total cost instead of the integral one and obtaining the so-called normative flow assignment. The solution of this procedure is optimal because the same objective function is adopted in the two stages.

Generally this solution (flow pattern and signal setting) is not a feasible solution of the original model built for equilibrium network traffic signal setting, since the set of user behaviour constraints has been removed.

On the other side, an upper bound is easily defined by finding a feasible solution of the problem. This task can be accomplished by performing an equilibrium assignment in which the cost-flow functions are derived from the signal setting relative to the previous defined lower bound.

The above considerations furnish useful indications to approach the problem of choosing the starting solution of an iterative procedure, because as noted, the final solution strongly depends on it. Therefore it seems useful the starting solution be univocally defined. Flow pattern and signal setting relative to the previous defined upper bound seems to be a good choice as well as an equilibrium flow pattern obtained assuming no delay at the junctions. At this aim experimental tests are needed to furnish better indications.

Some interesting theoretical works about the existence of equilibrium solutions for signalized networks appeared in recent literature. Particularly M.J. Smith produced many papers in this field, based on his definition of equilibrium and stability (Smith, 1979 a,b) and on the conditions which guarantee existence, unicity and stability of the equilibrium solution.

Smith's theoretical conclusions are important in order to estimate flow interaction at the junctions of an urban network. He assumes that junctions are non interacting and therefore does not consider offsets as decisional variables. On the other hand he assumes that delay at an approach depends on the flows at the other approaches of the same junction. This surely happens when the junction is not signalized, and can also happen for a signalized junction because delays depend on green times which depend on flows at different approaches. This assumption determines the need for an asymmetric equilibrium model. In Smith (1982) the set of solutions for the assignment problem is analyzed and demonstrated to be convex under some conditions related to delay functions. In Smith (1981a, 1981b) the properties of a control policy which guarantees the existence of an equilibrium pattern consistent with this policy are defined. The knowledge of these equilibrium and stability properties can be useful for designing an efficient control system. From this viewpoint he shows that a signal setting based on Webster's method or delay minimization does not satisfy the above mentioned conditions. He defines a theoretical control policy which meets these conditions but it appears to be a not "natural" policy. Smith (1985) is the most recent paper about these topics.

Heydecker (1983) develops analogous considerations based on the existence, unicity and stability conditions formulated by Dafermos (1980). He analyzes in detail the commonly used control strategies such as total delay minimization and capacity factor maximization.

CONCLUSIONS

The solution of the equilibrium network traffic signal setting problem can be approached using a global optimization model or an iterative procedure.

It has been noted that the proposed global optimization models assumes as known parameters cycle time, stage matrix and link offsets and therefore the actual decisional variables are only the green times. In particular it is worthwhile to underline that all these contributions to the equilibrium network traffic signal setting assume that there is no interaction between adjacent junctions, thereby omitting the influence of signal coordination. This is a strong assumption for an urban traffic network where signal coordination plays a very important role in link cost definition and can greatly improve network performance.

Iterative procedures, instead, allow to assume a greater number of control parameters as decisional variables without constraining some of them to be input data. They also allow to solve medium-scale real problems with an acceptable computational effort.

For these reasons real problems are usually solved via iterative procedures, in spite of their known limitations in finding the optimal solution.

Previous considerations indicate that equilibrium network traffic signal setting is an open research field both from theoretical and practical viewpoint. Further researches are needed to remove some simplifying assumptions of global optimization methods, so extending their field of application. On the other side, further experimentations are needed to test performance of iterative procedures, deeply investigating their mathematical behaviour.

REFERENCES

Aashtiani H.Z. (1979). The multimodal traffic assignment problem. Ph. D. Dissertation, M.I.T., Sloan School of Management.

Aashtiani H.Z. and Magnanti T.L. (1981). Equilibria on a Congested Transportation Network. Siam J. on Alg. and Discr. Math., 2: 213-226.

Allsop R.E. (1968). Choice of Offsets in Linking Traffic Signals. Traffic Engng. Control, 9: 73-75.

Allsop R.E. (1971). Delay Minimizing Setting for Fixed Time Traffic Signals at a Single Road Junction. J. Inst. Maths. Applics., 8: 164-185.

Allsop R.E. (1972). Estimating the Traffic Capacity of a Signalized Road Junction. Transpn. Res. , 6: 345-355.

Allsop R.E. (1974). Some Possibilities for Using Traffic
    Control to Influence Trip Destination and Route Choice.
    Proc. of the Sixth Int. Symp. on Transportation and Traffic
    Theory: 345-374. Elsevier, Amsterdam

Allsop R.E. (1983a). Optimization of Timing of Traffic Signals.
    Atti delle Giornate di Lavoro AIRO 1983: 103-120, Guida
    Editori, Napoli.

Allsop R.E. (1983b). Network Models in Traffic Management and
    Control. Transport Reviews, 3, 2: 157-182.

Allsop R.E. and Charlesworth J.A. (1977). Traffic in a Signal
    Controlled Road Network : An Example of Different Signal
    Timings Inducing Different Routing. Traffic Engng. Control,
    18: 262-264.

Beckman M.J., McGuire C.B. and Winsten C.B. (1956). Studies in
    the Economics of Transportation. Yale University Press, New
    Haven, CT.

Cantarella G.E. e Improta G. (1981). Una Metodologia per il
    Progetto Globale di Intersezioni Semaforizzate. Atti delle
    Giornate di lavoro AIRO 1981, Torino: 33-48.

Cantarella G.E. and Improta G. (1985). Capacity Factor
    Maximization and Cycle Time Minimization for an Individual
    Signalized Junction. Dipartimento di Informatica e
    Sistemistica, Universita' di Napoli. Rapporto interno n. 4.

Cantarella G.E. and Improta G. (1987). Capacity Factor and
    Cycle Time Optimization: A Graph Theory Approach, to appear
    on Transpn. Res. B.

Cantarella G.E., Improta G. e Sforza A. (1985). Progetto del
    Sistema di Controllo Semaforico su Reti in Equilibrio.
    Giornate di Studio su : "Mobilita' e Trasporti in un'area
    urbana. Problemi, Esperienze, Prospettive". Centro
    Scientifico I.B.M. Pisa, 17-18 ottobre 1985.

Cantarella G.E., Cosentino M., Improta G. and Sforza A. (1986).
    Equilibrium Network Control System Design. Preprints 5th
    IFAC - IFIP - IFORS Int. Conf. on Control in Transportation
    Systems,Wien, 8-11 July, 1986: 395-403.

Cantor D.G. and Gerla M. (1974). Optimal Routing in a packed
    switched computer network. IEEE Trans. on Computers, C-23
    : 1062-1069

Charlesworth J.A. (1975). Relations between Travel-Time and Traffic Flow for the Links of Road Networks Controlled by Fixed-Time Signals. Transport Oper. Res. Group, Research Report n. 13. Newcastle upon-Tyne.

Charlesworth J.A. (1977). Mutually Consistent Traffic Assignment and Signal Timings for a Signal-Controlled Road Network. Transport Oper. Res. Group, Research Report n.24. Newcastle upon-Tyne.

Dafermos S.C. (1971). An Extended Traffic Assignment Model with Applications to Two-Way Traffic. Transpn. Sci., 5: 366-389.

Dafermos S.C. (1980). Traffic Equilibrium and Variational Inequalities. Transpn. Sci., 14: 42-54.

Dafermos S.C. and Sparrow F.T. (1968). The Traffic Assignment Problem for a General Network. J. Res., Nat. Bur. Standards B. 73 B, 2: 91-117.

Dickson T.J. (1981). A Note on Traffic Assignment and Signal Timings in a Signal Controlled Road Network. Transpn. Res., 15B: 267-271.

Fisk C.S. (1984). Game Theory and Transportation Systems Modelling. Transpn. Res., 18B, 4/5: 301-313.

Fisk C.S. and Nguyen S. (1982). Solution Algorithm for Network Equilibrium Models with Asymmetric User Costs. Transpn. Sci., 16: 361-381.

Florian M. and Spiess H. (1982). Convergence of Diagonalization Algorithm for Fixed Demand Asymmetric Equilibrium Problems. Transpn. Res., 16B: 477-483.

Frank M. and Wolfe P. (1956). An Algorithm of Quadratic Programming. Nav. Res. Log. Quart, 3: 95-110.

Friesz T.L. (1985). Network Equilibrium, Design and Aggregation: Key Developments and Research Opportunities, Transpn. Res., 19A, 5/6: 413-427.

Gartner N.H. (1972a). Constraining Relations among Offsets in Synchronized Networks. Transpn. Sci. 6: 88-93.

Gartner N.H. (1972b). Optimal Synchronization of Traffic Signal Networks by Dynamic Programming. Proc.5th.Int.Symp. on the Theory of Traffic Flow and Transportation. (Edited by G.F. Newell): 281-295. American Elsevier, New York

Gartner N.H. (1976). Area Traffic Control and Network

Equilibrium. Traffic Equilibrium Methods (Edited by
M.Florian) : 274-297.Springer-Verlag, Berlin

Gartner N.H. and Little J.D.C. (1975). Generalized
Combination Method for Area Traffic Control. Transpn. Res.
Rec. 531: 58-59.

Gartner N.H., Little J.D.C., and Gabbay H. (1975). Optimization
of Traffic Signal Settings by Mixed-Integer Linear
Programming. Part I: The Network Coordination Problem; Part
II: The Network Synchronization Problem, Trans. Sci., 9:
321-363.

Gartner N.H., Little J.D.C. and Gabbay H. (1976). Simultaneous
Optimization of Offsets, Splits and Cycle Time. Transpn.
Res. Rec. 596: 6-15.

Gartner N.H., Gershwin S.B., Little J.D.C. and Ross P. (1980).
Pilot Study of Computer-Based Urban Traffic Management.
Transpn. Res., 14B: 203-217.

Hearn D.W., Lawphongpanich S. and Nguyen S. (1984). Convex
Programming Formulations of the Asymmetric Traffic
Assignment Problem. Transpn. Res. 18b: 357-365.

Heydecker B.G. (1983). Some Consequences of Detailed Junction
Modelling in Road Traffic Assignment. Transpn. Sci., 17:
263-281.

Hillier J.A. (1966). Appendix to Glasgow's Experiment in Area
Traffic Control. Traffic Engng. Control, 7: 569-571.

Improta G. (1978). Un Contributo sul Problema della Capacita'
per le Intersezioni a Raso Semaforizzate con Sistema Fixed-
Time. Atti del XVIII Cong.Naz. Stradale, Taormina, Italy.
: 349-356.

Improta G. (1986). Mathematical Programming Methods for Network
Control. Proc. of Advanced Workshop on Flow Control of
Congested Networks (A. Odoni, G. Szego Eds.). Springer-
Verlag, Berlin.

Improta G. and Cantarella G.E. (1984). Control System Design
for an Individual Signalized Junction. Transpn. Res., 18B
: 147-167.

Improta G. and Sforza A. (1982). Optimal Offsets for Traffic
Signal Systems in Urban Networks. Transpn. Res.,16B: 143-
161.

Le Blanc L., Morlok E. and Pierskalla W. (1975). An Efficient Approach to Solving the Road Network Equilibrium Traffic Assignment Problem. Transpn. Res., 9: 309-318.

Leventhal T.L., Nemhauser G.L. and Trotter L.E. (1973). A Column Generation Algorithm for Optimal Traffic Assignment. Transpn. Sci., 9: 168-176.

Magnanti T. and Wong R. (1983). Network Design and Traesportation Planning: Models and Algorithms. OR 125-83, MIT.

Marcotte P. (1983). Network Optimization with Continuous Control Parameters. Transpn. Sci., 17: 181-197.

Nguyen S. (1976). A Unified Approach to Equilibrium Methods for Traffic Assignment. Traffic Equilibrium Methods (Edited by M. Florian): 148-182. Springer-Verlag, Berlin

Nguyen S. and James-Lefebre L. (1975): TRAFFIC: An Equilibrium Assignment program. C.R.T. Montreal, pubbl. no. 17.

Robertson D.I. (1969). TRANSYT Method for Area Traffic Control. Traffic Engng. Control, 11: 276-281.

Sheffi Y. (1985). Urban Transportation Networks. Prentice Hall Inc. Englewood Cliffs, N.J.

Smith M.J. (1979a). Traffic Control and Route-Choice: A Simple Example. Transpn. Res., 13B: 289-294.

Smith M.J. (1979b). The Existence, Uniqueness and Stability of Traffic Equilibria. Transpn. Res., 13B: 295-304.

Smith M.J. (1981a). The Existence of an Equilibrium Solution of the Traffic Assignment Problem when there are Junction Interactions. Transpn. Res., 15B: 443-451.

Smith M.J. (1981b). Properties of a Traffic Control Policy which Ensure the Existence of a Traffic Equilibrium Consistent with the Policy. Transpn. Res., 15B: 453-462.

Smith M.J. (1982). Junction Interaction and Monotonicity in Traffic Assignment. Transpn. Res., 18B: 1-3.

Smith M.J. (1983). The Existence and Calculation of Traffic Equilibria. Transpn. Res., 17B: 291-303.

Smith M.J. (1985). Traffic Signals in Assignment Transpn.Res. 19B: 155-160.

Tan H., Gershwin S.B. and Athans M. (1979). Hybrid Optimization in Urban Traffic Networks. Report No. DOT-TSC-RSP- 79-7,MIT

Cambridge, MA.

Vincent R.A., Mitchell A.I. and Robertson D.I. (1980). TRANSYT Version 8. Road Res. Lab. Rep.,LR 80, Crothorne, Berkshire.

Wardrop J.G. (1952). Some Theoretical Aspects of Road Traffic Research. Proc. Inst. Civ. Eng., Part. II, 1: 325-378.

Webster F.V. (1958). Traffic Signal Setting. Road Research Technical Paper n.39 HMSO, London.

Webster F.V. and Cobbe B.M. (1966). Traffic Signals. Road Research Technical Paper n.56 HMSO, London.

Zangwill W.I. (1967). The Convex Simplex Method, Man. Sci. 14, 3: 221-238.

Zuzarte Tully I.M. and Murchland J.D. (1978). Calculation and Use of the Critical Cycle Time for a Single Traffic Controller. Proc. of PTRC Summer Annual Meet.. PTRC-P152 :96-112.

Static and Dynamic Models of Stochastic Assignment to Transportation Networks.

Ennio Cascetta
Department of Transportation Engineering
University of Neaples
Via Claudio, 21
80125 - Napoli
Italy

Introduction

Assignment models can be seen as correspondences or mappings associating link flows to a given (constant) travel demand and transportation network through a model of users'behaviour.

Assignment models can be classified as deterministic or stochastic depending on the assumptions made on users' perception of costs.

Both type of models assume that travelers are rational decision-makers, i.e. they choose the path of minimum perceived cost among those connecting the origin to the destination of their trip.

Deterministic models assume users endowed with an exact knowledge of alternative path costs so that only "shortest" or minimum cost paths are used.

Stochastic models assume, perhaps more realistically, that travelers choose on the basis of "perceived" costs, which are random variables distributed across the users' population, and possibly time, with true costs as mean values.

In the context of stochastic assignment models all paths have a theoretical probability of being chosen.

Stochastic models can be further subdivided on the basis of

NATO ASI Series, Vol. F38
Flow Control of Congested Networks
Edited by A. R. Odoni et al.
© Springer-Verlag Berlin Heidelberg 1987

the type of cost functions (constant or flow dependent) and of the type of system description (static or dynamic) assumed.

In this paper theoretical and computational aspects of different stochastic assignment models will be discussed.

1 - Notation and Terminology

Let us consider a transportation network $T(N,L,C)$ consisting of a set N of nodes, a set L of directed links and a set C of link travel cost functions.

Let N_c be the subset of centroid nodes, i.e. nodes in which trip can originate and/or terminate.

Let d be the Origin-Destination (O-D) demand vector whose components d_{ij} are the number of travelers moving between the centroid pair (i,j) in a given reference period.

The number of a-ciclic paths connecting centroid nodes is finite so that individual paths can be enumerated. Let us denote by I_{ij} the set of indexes relative to paths connecting the centroid pair (i,j).

Let F be the path flow vector whose components F_k are the number of trips between centroids (i,j) using path k in the reference period, with $k \in I_{ij}$.

Let P be the path choice proportion matrix of dimensions (nr. of paths) x (nr. of O/D pairs) whose elements p_{kij} are the proportions of users choosing path k between the O/D pair (i,j):

$$p_{kij} = F_k/d_{ij} \qquad \forall \, k \in I_{ij} \qquad (1.1)$$

the p_{kij} 's must obviously satisfy non negativity and conservation conditions:

$$p_{kij} \geq 0 \qquad \sum_{k \in I_{ij}} p_{kij} = 1 \qquad\qquad (1.2)$$

Expressions (1.1) and (1.2) can be combined together as

$$F = Pd \qquad\qquad (1.3)$$

Let v be link flow vector whose components v_l are the number of trips using link l in the reference time period.

Let us denote by A the link-path incidence matrix whose components a_{lk} are either 1 (link l belong to path k) or zero (link l doesn't belong to path k).

The relationship between path and link flows can be expressed via the above incidence matrix as:

$$v = A F \qquad\qquad (1.4)$$

In the following the usual assumption will be made of demand flows constant over a period long enough to allow a stable link and path flow pattern to take place.

Let c be the link cost vector whose components c_l are the average generalized travel cost (time, money, discomfort, etc) incurred by a traveler using link l. Because of congestion average link costs usually depend on link flows and link cost functions express such a dependence.

In general travel cost (time) on a given link depends on flows on many links so that the argument of cost function is the whole vector of link flows: $c_l = c_l(v)$. In the case in which cost on link l depends only on the flow on that link, $c_l = c_l(v_l)$, cost functions are called separable.

As we will see later on separable cost functions are particularly convenient for theoretical as well as computational reasons.

Let C denote the path cost vector whose components C_k are the

average generalized cost of path k. In transportation networks
is usually assumed that path cost is the sum of costs of
component links (additive cost structure).

Path costs can thus be expressed in terms of link costs as:

$$C = A^T c \qquad (1.5)$$

where A^T denotes the transpose of the link-path incidence
matrix.

Path costs can also be expressed as function of path and link
flows via expressions (1.4) and (1.5) as:

$$C = A^T c(v) = A^T c(AF) \qquad (1.6)$$

By using notation and terminology introduced in this section
assignment models can be defined as models giving
approximations of path choice fractions p_{kij} as function of
path cost based on assumptions concerning trip maker's route
choice behaviour. The assignment mapping can thus be formally
expressed as:

$$F = P(C) d \qquad (1.7)$$

$$v = A P(C) d \qquad (1.8)$$

2 - Random utility models of path choice

Stochastic assignment models assume that travelers choose
their path following a "random utility" choice model.

It is assumed that each user associates a perceived cost C_k to
each path k connecting his O-D pair and choose as to minimize
this cost. Because of a number of causes, such as random
fluctuation of travel costs, omitted and/or inexact evaluation
of some attributes making up the generalized cost, limited
information, variation of tastes across the population and,

for the same user, over time etc, it is assumed that perceived path costs are random variables with mean value given by the "true" average cost C_k :

$$\hat{C}_k = C_k + \epsilon_k \qquad (2.1)$$

Under the above assumptions path choice probabilities p_{kij} can be expressed as

$$p_{kij} = \text{Prob} [C_k - C_h < \epsilon_h - \epsilon_k] \quad \forall\ h \neq k, \quad h,k \in I_{ij} \qquad (2.2)$$

The functional form of path choice probabilities (2.2) depends on the joint probability law assumed far random terms ϵ_k.

If error terms are assumed to be independent, identical Weibull variates: $\epsilon_k \approx W(0,\sigma)$ with zero mean and standard deviation σ, equation (2.2) reduces to the well-known multinomial logit model (see Domencich and McFadden (1974), Sheffi (1985)):

$$p_{kij} = \exp(-\delta C_k) / \sum_{h \in I_{ij}} \exp(-\delta C_h) \qquad (2.3)$$

where the parameter δ is inversely related to the standard deviation of residuals through the expression:

$$\delta = \pi / \sqrt{6} * \sigma$$

The above closed form is a significant advantage of logit model over other path choice models while the assumption of independent perception errors for often heavily overlapping paths is its major theoretical shortcaming; see Daganzo and Sheffi (1977), Florian and Fox (1976) among others for a critique to the logit path choice model.

On the other hand if residuals are assumed to follow a joint multivariate normal distribution: $\epsilon_k \approx \text{MVN}(0,\Sigma)$ with moments:

$$E(\epsilon_k) \quad = 0$$

$$Var\ (\epsilon_k) \quad = \delta\ C_k \tag{2.4}$$

$$Cov\ (\epsilon_k\ \epsilon_h) = \delta\ C_{kh} \qquad if\ h,k \in I_{ij}$$

$$= 0 \qquad\qquad otherwise$$

where C_{kh} is the cost of the overlapping parts of paths k and h, path choice probabilities are given by the probit model (see Daganzo and Sheffi (1977), Daganzo (1979):

$$p_{kij} = \int_{C_h < S_k} \emptyset\ (x/C,\Sigma)\ dx \tag{2.5}$$

where S_k is the set of values $\hat{C}_h > \hat{C}_k$ for any value of $\hat{C}_k$ and $\emptyset(.)$ is the MVN probability density function.

The probit model rests on theoretically sounder assumptions but computing path choice probabilities is a more demanding task as no closed form exists for (2.5).

In practice application of the probit model requires the use of approximate Monte Carlo simulation methods. We will come to this point again in section 7. A quantity related to any random utility path choice model is the satisfaction S_{ij} relative to users moving between O/D pair (i,j) and defined as the expected value of the r.v. perceived travel cost between (i,j):

$$S_{ij} = E\ [\ \min_{k \in I_{ij}}\ \{C_k\}\ /\ C\ ] \tag{2.6}$$

Note that the satisfaction (2.6) depends on the vector of "true" costs C via equation (2.1). It can be shown under some assumptions that the partial derivative of S_{ij} with respect to C_k is:

$$\frac{S_{ij}}{C_k} = p_{kij}\ (C) \qquad if\ k \in I_{ij}$$

$$= 0 \qquad\qquad otherwise \tag{2.7}$$

The above property can be immediately verified for the logit model by differentiating expression (2.2).

3 - Stochastic network loading (SNL) models.

Stochastic network loading models assume that link average travel costs are constant. This is a reasonable assumption for rather uncongested networks where cost functions are almost "horizontal" for low flow/capacity ratios.

Constanty of link costs implies that of path costs so that path choice probabilities can be computed independently from link flows via expressions (2.2) and (2.5).

Actual path and link flows taking place in the network are random variables because such is the number of travelers choosing each path in a given reference period. In particular under the assumption that users choose their path independently from each other, flows on paths connecting tha same O-D pair are multinomial random variables. The expected path flow vector is:

$$E(F) = P(C) d \qquad (3.1)$$

and variances and covariances between flows are given by

$$Var(F_k) = d_{ij} p_{kij}(C) (1-p_{kij}(C)) \quad k \in I_{ij} \qquad (3.2)$$

$$Cov(F_k F_h) = -d_{ij} p_{kij}(C) p_{hij}(C) \quad h,k \in I_{ij} \qquad (3.3)$$

$$= 0 \qquad otherwise$$

The link flow vector is a linear function of the path flow vector, see equation (1.5), so that its expected value can be computed as:

$$E(v) = A P(C) d \qquad (3.4)$$

and its variance-covariance matrix Σ_v can be obtained by:

$$\Sigma_v = A \Sigma_F A^T \tag{3.5}$$

where Σ_F is the variance-covariance matrix of path flows with elements given by (3.2) and (3.3).

4 - Stochastic user equilibrium (SUE) models.

For congested networks travel costs vary with traffic flows; the path choice probabilities, however computed, are function of the flow vector: $p_{kij}(C(v))$.

Equilibrium flow vectors are defined in terms of the following fixed point conditions:

$$F^* = P(C(AF^*)) d \tag{4.1}$$

$$v^* = A P(C(v^*)) d \tag{4.2}$$

Mathematical properties of SUE flows parallel closely those of deterministic equilibrium flows.

Existence of SUE vectors is assured under mild conditions by Brower's fixed point theorem; unicity of equilibrium link flows requires the positive definiteness of the link cost function Jacobian $J[c(v)]$, Daganzo (1982).

If the Jacobian matrix of $c(v)$ is symmetric the equilibrium vector v^* is solution of the following unconstrained optimization problem:

$$v^* = \arg \min Z(v) = c(v)^T v - S(A^T c(v))^T d - \int_0^v c(x) \, dx \tag{4.3}$$

where $S(.)$ is the vector of satisfaction functions (2.6) for all O/D pairs. The satisfaction function Jacobian with respect to link flows is

$$J[S(v)] = P(C(v))^T A^T J[c(v)] \tag{4.4}$$

because of (2.7) so that the gradient of $Z(v)$ results:

$$\nabla Z(v) = (v - A P(C(v)d))^\tau J[c(v)] \qquad (4.5)$$

The nullity of gradient computed for the optimum point v^* requires that:

$$v^* = A P(C(v^*)) d$$

which coincides with (4.2).

The symmetry condition holds trivially for separable cost functions; in that case equation (4.3) becames:

$$v^* = \text{arg min } \Sigma_i \ v_i \ c_i(v_i) - \Sigma_{ij} \ d_{ij} \ S_{ij}[C(v)] - \Sigma_i \int_0^{v_i} c_i(x)dx \qquad (4.6)$$

Actual flows are also in this case random variables; it is assumed that $C(v^*)$ is the expected perceived cost vector, means, variances and covariances of path flows can be obtained as:

$$E(F) \qquad = P(C(AF^*))d = F^* \qquad (4.6)$$

$$Var(F_k) \qquad = d_{ij} \ p_{kij}(C(F^*)) \ [1-p_{kij}(C(F^*))] \qquad (4.7)$$

$$Cov(F_h \ F_h) = -d_{ij} \ p_{kij}(C(F^*)) \ p_{hij}(C(F^*)) \quad \text{if } h,k \in I_{ij} \qquad (4.8)$$

$$\qquad = 0 \qquad\qquad\qquad\qquad \text{otherwise}$$

and for link flows:

$$E(v) = A P (C(v^*))d = v^* \qquad (4.9)$$

$$\Sigma_v \quad = A\Sigma_F A^\tau \qquad (4.10)$$

Here and in the following path costs will be expressed directly as function of both path and link flows for notational convenience.

The interpretation of SUE flows as expected values of the actual flow probability distribution requires some approximate assumptions.

In fact if actual flows taking place in a given network are random variables so are travel link and path costs. Expected travel costs differ from cost computed for expected flows:

$$E[C(v)] \neq C(E(v)) = C(v^*) \qquad\qquad (4.11)$$

apart from the rather uncommon case of linear (affine) cost functions.

Equilibrium models justify the hypothesys of $C(v^*)$ as the expected perceived cost vector underlying the interpretation of v^* and F^* as expected vectors, (4.6) and (4.9), by assuming that actual flows v are close enough to their expected values so that (4.11) holds as an equality.

In real cases the link flow variance-to-mean ratios are very small, it should however be noted that small deviations in flows do not necessarily imply small deviations in costs, expecially for very congested networs, see fig. 1.

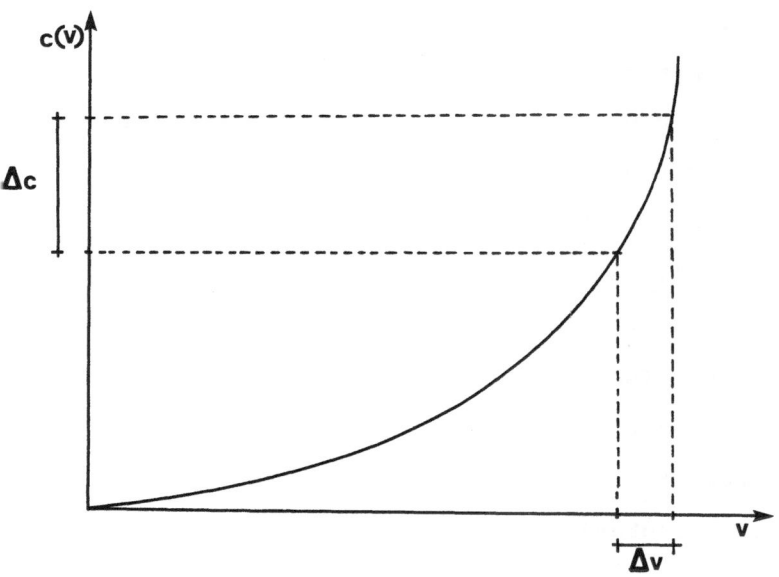

Fig. 1 - Cost-flow relationship

5 – Network dynamics as a stochastic process

A possible alternative to the "static" equilibrium analysis of transportation systems described in section 4 is that of modeling its evolution over a sequence of "times" or "epochs"...t-1, t, t+1,...

Times or epochs can have either a chronological interpretation as successive reference periods of similar characteristics (e.g. the morning peack period of successive working days) or they can be seen as "ficticious" moments in which the system changes its state.

The dynamical approach is based on the following considerations:

i) no transportation system remains in the same state, however defined, over successive time periods

ii) the state occupied by the system at any time is not predictable "a priori" because of several causes such as the random fluctation of attributes, random event in the network determining actual path choices variation, in the composition of the total (fixed) demand etc. Furthermore the state occupied at any time "depends" on previous states through the dependence of traveler's choices at that time on costs incurred in the past.

System states can be defined at different level of aggregation according to their use; to fix ideas they can be throught as path or link flows taking place at each time.

The above assumptions imply that the system evolves among different states over successive time periods as a stochastic process the type of which depends on the particular choice mechanism followed by travelers.

In this framework even if equilibrium flows took place at one point in time they would not necessarily reproduce themselves in successive epochs.

Stationariety is a very important property of the stochastic process insomuch it ensures a stable evolution law and allows to associate a flow distribution to each demand-network system independently of its starting configuration and elapsed time.

It can be shown, Cascetta (1986), that the stochastic process is "dynamically stable", i.e. admits an unique stationary distribution of probabilities of occupying any feasible state, if the following sufficient conditions hold:

i) at each time users base their choices on informations acquired in a finite number of previous times (learning mechanism)

ii) travelers' choice mechanism is time-invariant

$p_k{}^t(u)[S_{t-1}=S_a, S_{t-2}=S_b...]=p_k{}^{t'}(u)[S_{t-1}=S_a, S_{t-2}=S_b...]$ $k \in I(u)$

where $p_k{}^t(u)$ is the probability, however computed, that the u-th traveler chooces path k at time t, S_{t-1} is the state occupied by the system at time t-1 (e.g. the path flow vector occurring at that time) and I(u) is the set of alternative paths considered by the u-th user.

iii) path choice probabilities are positive for all paths considered by the user

$p_k{}^t(u) > 0$ $\quad\quad \forall \; k \; \in \; I(u)$

Various stationary stochastic processes can be obtained by assuming different choice and learning mechanisms satisfying the above conditions.

6 - A dynamic model of stochastic assignment (STODYN).

Any stationary model of system dynamics can be considered as a stochastic assignment model insofar it allows a probability distribution of flows with its moments (means, variances, covariances, etc.) to be associated to a demand-network system.

The STODYN model is based upon a set of hypotheses similar to those underlying other stochastic assignment models, Cascetta (1986):

i) Users moving between the same O-D pair are "indistinguishable", i.e. they have the same set of alternative paths ($I(u) = I_{ij}$) and follow the same behavioural mechanism.

ii) Users moving at each time t follow a "random utility" model of path choice:

$$p_{kij}{}^t = \text{Prob} [\hat{C}_k{}^t < \hat{C}_h{}^t] \quad \forall h \; k \quad h,k \in I_{ij} \qquad (6.1)$$

The perceived cost of path k at time t, $\hat{C}_k$ is a random variable with expected value C_k:

$$\hat{C}_k{}^t = \bar{C}_k{}^t + \epsilon_k{}^t \qquad (6.2)$$

Logit and Probit models of path choice probabilities can be obtained by assuming i.i.d. Weibull or MVN residuals.

iii) Users have a learning mechanism based on m previous periods, i.e. they base their choices at time t on cost occurred in m previous times:

$$\hat{C}_k{}^t = \Sigma_i \; w_i \; C_k{}^{t-i} + \epsilon_k{}^t \qquad (6.3)$$

and, because of congestion:

$$\hat{C}_k{}^t = \Sigma_i \; w_i \; C_k(F^{t-i}) + \epsilon_k{}^t \qquad (6.4)$$

where $\{w_i\}$ is a set of weights taking into account the

possibly different impact of most recent experiences.

Under the above assumptions system states can be conveniently identified as path flows occurring at each time ($S^t = F^t$).

The number of "feasible" path flow vectors with integer components is finite so that they can be enumerated; furthermore the probability of observing any given state F_i at time t depends on the states occurred in the m previous times and is time-invariant.

This implies that the stochastic process in the path flow space is an m-dependent Markov Chain with the following properties:

i) it exists an unique stationary distribution of feasible path flow vector probabilities

$$\tau = \{\pi_i\}$$

with π_i expressing the probability of finding the system in state F_i at any time.

ii) the process is ergodic:

$$\pi_i = \lim_{t \to \infty} n_i(t)/t \tag{6.5}$$

where $n_i(t)$ is the number of times that state F_i occurs in t epochs.

First and second order moments of path and link flows can be expressed as:

$$\bar{F} = E(F) = \Sigma_i \ \pi_i \ Fi \tag{6.6}$$

$$\Sigma_F = \Sigma_i (F_i - \bar{F})(F_i - \bar{F})^\tau \ \pi_i \tag{6.7}$$

$$\bar{v} = E(v) = A \ \bar{F} \tag{6.8}$$

$$\Sigma_v = A \ \Sigma_F \ A^t \tag{6.9}$$

7 - Relationship between STODYN and SUE expected flows

In order to compare STODYN and SUE expected flows the case of a one step (m=1) learing mechanism will be considered.

By using the fixed point property of the steady state probability vector π for a Markovian process, equations (6.7) and (6.9) can be reformulated as:

$$\bar{F} = \Sigma_i \ \pi_i \ P(C(F_i)) \ d = E[P(C(F))] \ d \qquad (7.1)$$

$$\bar{v} = A \ E[P(C(F))] \ d \qquad (7.2)$$

Comparaison of equations (7.1) and (7.2) with equations (4.6) and (4.9) defining F* and v* shows that STODYN and SUE expected flows coincide if:

$$E[P(C(F))] = P(C(E(F))) = P(C(\bar{F})) \qquad (7.3)$$

Expression (7.3) holds for constant costs or for linear path choice probabilities and link cost functions, i.e. for uncongested networks or for linear systems.

In general, however, $\bar{F}$ and $\bar{v}$ are approximately equal to F* and v* within the limits of a first order approximation of the vectorial function P(C(F)). It can be shown by expanding the above function in Taylor's series up to the second term that the approximation is acceptable when the elements of dispersion matrices Σ_F and/or Σ_v are "small" enough.

Similar results hold for a m-step learning mechanism.

In the limiting case of m tending to infinity with uniform weights, users tend to base their choices on average costs E[C(F)] which, as already remarked in section 4, are still theoretically different from costs computed with average flows for non-linear cost functions, see expression (4.11).

It should finally be noted that the above approximate

coincidence of STODYN with SUE expected flows doesn't extend to second order moments; higher variances and an autocorrelation structure should be expected for STODYN link flows.

8 - Algorithms for stochastic assignment.

Computation of expected flows for a logit-based SNL model can be efficiently carried out by the Dial's algorithm, Dial (1971), which obviates the explicit enumeration of elementary paths. The above algorithm assumes that only efficient paths are used, i.e. paths taking the traveler farther away from his origin and/or closer to his destination. A detailed discussion of the Dial's algorithm is beyond the scope of this paper and can be found in the original paper and in the book of Sheffi, Sheffi (1985), among others.

Algorithms for the probit-based SNL model use a Monte Carlo simulation tecnique. They sample repeatedly perceived link costs from independent normal variates with mean value equal to the average link cost and variance proportional by a parameters δ, equation (2.4). For each replication the O/D demand is assigned to shortest paths computed with the perceived travel costs and obtained flows are averaged over all replications untill a convergence criterion is met.

The algorithmic scheme used for computing SUE expected link flows is known as Method of Successive Averages (MSA), Powell and Sheffi (1982). It can be shown that MSA for separable cost functions is a gradient algorithm for the unconstrained optimization problem (4.6). The main steps of the algorithm are:

Step 0: Initialization. Perform a (logit or probit) SNL based on a vector of initial link travel costs c^o. This generates a vector of link flows v^1. Set $K = 1$

Step 1: Update the link cost vector $c^k = c(v^k)$.

Step 2: Perform a SNL based on the current cost vector c^k. This yelds an auxiliary link flow vector y^k.

Step 3: Average the current flow vector over all previous iterations:

$$v^{k+1} = v^k + 1/k \ (y^k - v^k) = 1/k \ \Sigma_i \ y^i$$

Step 4: Convergence test. If a convergence criterion, e.g. maximum percentual variation of flows n iteration apart, is not met set k = k+1 and go to Step 1.

Algorithmic schemes for computing STODYN expected flows are based upon the ergodic property of the stochastic process. In other words they simulate a stationary realization of the stochastic process and use it for computing moments of the variables of interest. The general structure of the algorithm is:

Step 0: Inizialization. As in the MSA, furthermore a number of cycles Step 1 - Step 2 are repeated untill the stationariety of the link flows sequence is not accepted by a suitable test. Set k = 1 and keep last m previous flow vectors.

Step 1: Update the average perceived link cost vector over last m iterations

$$c^k = \Sigma_i \ w_i \ c \ (y^{k-i})$$

Step 2: Perform a SNL based on the current cost vector c^k.

Denote by y^k the resulting link flow vector.

Step 3: Average the current flow vector over all previous iterations.

$$v^{k+1} = v^k + 1/k \; (y^k - v^k)$$

Step 4: Convergence test. If a convergence criterion is not met set k = k+1 and go to Step 1.

Observe that the sequence of link flow is a pseudo-realization of a stochastic process because the y^k's result from a SNL rather than from a simulation of individual users' choices. Some numerical tests however show that the influence of this approximation over final flows is completely negligible, Cascetta (1986).

The STODYN algorithm doesn't require separable cost functions for its convergence.

9 - Empirical validation of stochastic assignment models.

Very few studies about empirical validation and comparaison of stochastic assignment models are reported in the literature. In this Author knowledge the only piece of work carrying out a systematic analysis and comparaison of different assignment models is the one produced by A. Nuzzolo and himself, Cascetta and Nuzzolo (1986).

In that paper all stochastic assignment models dealt with in this paper were applied to the car netoworks of two medium-sized italian towns (Parma and Foggia) and compared on the basis of their ability to reproduce traffic flows counted on a number of links.

The main results were that SUE and STODYN expected flows were in the whole largely comparable (as expected by the results of section 7) even if some differences appeared at the level of individual links. Parameters δ both for logit, expression (2.3), and probit, expression (2.4), path choice models were very stable in the two areas. Furthermore, in spite of their theoretical difference, logit and probit choice models resulted in globally equivalent equilibrium flows.

SNL models applied with free-flow times performed systematically worse that SUE and STODYN models, as it was to be expected because the two networks were rather congested at least in their central areas.

All tested models predicted higher flows more precisely and, at least for SUE and STODYN, with an accuracy level acceptable for most engineering applications.

Further detail relative to data bases, implemented algorithms, statistical analyses of results for stochastic as well as deterministic models can be found in the original paper.

Conclusion

In this paper different stochastic assignment models were discussed under "standard" hypotheses. A number of generalizations are possible; some are listed in the following:

- Multiple users'types (e.g. by trip purpose) with different perceived cost distributions

- Multiple vehicle types (interactions among different models using the same network for fixed modal O/D demands)

- Models with elastic demand considering generation / distribution / modal choice / assignment models in an unified and internally consistent framework. The random utility path choice model make stochastioc assignment models "homogeneous" with most popular demand models so that a model system based on coherent assumptions can be built.

- Different choice mechanism (e.g. "adaptive" models of path choice in transyt networks).

- Different learing mechanism (e.g. the presence of habit in route choice).

Stochastic assignment models are based on assumptions on users' behaviour more realistic than their deterministic counterparts. Furthermore they are more convenient for estimating O/D demand from traffic flows, Nguyen and Cascetta (1986).

In spite of their theoretical advantages stochastic models,and expecially equilibrium ones, still find only limited use in practical application. This is probably due to their relatively recent introduction as well as to the greater computational effort required. The latter inconvenient should however be reduced by current improvements in computer technology.

References

Cascetta, E. (1986) A Stochastic process approach to the analysis of temporal dynamics in transportation networks, Dept. of Transp. Engineering Univ. of Neaples, Intern. Report and sent for pubblication to Transp. Res.,

Cascetta, E. and Nuzzolo, A. (1986) An empirical analysis of assignment models for urban car networks, Dept. of Transp. Eng. University of Neaples, Internal Report, sent to the IX Int. Symp. on Transportation and Traffic Theory, Cambridge MA,

Daganzo, C.F. and Sheffi, Y. (1977) On stochastic models of traffic assignment, Transp. Sci. 11, 253-274,

Daganzo, C.F. (1979) Multinomial Probit: the theory and its application to demand forecasting. Academic Press, New York,

Daganzo, C.F. (1982) Unconstrained extremal formulation of some transportation equilibrium problems. Transp. Sci. 16, 332-360,

Dial, R.B. (1971) A probabilistic multipath traffic assignment model which obviates path enumeration. Transp. Sci. 5, 83-111,

Domencich, T.A. and Mc Fadden, D. (1975) Urban travel demand: a behavioural analysis, American Elsevier, New York,

Fisk, C. (1980) Some developments in equilibrium traffic assignment, Transp. Res. 14(B), 243-255,

Florian, M. and Fox, B. (1976) On the probabilistic origin of Dial's multipath traffic assignment model. Transp. Res. 10, 339-341,

Nguyen, S. and Cascetta, E. (1986) Estimating Trip matrices from traffic counts: theory and applications, Proc. of the 1st International Course in Planning and Management of urban transp. systems, Montreal,

Powell, W.B. and Sheffi, Y. (1982) The convergence of equilibrium algorithms with predetermined step sizes, Transp. Sci. 16, 45-55,

Sheffi, Y. (1985) Urban transportation networks, Prentice Hall, Englewood Cliff. N. 5.

CONGESTION CONTROL IN FREEWAY CORRIDORS:
THE IMIS SYSTEM

N. H. Gartner
Department of Civil Engineering
University of Lowell
Lowell, Massachusetts, U.S.A.

R. A. Reiss
Sperry Systems Management
Great Neck, New York, U.S.A.

ABSTRACT

This paper describes a traffic control system design for congestion control in freeway corridors. The system has a hierarchical structure. A corridor level control acts in a supervisory capacity dynamically allocating traffic among alternative corridor facilities, such as freeways, frontage roads and signalized arterials. A local level control then selects control parameters for the individual facilities based on the predicted usage at the corridor level. A user specified performance function is optimized in the process.

Extensive simulation testing was conducted to verify algorithm performance and evaluate potential system benefits. A prototype system named IMIS (Integrated Motorist Information System), which is based on this design, is now being implemented in Long Island, New York.

INTRODUCTION

Urban traffic congestion is a serious and worsening problem, drawing increasing attention from transportation professionals and government officials. U.S. urban freeways, which carry nearly 30 percent of all traffic in urban areas, are particularly affected by the growing urban congestion. According to a recent report [1], urban freeway congestion annually causes 1.2 billion vehicle-hours of delay, 1.3 billion gallons of wasted fuel and $9 billion in excess user costs per year. Since urban freeway travel increases at a rate of 1.9 percent per year, and no significant additional physical capacity is contemplated, the problem will continue to increase in severity.

One of the most effective remedial measures is the installation of a surveillance and control system for the congested segments of a freeway. Such a system would contain mainline and ramp surveillance through loop detectors, a traffic responsive ramp metering system and an incident management program. In this paper we describe the logic structure of a traffic control system whose purpose is congestion control in a freeway corridor. This structure serves as a basis for the design of the IMIS system [2].

THE IMIS SYSTEMS

The Integrated Motorist Information System (IMIS) is a freeway and arterial-street traffic management project on 128 miles of heavily traveled highways in a 35-mile-long east-west corridor. It is located in the densely developed northwestern quadrant of Long Island, New York. IMIS is a remedial system, effective at levels of service (LOS) D and E. It is designed to reduce annoying and expensive recurring congestion and traffic flow instabilities.

Traffic flow instabilities result from a variety of causes, chief of which is an excessive ratio of traffic volume to highway capacity. Secondary causes are: disabled motorists on the shoulders, distracting roadside events, in-roadway accidents, poor weather, construction projects, routine maintenance, high levels of defensive driving by vehicle operators, and the mix of vehicle types in the traffic flow. With the onset of unstable stop-and-go operation, capacity per lane drops suddenly and dramatically. The goal for this corridor is to actively manage traffic demand volumes to sustain stable, efficient, low-cost rush hour operation at LOS D and E, and avoid the onset of stop-and-go jam conditions.

The system's operation is based on routing traffic past traffic jams and lane closures, via existing alternative routes that are not fully used. During incidents, route diversion information provides motorists with knowledge that a better alternate route exists past a congested highway section ahead. Alternate route traffic control, during diversions, based on real-time surveillance, reduces the instabilities of high density traffic flow on that route. During non-incident conditions providing high quality real-time operating condition information to motorists also will reduce instabilities. The project is being pursued on the basis that the expected benefits to the motoring public are 2.3 times government costs for building, operating, and sustaining the system, according to reports from project planning stages.

FREEWAY CORRIDOR TRAFFIC CONTROL

A corridor is a roadway system consisting of a few primary longitudinal
roadways (freeways or major arterials) carrying a major traffic movement,
with interconnecting roads which offer the motorist alternative paths to his
destination. Examples of such corridors include the IMIS corridor in Long
Island, New York (Figure 1) and the I-10 corridor in Los Angeles (Figure 2).
A corridor which can benefit from a corridor traffic control system is one
in which one or more routes can become congested even though the corridor as
a whole has sufficient capacity to provide travellers with a reasonable
level of service. In such corridors, a traffic control system serves to
rapidly detect congestion and implement controls to minimize traffic disruption. These controls can take the form of diversion, ramp metering and signal timing.

In diversion control, traffic is dynamically allocated among the various corridor facilities such as freeways, freeway-frontage roads and signalized arterials. This control is generally implemented by variable message
signs or highway advisory radio which convey diversion information to motorists. Ramp metering uses traffic signals on entrance ramps to control the
volume of traffic entering the freeway mainline. Metering rates are selected
to respond to present and predicted traffic demands on the mainline. Signal
timing control enables timing plans on corridor arterials to be responsive
to changes in demand due to diversion as well as regular time-of-day changes.

The key ingredient in the corridor traffic control system is the
traffic diversion software. The algorithm contained in the traffic diversion software selects values of diversion fractions, ramp metering rates
and signal timing parameters (cycles, splits, offsets) which optimize a
measure, or measures, of traffic movement through the corridor.

TRAFFIC DIVERSION SOFTWARE

The diversion algorithm is structured as a control hierarchy as shown
in Figure 3. The corridor level acts in a supervisory capacity, dynamically
allocating traffic optimally among the various corridor facilities such as
freeways, freeway-frontage roads and signalized arterials. Then the local
level optimizes flow over the individual facilities, based on predicted
usage as determined at the corridor level.

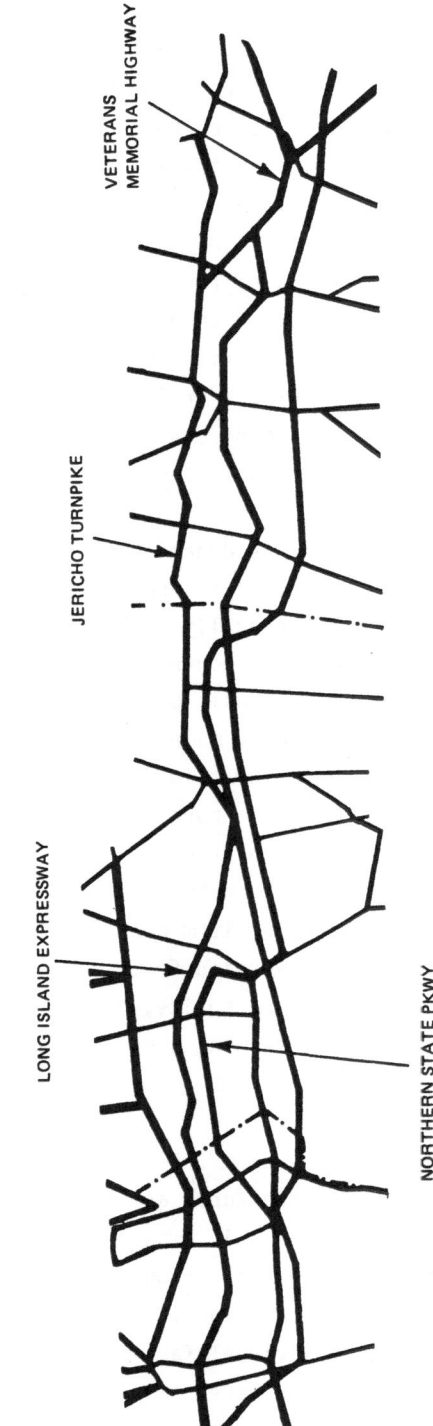

FIGURE 1. INTEGRATED MOTORIST INFORMATION SYSTEM (IMIS) CORRIDOR

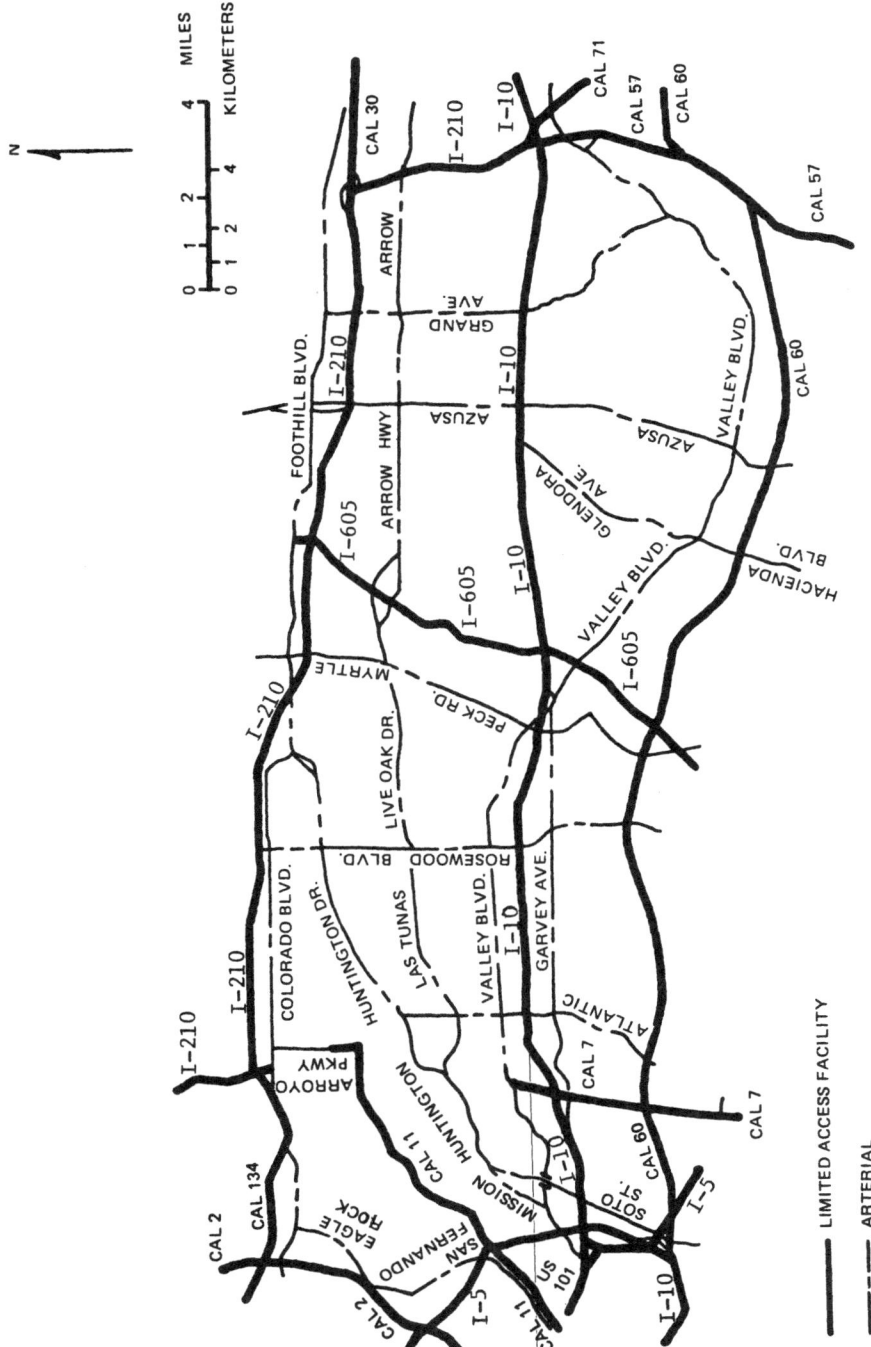

FIGURE 2. LOS ANGELES I-10 CORRIDOR

Corridor Level Control Algorithm

The corridor level control algorithm dynamically assigns traffic to corridor routes in such a way as to optimize a selected performance criterion or objective function. The corridor-level control process is performed at periodic intervals, typically, 10 to 15 minutes, or whenever measured corridor conditions appear to warrant immediate re-optimization, such as when a major incident is detected. The algorithm generates a <u>diversion policy</u> (consisting of optimal diversion fractions at each control node) which satisfies the origin-destination desires of all corridor users. This policy accounts for all predicted traffic effects during the optimization level (typically 1/2 hour). The algorithm is iterative in nature and has the attribute that any intermediate solution constitutes an acceptable assignment which can be implemented if for any reason the optimization process is terminated.

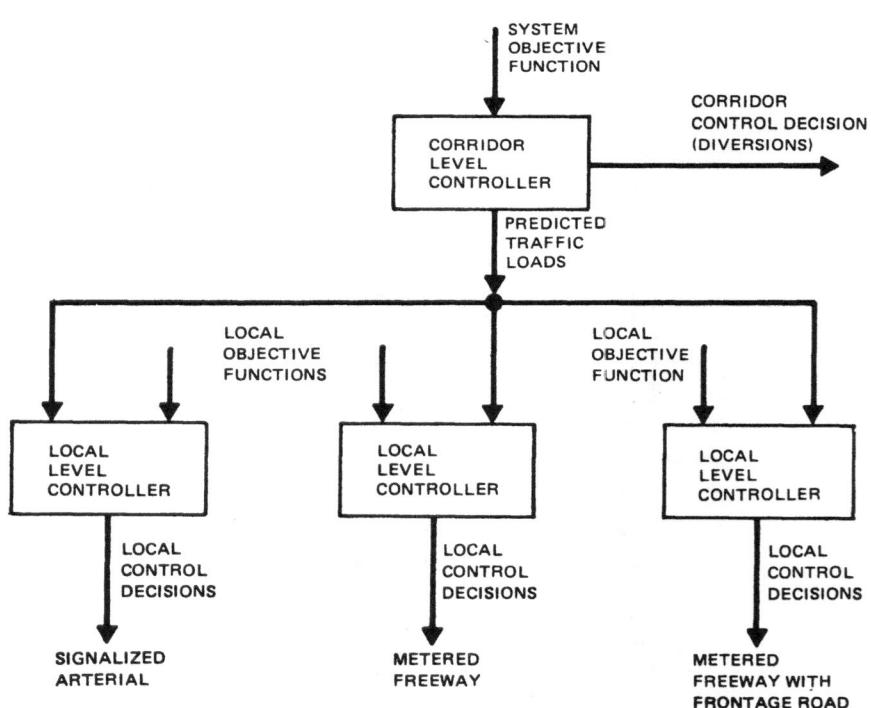

FIGURE 3. DIVERSION ALGORITHM HIERARCHY

Thus, the major components in corridor level control are:
- Selection of an objective function
- Specification of the corridor origin-destination matrix
- Computation of an optimal assignment (diversion fractions)

A wide choice of system objective functions is available to the algorithm user. Selectable by user option are: travel time, speed, throughput, delay, fuel consumption and pollutant emissions or any linear combination of these. These functions can be computed from conventional sensor outputs with reasonable processing requirements.

The second component listed above is the requirement for a corridor origin-destination (O-D) matrix. This is to ensure that optimal assignments are consistent with the origin-destination (O-D) desires of corridor users. Because these desires are continually changing, a method is included in the algorithm which automatically synthesizes the corridor O-D matrix from volume data collectable by a real-time surveillance system. With this technique, costly manual O-D surveys are unnecessary for operation of the algorithm.

Finally, the heart of the corridor level control is the flow optimization, i.e., the computation of optimal diversion fractions at each diversion node in the network. This computation has the following aspects:
- least cost path calculation
- α-search process
- traffic prediction model

The general structure of the flow optimization algorithm is shown in Figure 4.

Based on an initial set of link costs, least cost routes are computed for all trips specified by the present and predicted origin-destination matrix. Since link costs are flow dependent, the actual costs incurred by the least cost routing will be different from those which are assumed. Hence, a convergent iterative process develops: assume link costs, assign traffic over least cost routes, calculate new costs, perform new assignments, etc.

The α-search process refers to a technique which stabilizes the iteration by assuming the next iterated traffic assignment to lie between the previous computed "best" assignment and the newly computed "best" assignment. This results in a smooth convergence to the optimal set of assigned flows. The process is based on the Frank-Wolfe [3] method for solving linearly constrained mathematical programming problems.

Finally, the traffic prediction model calculates system performance as

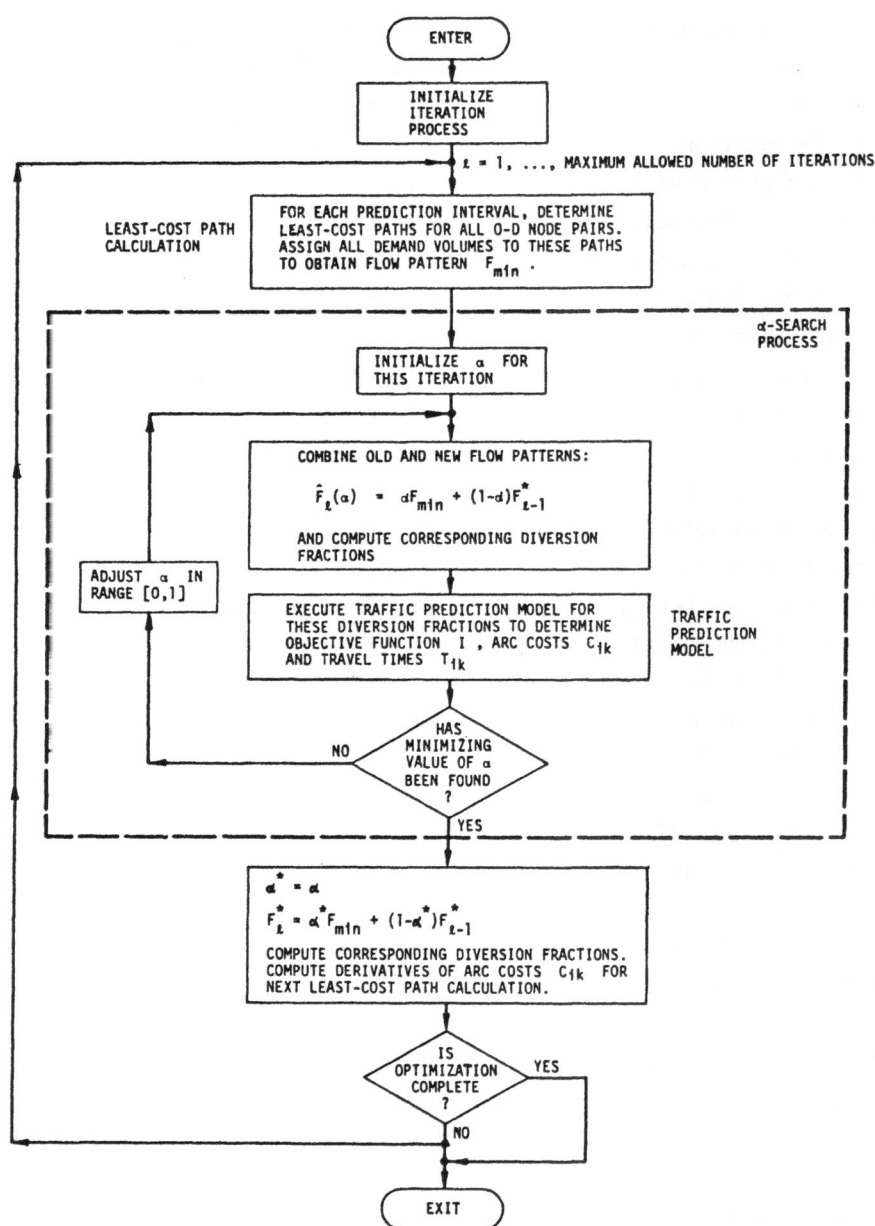

Figure 4.   Flow Diagram of Flow Optimization Process

a function of current and predicted traffic states on each link as well as the cost and travel time to be experienced in traversing each link. Traffic prediction on limited access routes is accomplished using a hydrodynamic flow model with provisions for queue generation and spillback onto upstream links. Signalized arterial prediction is done by superimposing changes in flow as a result of diversions onto "normal" time-of-day traffic changes. The link costs associated with these flows are computed by retrieving previously run TRANSYT [4] evaluations for various flow levels.

Local Level Control Algorithm

After the corridor level control has dynamically assigned the flows, traffic control on individual freeways, freeway-frontage road pairs and signalized arterials must be optimized. To accomplish this, the local level algorithm selects ramp metering rates at freeway on-ramp controllers and controller timing parameters along arterials and frontage roads based on the flows predicted to take place after diversion.

The ramp metering strategy is one which adjusts metering rates to prevent flow from exceeding capacity at any point along the freeway.

Control along signalized arterials and frontage roads is performed by selecting one of a set of patterns previously generated by the TRANSYT signal optimization program.

DIVERSION PROGRAM STRUCTURE

The diversion algorithm consists of three levels of control: (1) the corridor executive, which supervises the overall control process by allocating and scheduling computer resources to the various traffic control functions; (2) the corridor level control, which optimizes traffic flow "in the large" by allocating traffic among the various corridor facilities such as freeways and signalized arterials; and (3) the local level controls, which operate to optimize the use of individual facilities independently from one another, but based on the predicted demand as determined at the corridor level.

Based on this three-tiered hierarchical control structure, the control program has been configured into nine distinct modules, or functions, as follows:

• Corridor Executive Control

Corridor-Level Control
- Demand Estimation
- Corridor Control
- Corridor Control Message Selection
- Performance Evaluation

Local-Level Control
- Ramp Metering Control
- Arterial Signalization
- Incident Detection
- State and Parameter Estimation

The key interactions among these nine functions and between these functions and the corridor surveillance network are shown in Figure 5. Of interest here are the major and minor control loops by which traffic surveillance information gathered by roadway sensors is fed back to the various control functions.

Corridor Executive Control

The Corridor Executive function allocates available computer resources among the other control functions so as to maximize the effectiveness of the corridor control process under widely varying conditions of traffic load and in face of unpredicted incidents of varying severity. It establishes the fundamental rates at which the lower-level functions are to be executed normally, and it maintains a priority-ordered list of these functions for the purpose of resolving potential conflicts. Further, it assures timely response to rapidly changing conditions by enabling specific functions for execution out of their normal sequence when incidents are identified in the system.

Corridor Level Control

1. Demand Estimation

A key ingredient of the corridor-level control process is the estimation of the corridor users' time-varying origin-destination demand pattern. The Demand Estimation function calculates estimates of current and projected origin-destination volume flows between every feasible origin-destination

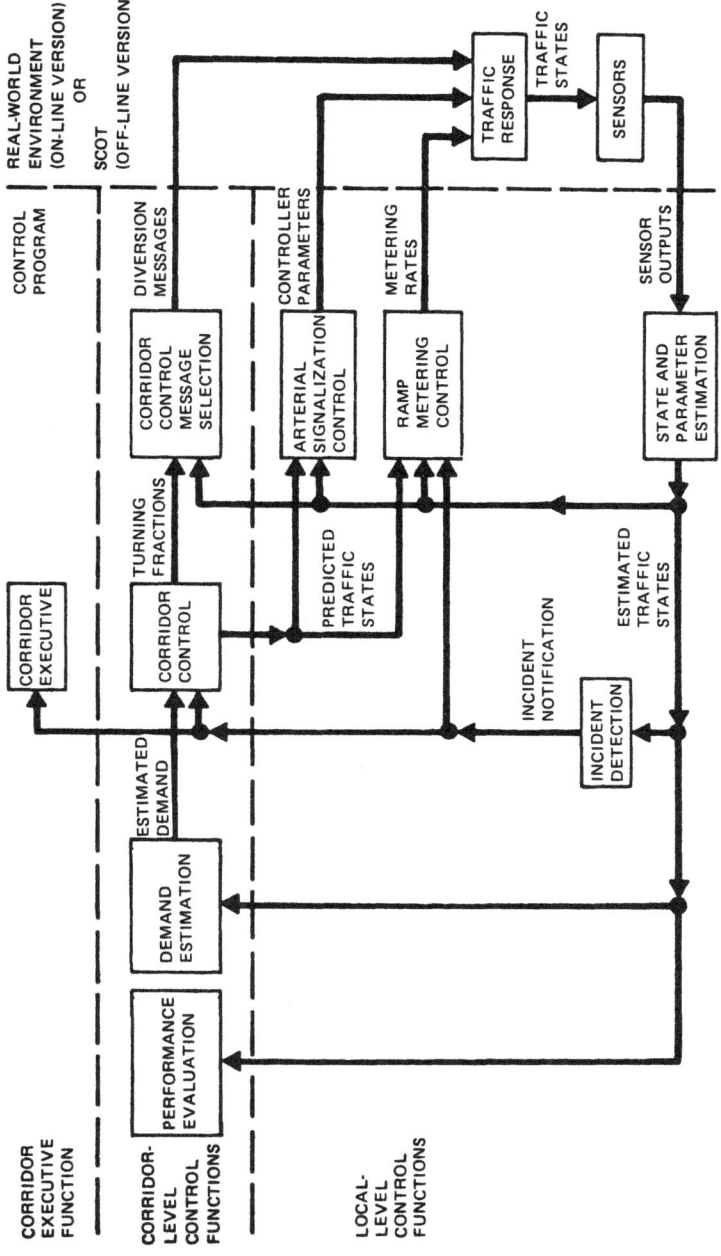

FIGURE 5. CONTROL PROGRAM FUNCTIONAL INTERACTIONS

node pair in the corridor network. The estimates are based on a combination of (1) historical demand data as obtained from O-D surveys, and (2) synthesized O-D data, generated from estimated on- and off-flows at system entrance and exist ramps and estimated link volumes. The future period over which predicted demand data are required is the "look-ahead" interval of the Corridor Control algorithm.

2. Corridor Control Message Selection

The Corridor Control Message Selection function operates the corridor control devices (i.e., message signs) to attempt to shape the traffic flow into the optimized pattern determined by the Corridor Control function. A finite repertoire of candidate messages is assumed, each quantifiable in terms of the amount of diversion that can be expected when that message is displayed. When a change in diversion policy at any diversion node is dictated by the Corridor Control function, a scheme of sequenced message changes is implemented so that a smooth transition from the old policy to the new one is made.

3. Performance Evaluation

The Performance Evaluation function provides estimates of the current performance being achieved, both on a link-by-link basis and on a system-wide level. This function utilizes the current link state estimates to compute measures of performance for each link and for the system as a whole. Generally, these performance measures include the objective function which the Corridor Control function is attempting to optimize, as well as the other link and corridor measures which the control function evaluates on a predictive basis.

Local Level Control

1. Ramp Metering Control

The Ramp Metering Control function calculates metering rates for all metered ramps along a freeway section. The function may be executed independently for each of several different freeway sections in the network. The metering rates calculated by the function are based on the predicted

demands at the on-ramp entry points and the downstream reserve capacities.
In calculating the ramp metering rates, the function explicitly takes into
account expected traffic demands resulting from traffic diversions instituted
by the Corridor Control function.

2. Arterial Signalization Control

The Arterial Signalization Control function establishes cycle lengths,
offsets, and splits for a set of signal controllers along a section of arterial roadway. This function may be executed independently for each of several
such arterial sections within the network.

When exercised for a particular section, this function first makes a
near-term (e.g., ten-minute) prediction of the volumes to be seen at each of
the detector station sites in the section, based on currently measured volumes,
historical values of local demand (or demand changes) and possible diversion
flows resulting from corridor-level control decisions. This predicted flow
"pattern" is then compared with a number of stored reference patterns appropriate to the section, the time of day and day of week. Identification of
the reference pattern most closely matching the predicted pattern also identifies the stored set of controller parameters which is associated with the
chosen pattern, both having been generated together offline. These parameters
include the control variables for each controller in the section, together
with additional parameters which govern the transition process from the signal pattern currently in effect to the new one.

3. Incident Detection

The Incident Detection function identifies freeway incidents (e.g.,
accidents or severe congestion) of a magnitude sufficient to disrupt traffic
flow. It also estimates the severity of such incidents and calls the Corridor Control function to reoptimize the network. Incidents are detected by
examining occupancy data for adjacent pairs of detector stations and util-
izing the California incident detection algorithm to test for incidents.
When an incident is identified, its approximate location is determined, together with the remaining (incident) capacity of the affected link for use
by the Corridor Control function in reoptimizing the network traffic flow,
taking into account the reduced capacity of the incident link.

4. State and Parameter Estimation

The State and Parameter Estimation function provides up-to-date estimates of the current traffic states at each detector station and on each
network link. This function processes real-time roadway sensor data to obtain estimates of the traffic states in the network required by the other
control functions. Specifically, it uses smoothing techniques to calculate
volumes, occupancies and speeds for all detector stations in the network.
For stations or links where detectors are inoperative, estimates are made
based on historical data. Data from individual freeway sensors are then
combined to produce a volume and density on each freeway link for use by
the Corridor Control function. Incident capacities calculated by the Incident Detection function are used in producing state estimates for links and
stations affected by incidents.

Details of the corridor traffic control logic and computer program documentation are given in a series of reports that were prepared by Sperry
Systems Corporation for the U.S. Department of Transportation and are listed
in the References [5,6].

SIMULATION TESTING AND EVALUATION

A series of corridor scenarios was simulated by SCOT [7] in order to
verify the performance of the control algorithm [8]. The results of these
tests demonstrated significant benefits, particularly during incident envir-
onments. Delay improvement of up to 66 percent was observed. During incidents, the extent of congestion decreased by an average of 52 percent, while
duration of congestion decreased by 56 percent on average.

The scenarios were defined to be representative of traffic flow in the
Integrated Motorist Information System (IMIS) corridor located in northern
Long Island, New York. The networks ranged in length from about 3 to 14
miles (4.8 to 22.5 km), including two limited-access facilities and, in some
cases, a parallel arterial. They contained up to four diversion nodes.
Each scenario was executed with and without control. A summary of the algorithm performance for each scenario is given in Table 1. The following
conditions were included in the nine scenarios which were run: typical weekday morning and evening rush-hour peaks; typical recreation peak; off-peak
period; and traffic incidents of varying duration and severity, at critical
corridor locations during both peak and off-peak periods and during recreation

TABLE 1. SUMMARY OF PERFORMANCE BY SCENARIO

Scenario No.	Excess Demand (veh/hr)	Reserve Capacity on Alternate Route (veh/hr)	Duration of Incident (minutes)	Total Delay without Control (min)	Total Delay with Control (min)	Delay Reduction (% Improvement)	Congestion Clearance (% Time Reduction)	Max. Queue Extension (% Reduction)
1	1038	1112	20	13414	12981	3.2	16.7	12.5
2	820	1394	15	11391	9361	17.8	66.7	63.6
3	1235	1530	15	16527	12385	25.1	54.5	74.1
4	587	380	25	11398	11490	(0.8)	50	57.1
5	1385	3029	15	13928	4611	66.9	91.7	81.3
6	403	757	25	4074	3847	5.6	0	0
7	–	–	No incid	24550	23791	3.1	–	–
8	1576	627	20	53847	46227	14.2	50	34.1
9	(132) No excess	1587	30	2093	2022	3.4	No backup	–

peak. A summary of a typical scenario and its results are given in Table 2.

The overall effectiveness of the algorithm can be judged by comparing the baseline case (no control) with the control case using the various measures of effectiveness (MOE) computed by SCOT. The first MOE compared is delay, which was selected as the objective function for all the scenarios.

As seen in Table 1, reduction in delay is greatest in the case of a "hard" incident (i.e., one in which the excess of demand over capacity at the incident site is at least 750 vehicles/hour) when there is sufficient reserve capacity on the alternative route to accommodate this excess. These conditions are met in Scenarios 2, 3, and 5. Generally, as the excess of demand over capacity increases, the benefits in terms of delay will increase provided reserve capacity on the alternative route is sufficient to accommodate the excess. Expected benefits decrease as reserve capacity on the alternative route decreases below the value of excess demand at the incident site.

Expected benefits also decrease as the incident becomes "softer" (less excess of demand over capacity). For example, the no-incident case simulated in Scenario 7 shows only a 3.1 percent improvement. Scenario 9, an incident case where the demand was less than the incident capacity, shows a 3.4 percent improvement. Scenario 4, a soft incident, actually shows a small (less than 1 percent) disadvantage with control. This is not considered significant as it represents only 92 vehicle-minutes of delay out of 11,398 vehicle-minutes.

Another factor to be considered is that the delay improvement values tend to be diluted, for two reasons. First, the delay computed by SCOT includes that experienced by vehicles at signalized intersections. Even with optimum timing, there is some irreducible minimum value of this delay which occurs in both the no-control and control cases. This tends to reduce the percentage improvement achievable by the algorithm. Second, the time span of the scenarios ordinarily included normal periods both before and after the period of congestion. During these periods, only small delay improvements are possible, but the irreducible surface street delay just described continues to accumulate. As a result, the overall percentage improvement for the scenario is less than that realized during the incident. Shortening the duration of the scenarios would thus increase the apparent delay improvement.

Measures of effectiveness which apply only during the incident show far greater improvement. Referring to Table 3, which presents a summary of

TABLE 2. TYPICAL SCENARIO RESULTS

SCENARIO 2

• Typical westbound weekday AM peak

• Single lane blocked by incident on LIE W/B at Willis Ave. (Between Willis Ave. off-ramp and on-ramp)

• EVENT   TIME (MINUTES FROM BEGINNING OF RUN)

Start of run	0
1st control period	4
Incident starts	10
Incident detected	18
Incident cleared	25
Clearance detected	28
Congestion cleared	37 W/O Control 29 W/ Control
End of run	75

• CUMULATIVE MEASURES OF EFFECTIVENESS

MOE	W/O CONTROL	W/CONTROL	CHANGE (%)
Veh trips	18526	18533	—
Veh miles (veh km)	49536 (79718)	49445 (79572)	0.2
Total delay (min)	11391	9361	17.8
Avg delay/veh (sec)	36.9	30.3	17.9
Delay (min/veh mi)	0.23 (0.14 min/ veh km)	0.19 (0.12 min/ veh km)	17.4
Travel time/veh (min)	3.26	3.15	3.4
Congestion clearance time after incident removed (min)	12	4	66.7
Max extension of queue upstream of incident (mi)	1.1 (1.8 km)	0.4 (0.6 km)	63.6

TABLE 2. TYPICAL SCENARIO RESULTS (continued)

INCREMENTAL DELAY
DURING RUN (MIN)

TIME PERIOD		W/O CONTROL	W/CONTROL	IMPROVEMENT (%)
FROM	TO			
7:00	7:04	421	440	(4.5)
7:04	7:14	1217	1151	5.4
7:14	7:18	688	542	21.2
7:18	7:29	2860	1917	33.0
7:29	7:39	1886	1232	34.7
7:39	7:49	1201	1095	8.8
7:49	7:59	1254	1126	10.2
7:59	8:09	1173	1164	0.8
8:09	8:15	691	694	(0.4)
	Total	11391	9361	(17.8)

ALGORITHM COMPUTATIONAL TIME

TIME IN (MIN)	TIME OUT (MIN)	SEC
424	424.3	18
434	434.4	24
438	438.2	12
448	448.4	24
458	458.4	24
468	468.6	36
478	478.4	24
488	488.4	24

AV = 23.24 SEC

SIGNAL TIMING

TIME PERIOD	PATTERN
7:00 – 7:18.2	55 sec Cycle
7:18.2 – 7:30.2	65 sec Cycle
7:30.2 – End of Run	55 sec Cycle

INCIDENT FALSE ALARMS: NONE

MISSED INCIDENT DETECTIONS: NONE

TABLE 3. OVERALL PERFORMANCE SUMMARY

MOE	WITHOUT CONTROL	WITH CONTROL	PERCENTAGE IMPROVEMENT
Average total delay per scenario	16802 veh min	14080 veh min	16.2
Average congestion clearance time after incident removed (per scenario)	16.3 min	7.1 min	56.4
Average extension of queue upstream of incident (per scenario)	1.74 mi (2.80 km)	0.84 mi (1.35 km)	51.7

all the simulation tests, we see that the average time required to clear congestion after an incident is removed shows a 56 percent improvement with control. The length of queue generated by an incident was reduced on average by about 52 percent. These magnitudes indicate the dramatic improvements potentially available during incident conditions.

It can be seen in Table 1 that the incident durations average about 20 minutes. These durations had to be limited so that the SCOT simulation would not be overloaded with vehicles and platoons. Since delay due to an incident can be shown to be proportional to the square of its duration, benefits would have been greater had longer incidents been simulated.

ACKNOWLEDGMENT

This paper is based on research sponsored by the Offices of Research and Development, Federal Highway Administration, under Contract DOT-FH-11-9557 with Sperry Systems Management.

REFERENCES

1. Lindley, J. A., "Urban Freeway Congestion: Quantification of the Problem and Effectiveness of Potential Solutions," ITE Journal, January 1987.
2. Adams, L. H., "IMIS: Computer Control of a Freeway Corridor," NYSDOT, 1982.
3. Frank, M. and Wolfe, P., "An Algorithm of Quadratic Programming," Naval Research Logistics Quarterly. Vol. 3, 1956.
4. Robertson, D. I., TRANSYT: A Traffic Network Study Tool, Road Research

Laboratory Report LR253, 1969.

5. Reiss, R. A., et al. Development of Traffic Logic for Optimizing Traffic Flow in an Intercity Corridor (4 volumes). Final Report under Contract DOT-FH-11-8738. U.S. Department of Transportation, Washington, D.C., January 1978.

6. Gragg, B. B. and Hambly, L. W. Corridor Control Program for Optimizing Traffic Flow in an Intercity Corridor (Program Documentation Reports). Contract DOT-FH-11-8738. U.S. Department of Transportation, Washington, D.C., January 1978.

7. SCOT Model--User's Manual and Program Documentation. Office of Research, Federal Highway Administration, May 1975.

8. Reiss, R. A. Traffic Diversion Software--Applications Summary. Report No. FHWA/RD-80/100, Federal Highway Administration, September 1981.

On integrating public and private traffic control systems

A. BOLELLI, M. CHIFARI(*), A. MARGARIA(**)
R & D Department
MIZAR Automazione S.p.A.
V. Monti, 48 - 10126 TORINO
ITALY

Abstract

In the present paper we discuss the problem of the creation of automated managing and control systems to increase the efficiency of transport on urban areas.
Particularly we consider here the problem of simultaneously controlling private and public traffic when both absolute and controlled priority are required for public means of transportation.
A general description of a feasible architecture for dynamical traffic control systems is given, including some mathematical problems involved.

Foreword

The creation of automated management systems to increase the efficiency of urban areas transportation systems in not an ordinary problem. Frequently it is even difficult to define systems operation quality by a global point of view, because of the strong interactions between urban traffic subsystems such as multimodal public traffic, private traffic, parking areas and so on.
When we refer to automated managing traffic systems these links become requirements for a strong cooperation between all the elements.
Let's consider here, the problem of simultaneously managing and controlling urban private and public traffic on urban areas.
The importance of this integrated management is evident: both kinds of traffic are affected by low average speed due to stops and congestions, high energy consumptions, pollutions etc. Furthermore the strong irregularity in public means motion, also due to stops and congestions makes public transportation less attractive.
In most cases structural interventions are not strictly required. It would be rather necessary to correctly manage main reason of the above mentioned problems, that is traffic light regulatio. at the intersections.
During the last five years two automated control systems that follow

---

(*) C.S.S.T. - Via Giolitti, 48 - TORINO
(**) Area Dipartimentale IX Circoscrizione e Traffico - TORINO

NATO ASI Series, Vol. F38
Flow Control of Congested Networks
Edited by A. R. Odoni et al.
© Springer-Verlag Berlin Heidelberg 1987

this statement have been designed, developed and integrated in Turin: the first is an innovative centralized Traffic Light Control System and the second is a Monitoring and Regulating Mass Transit System. As a result of this experience it's now possible for us to single out main features required for a correct integration between Private and Public Traffic Control Systems.

Main features of the System Architecture

The philosophy of the above mentioned integration is to perform a coordinate traffic control minimizing the total time lost by private vehicles during their trips, subject to the constraint that public vehicles of some selected lines (we say with absolute priority) should not be stopped at the intersections with traffic lights. Also a "controlled priority" can be obtained by determining which trips and on what intersections should be privileged.

To set this goal, we could expect that an Integrated Traffic Control System should be provided with:

- a localization subsystem to update continuously public means positions on the network (with a few meters standard deviation) and to make them available.

- a "public means forecasting module" making use of the knowledge of current positions on the network and the "public means motion model", to point-wise forecasting the motion of every vehicle at traffic light intersections with an increasing accuracy. Here the following hypothesis is made: the forecasting module evaluates arrival time at the intersections on a "time horizon" depending on the traffic light control; to improve the quality of these evaluations, the module is free to "ask" the localization subsystem for new localizations on a given time (with a one-or-two seconds standard deviation).

- a "regularizing" module making use of the departure time-table and the forecasting module evaluations to state where and on what trips, priority must be requested. For those public means of selected lines where an absolute priority is required, this module provides request to every controlled intersection.

- the "traffic light control system" which performs a two level traffic control: area and local control. At the higher level an "observer" making us of the aggregate traffic conditions attempts a private traffic O/D prediction (based on past/present informations). The outputs of the Area Control are desired speed and flows on the network translated in a "reference plan" with boundaries and weights to be assigned to the individual components of local cost functionals.

The local controllers act on their own local model using at their best the information coming from the higher level. At every decision instant their goal is to minimize a cost functional extending the considered time over an interval of the order of a few minutes.

In the functional are taken into account (with weights dynamically assigned by the area controller) both time lost by vehicles at the controlled intersections and the correspondence of local decisions and policy with decisions given as reference by the area controller.
To perform a correct integration of public and private traffic control, the local controller should be able to accept forecasting messages on means motion and to adequate time by time its policy to obtain public vehicles arrival at the intersections during the "green phase".
How can a traffic light control system do this without damaging private traffic control is clearly explained in [1] and proved by the results of the above mentioned first realization in Turin[2].
We should like to introduce here some mathematical models developed for the prediction of public vehicles motion and a generalized algorithm for the regularization of public transportation network.

The public traffic "moving forward" model and the forecasting algorithm.

The required accuracy in evaluating the prediction messages set the opportunity of modelling the slow variation of travel times along the day. The public traffic model we consider is deterministic as regards the vehicle generation (the departure time table of each vehicle is assumed to be known and enforced) and the routes of each trip. It is stochastic instead as to travel time. Hypothesis is made that the travel time of a vehicle can be subdivided into three components:

- free travel time
- waiting time at stop or station
- lost time at the intersection with traffic lights.

The third component could be subject to the system control or not; in the first case we can assume it known, in the second one we could model this component as a noise on free travel time. The first two components can be described by a stochastic process non stationary in the expected value, whose expected value has a slow variation with respect both to the absolute time (time-of-day) as well as to the trip k. This model is employed on-line to forecast the individual time components of each section i:

$$t^i_{K+1} = \alpha\, t^i_K + (1-\alpha)T^i_{K+1} + (1-\alpha)g^i\, R^i_{K+1} + w^i_{K+1} \qquad\qquad (I.a)$$

$$T^i_{K+1} = T^i_K + s^i_{K+1} \qquad\qquad (I.b)$$

$$R^i_{K+1} = \beta R^{i-1}_{K+1} + v_{K+1} \qquad\qquad (I.c)$$

Where

t^i_K is the actual travel time in the section i during the trip k.

T^i_K is the expected value (non stationary) for the travel time in the section i during the trip k.

R^i_K is the expected time component due to the drift for the trip k in the section i

w^i_K , s^i_K , v^i_K are independent noises.

g^i is a gain parameter for the drift time component depending upon the tipology of the section i (i.e. it could be set to zero for a free travel section and non zero for a section including a station).

α,β are once for all fixed parameters.

The model shown here is linear in the states t, T and R.
We can assume the following:

$$m^i_{K+1} = t^i_{K+1} + \delta^i_{K+1} \qquad\qquad (I.d)$$

as the output equation of this model, where δ^i_K is a measuring corrupting noise.

Let p the number of considered sections of a route; an optimal solution to the problem of state estimation could be obtained by introducing the Kalman filter for the 2p+1 state equations (two for each section and one for the drift time component R).

A more feasible sub-optimal solution can be adopted in order to forecast state evolution, by measuring time components t^i_K as the vehicles move on. Let

$$x^i_{K+1} = A x^i_K + n^i_K \qquad\qquad (II.a)$$

where

$$
x^i_K = \begin{vmatrix} \dot{t}^i_K \\ \dot{T}^i_K \\ \dot{R}^i_K \end{vmatrix} \quad A = \begin{vmatrix} \alpha & (1-\alpha) & (1-\alpha)g^i \\ 0 & 1 & 0 \\ 0 & 0 & \beta \end{vmatrix} \quad n = \begin{vmatrix} \dot{w}^i_K \\ \dot{s}^i_K \\ \dot{v}^i_K \end{vmatrix} \qquad \text{(II.b)}
$$

For all the sections i crossed by the vehicle during the trip k.
At any trip k+1 the current state estimation $\hat{x}_{K+1/K}$ and its a priori variance in given by:

$$
\hat{x}_{K+1/K} = \hat{x}_{K/K} + (CR^{-1} C^T + \textstyle\sum_{K+1/K}) \cdot CR^{-1} \cdot (\dot{m}^i - C^T \cdot \hat{x}_{K/K}) \qquad \text{(II.c)}
$$

$$
\textstyle\sum_{K+1/K} = A \sum_{K/K} A^T + Q \qquad \text{(II.d)}
$$

where $C^T = (1\ 0\ 0)$ (remember that only $\dot{m}^i$ is available) Q is the co-variance matrix of the input noise n, R is the co-variance matrix of the output corrupting noise δ.

The advantage of this recursive estimation is that a simple computation can be used to provide state estimate, as each new data ($\dot{m}^i_K$) becomes available, and therefore it offers the realistic possibility of on-line computer implementation. The state estimates, can be used to predict the "arrival times" to "section goals", on which the regularizing algorithm is based.

Furthermore it can be shown that the discrete Riccati equation given by (II.d) grows to a steady state in a few trip k when

$$
\textstyle\sum_{K+1|K} = \sum_{K/K} = \sum \qquad \text{(IV)}
$$

Hence the equation (II.a) becomes

$$
\hat{x}_{K+1/K} = \hat{x}_{K/K} + \textstyle\sum CR^{-1} \cdot (\dot{m}^i - C^T \hat{x}_{K/K}) \qquad \text{(V)}
$$

The "Regularizing" Control algorithm.

This algorithm makes use of the available time-tables of each trip of a public transportation route forecasted to any significant section goal (according with the previously introduced model) to determine regularity conditions and adequate adjustments.
We describe here the regularizing method employed to perform these functions using general feasibility rules. These rules are defined by the following statements:
each vehicle is taken into account together with the N/2 followers and N/2 leaders. If the motion of the total N + 1 considered vehicles was "regular" they should be located on the route according to the nominal time-table and particularly their mutual distances should be coherent with the nominal frequency of service.
We aim to choose the optimal sequence of adjustments to reach a regularity condition setting the following goals:

- the sequence of adjustments should obtain intervals between vehicles as close as possible to those generated by nominal frequency of service.
- The adjustments should be as small as possible to make control "robust" and cheap; they should approach vehicles to the nominal time-table.

The previously introduced goals state that "intervals", as well as "time-tables" must be taken into account; furthermore a "robust" control must be obtained. This task can be accomplished by introducing a "generalized" delay as input variable for the control algorithm. In fact, the simple "delay" versus the time-table could not provide a robust control, because of the inherent difficulty of following fixed time-tables in high frequency conditions; on the other hand the service level, in those conditions, is related to "intervals" between vehicles (which can be easier controlled in high frequency routes).
In other conditions (e.g. in low frequency routes) a strict observance of the time-table is required by the users. Finally a useful control must take into account deviations versus "time-table" as well as versus "intervals"; moreover, it must be able to assign variable weights to both deviations.
In the following the "generalized" delay is introduced: it must be simply computed on the basis of the service route information.
Let R and R be the time delays of the examined vehicle and the j-th vehicle among the remaining (N/2 leaders a N/2 followers) versus the nominal time table, observed on a specific section goal along the route; let x and x be the ideal objective adjustments for the same vehicles; O and o respectively the nominal and the actual time interval between the vehicle j and the examined one; we assume the following conditions:

$$O_j' \cdot a = o_j' + x_j' - x \qquad \text{with } a > o \qquad \text{(VI)}$$

condition (VI) implies that a feasible control could accept only a small, constant variation in "service level", here represented by "a". The control task can now be defined as the one providing the smallest "a" and $(x_1, x_2, \ldots x_N)$ in the domain of the feasible ones where $X : (x_1, x_2, \ldots x_N)$ is the sequence of the desired adjustments. To find X and "a" taking into account the previously defined goals, we should minimize a cost functional depending on delayes versus the time table and differences intervals. Let us define:

$$F = b \cdot A + g \cdot E + L \qquad \text{(VII.a)}$$

with the constraints expressed in, where

$$A = (R - x)^2 + \sum_{i=1}^{N} (R_i - x_i)^2 \qquad \text{(VII.b)}$$

$$E = x^2 + \sum_{i=1}^{N} x_i^2 \qquad \text{(VII.c)}$$

$$L = (1 - a)^2 \qquad \text{(VII.d)}$$

and b, g $\geqslant$ 0

In the functional F it can be observed how the term A is introduced to take into account delays versus the nominal time-tables, the term E to weight the control signals, the term L (with the imposed constant) to pay for "service level", i.e. for differences in high frequencies. So, a feasible control con be obtained by minimizing the given functional:

$$\min_{a,X} F \qquad \text{(VIII)}$$

where

$$Tn = \sum_{j=1}^{N} O_j' \qquad \text{(IX.a)}$$

$$Te = \sum_{j=1}^{N} o_j' \qquad \text{(IX.b)}$$

$$B = \sum_{j=1}^{N} O_j'^2 \qquad \text{(IX.c)}$$

$$P = \sum_{j=1}^{N} (o_j' * O_j') \qquad \text{(IX.d)}$$

$$x = \frac{b}{(b+g)} \cdot R + \frac{g}{(b+g)} \cdot \frac{(Te-Tn) + (b+g) \cdot (Te \cdot B - Tn \cdot P)}{(N+1) + (N+1) \cdot (b+g) \cdot B - (b+g) \cdot Tn^2} \qquad (X.a)$$

$$a = \frac{b \cdot B}{(b-g) \cdot B + 1} - \frac{g \cdot Tn}{(b+g)+1} \cdot \frac{(Te-Tn) + (b+g) \cdot (Te \cdot B - Tn \cdot P)}{(N+1) + (N+1) \cdot (b+g) \cdot B - (b+g) \cdot Tn^2} \qquad (X.b)$$

The (X.a) expresses the "generalized" delay for the examined vehicle.
In the particular (but very common) case of equal nominal time between
each couple of the considered N vehicles, we have Tn = 0 and then the
equation (X.a) can be reduced to:

$$X = \frac{b}{b-g} \cdot R + \frac{g}{b+g} \cdot \frac{Te}{N+1} \qquad (XI)$$

expressing a weighted average between the delay of the vehicle and the
error on time between vehicles. This simplified equation allows to
consider how the adjustments can be reduced to the nominal delay R as
regard to the nominal time-table (when g=0) or to the error as regard
to the mean time between vehicles (when b=0).

Conclusion and acknowledgements.

The mathematical models here described have been developed, experimen-
ted and now successfully employed for on-line control through the
integration of public and private traffic control systems carried out
in Turin during the last year.
The Authors express their grateful appreciation to the Transportation,
Traffic and Road Division of the City Administration of Turin, to the
Mass Transit Authorities and to all participants to this project.

REFERENCES

[1]. Dorati F., Mauro V., Roncolini G., Vallauri M. (1984) - A hierarchical decentralized traffic light control system. The first realization. IFAC 9th World Congress Vol. II 11.6/A-1.

[2]. Rif. IX "LL.PP. Ispettorato circolazione e traffico (1986) - Sistema di controllo del traffico pubblico e privato "Progetto Torino":
Valutazione delle prestazioni.

[3]. Bolelli A., Nepote F., Di Taranto C., Sapienza S. (1986). Inte-
grazione di sistemi semaforici e di monitoraggio per la regolariz-
zazione del servizio pubblico. Atti 4° Convegno Nazionale CNR Vol.
II 3-27.

AN EXPERT SYSTEMS APPROACH TO NONLINEAR OPTIMISATION

J.J. McKeown

Department of Civil Engineering

Queens University

Stranmillis Road

Belfast, N. Ireland

ABSTRACT

Existing methods for the numerical solution of optimisation problems are very powerful. However, this potential is often not realised in practice because few users of optimisation programs possess the specialist skills needed to use the available subroutines effectively. Difficulties very often arise with regard to problem formulation, algorithm selection, trouble-shooting and the intrepretation of solutions. In addition, many practical problems violate the strict assumptions underlying standard algorithms and must be used interactively to attempt to find solutions. This paper proposes an expert-systems approach to this problem and points a way to the development of systems which would be much more powerful problem-solving tools than the sum of the individual numerical subroutines contained within them.

NATO ASI Series, Vol. F38
Flow Control of Congested Networks
Edited by A. R. Odoni et al.
© Springer-Verlag Berlin Heidelberg 1987

SECTION 1 : INTRODUCTION

Nonlinear optimisation techniques are very powerful iterative tools for the solution cf minimisation problems either with or without constraints. Such techniques existed only in fairly rudimentary form (in most cases) prior to about 196C. The subsequent rapid development, particularly during the 70's, was certainly stimulated by the increasing power and availability of digital computers.

As a result of these developments, problems can now be solved reliably and accurately which previously could only be approached in simplified forms, or indeed which heretofore it would not have been considered practical to attempt. Examples of the former class can be found in the field of nonlinear parameter estimation and curve fitting, where simplified models or devices such as logarithmic transformations had to be used in order to achieve any approximation to a solution. Again, many engineering problems, from minimum weight structural design to the optimal design of nonlinear electronic circuits, are only now becoming feasible as routine operations because of the newly-developed methods. At the present time, however, a curious situation exists which has attracted surprisingly little comment or attention as yet; for while there is now a battery of powerful techniques, their availability seems to be virtually unknown to many of those who could make most use of them. Mary even of those who are aware of the new possibilities are not sufficiently skilled to make effective use of the algorithms available.

This situation exists despite the wide dissemination of excellent computer implementations of the best of the new algorithms; for example, the NAG

Subroutine Library and the OPTIMA package of the Numerical Optimisation Centre at Hatfield Polytechnic. Such libraries and packages, if intended for general use, invariably consist of a set of individual subroutines; if directed towards specialised use, eg for network analysis, their user interfaces make them inaccessible for other users. In the case of the subroutine libraries the onus is on the user to formulate his problem, to select a suitable algorithm, to analyse any initial failures (a crucial requirement) and to take corrective action, to interpret the results obtained and to assess the level of confidence which can be placed in them. Bearing in mind the fact that virtually all the current methods will find only local rather than global minima, and are effective only when certain assumptions about the behaviour of the objective function and constraints are satisfied; and given that in almost all practical cases such conditions are impossible to guarantee a-priori, the likelihood of a failure somewhere along this chain is high. It is therefore fair to say that at this time nonlinear optimisation techniques are accessible to general users only through the mediation of a numerical analyst. It is also true to say that most Computer Centres even at Universities lack such staff, with the result that users often reach the trouble-shoot stage only to conclude (wrongly) that the optimisation subroutines selected are unable to solve their problem.

This situation will sound familiar to many engineers, scientists and researchers. It does not reflect badly either upon the designers of the algorithms or upon the programmers who have implemented them. Rather, the problem lies in the neglect of the crucial interface between such professionals and the users on whose behalf, after all, the whole process of research, design and development has been carried out. It is tempting to place the blame instead upon a system which trains mathematicians who can design algorithms but underestimate the problems of making them accessible and computer scientists who understand software engineering but are often less than familiar with numerical techniques. However this may be, there is now a clear need for the development of bridging software, which will enable practitioners of any numerate discipline from economics to molecular biology to make full use of nonlinear optimisation methods without specialist training or advice and with complete confidence in (and reponsibility for) the results.

Such a project will be at least as demanding as the development of many of the algorithms themselves. With its help, a user will be able to formulate his problem and specify it in computer-executable form; to make an intelligent choice of optimisation algorithm (and perhaps to experiment with several); to diagnose initial faults and to have corrections made rapidly and easily, for example, by rescaling; to interpret the results and to make comprehensive, flexible sensitivity analyses. All the above steps must be possible within a context which allows backtracking, reformulation, requests for guidance and restarts, and it must lead ultimately either to a clear solution with well-defined measures of its credibility, or to the early detection of the intractability of the problem. Ideally, the word 'intractable' in this context would mean that there is no numerical solution to the problem, not that some arbitrary assumption has been violated. Such a system would be much more than a mere addendum to existing subroutines. It would be more powerful than any one of them because it could be used heuristically to solve problems outside the scope of its kernal numerical subroutines. It would thus allow the intuitive skills and the experience of the real expert to be brought to bear on a problem, while taking full advantage of good numerical techniques for the sub-problems.

A relatively few years ago, the development of such a system would have been infeasible for two, and possibly three, reasons. These were: lack of suitable optimisation .algorithms to make it worthwhile; lack of widely available (ie cheap) interactive computing power; and possibly, lack of understanding of techniques such as Artifical Intelligence, which, in this context, means Intelligent Knowledge-Based (IKBS) or Expert Systems. All of these prerequisites are now to hand, and the opportunity therefore exists to make a step forward in optimisation techniques which will at the same time both increase the power of these methods and ensure a much wider acceptance for them.

In this paper we shall consider in more detail the inadequacies of the present facilities, propose means by which these can be removed, and discuss the work already carried out towards this end.

SECTION 2 : THE DIFFICULTIES OF USING OPTIMISATION SUBROUTINES

The difficulties which many users experience when using or attempting to use nonlinear optimisation techniques for the first time are not merely incidental. They are fundamental and they stem from a particular model of the typical user which seems to be widely held by designers of optimisation software. This may be illustrated by considering the two following scenarios.

Scene 1 : The mathematician's tale (or how optimisation should be done)

Step 1   A potential user perceives the need for, or the advantage to be gained from, numerical optimisation techniques.

Step 2   He formulates his problem accordingly.

Step 3   He finds a suitable subroutine library and, with the help of the User Manual, selects an appropriate subroutine.

Step 4   He codes his problem, debugs and tests it.

Step 5   He runs his program, finds a solution and goes away satisfied.

This scenario, however implausible, seems to be the model underlying the provision of 'naked' subroutines to prospective users. A more likely scenario is offered next.

Scene 2 : The user's tale (or how optimisation fails to be done)

Step 1   The potential user, more often than not, fails to appreciate that optimisation techniques could be of use to him. We hear no more from him

Step 2   Or, the potential user does perceive a need, and proceeds (with misgivings)

Step 3   His Computer Centre directs him to a Subroutine Library user manual. From this he learns how to formulate his problem, select a subroutine and chose suitable values for the various parameters required by it.

Step 4   The user codes his problem, tests it and partially debugs it.

Step 5   He attempts to run his program, whereupon:

- the procedure stops after one or two iterations and prints an unhelpful message, or

- the procedure runs and runs, producing no useful results, or

- the procedure crashes, or simply goes quiet, or

- the procedure stops at a "solution" which the user knows to be wrong, or

- (fill in your own experience)

Step 6   The user goes away with his worst fears and prejudices confirmed and tells his friends that (a) the subroutine, (b) the Subroutine Library, and (c) optimisation in general are useless

Step 7   He devises his own algorithm (he always knew it was easy, anyway), which also fails, but in ways which are understandable if not correctible. He is now satisfied that numerical optimisation does not work and finds another way to approach his problem.

It will be noticed that the first of these scenarios is quite linear; there is not scope for initial mistakes, followed by feedback, learning and adjustment. It either works first time or, probably, not at all. The second, while perhaps a little whimsical, will probably sound familiar to most of those who have been concerned with the practical use of nonlinear optimisation methods; it could certainly be elaborated almost to the dimension of a tragic epic.

The difficulties encountered by non-specialist users of optimisation programs would seem to stem not from some incidental feature of the existing software but from the algorithm – rather than problem-orientated philosophy of program designers. For, while it is certainly fair to say that it is useful to have available a range of subroutines for finding local minima of smooth, continuous functions subject to smooth continuous constraints, it is not true that a collection of such subroutines constitutes an adequate facility for solving practical optimisation problems in the absence of a consultant numerical analyst. One is tempted to say that a set of subroutines is as useful to non-specialist users of optimisation methods as a box of transistors, diodes and resistors is to a typical hi-fi enthusiast. These are strong statements; how can they be justified?

Consider the difficulties which can be encountered at each of the main stages of an optimisation project.

1.   The decision to use numerical optimisation

Firstly, although optimisation is now taught in many undergraduate courses, few of those whose work might benefit from its use are aware of the real power as well as the limitations of optimisation methods. It is quite usual to find researchers using inefficient ad-hoc methods, unaware that numerical optimisation could probably solve their problem with little trouble. One particularly noticeable lack is of the awareness that optimisation methods can still be used even when the analytical form of the objective function is unknown. The first problem is therefore that of education. This can be addressed in two ways by means of an expert system of the type which we shall propose. Firstly, such a system would clearly have uses as an educational aid, allowing students to tackle a variety of real and simulated problems more quickly, and in a more controlled environment, than might otherwise be possible. This direct use of the system might, however turn out to be less important in the longer term than the change in the general perception of the potential benefits to be gained from optimisation methods which will come about once the user community comes to associate then with rapid, powerful problem solving rather than with potential frustration. At the present time those who advocate the more general use of these methods must be prepared to tackle queries and even complaints from those who take their advice. The aim of the development proposed would be to make users almost completely self-sufficient and indeed to make them advocates in their turn.

2. Problem formulation

Once it has been decided that numerical optimisation is to be used, the problem of formulation remains. There are several common difficulties associated with this. Inexperienced users often have a tendency towards over-elaborate manipulation of their problem. For example, in nonlinear least-squares problems to seek an expression for the residuals of the form: $y - F(X,Z)$, where the general form $G(y,X,Z) = 0$ may be more convenient. Such over-elaboration, while not serious in itself, is a potential source of potential mistakes in the coding of the problem. Once the user realises that he can use existing analysis models, with a little modification, as function evaluation routines he is coming closer to the spirit of the method. However, there are still possible pitfalls to avoid. For example, in the optimisation of an engineering structure, an engineer might use an existing routine for stress analysis as the basis for his constraint evaluation program. Such a routine would almost certainly have been developed for use in circumstances in which the input data were prepared, one run at a time, by an engineer and carefully checked for 'ridiculous' errors such as negative physical dimensions. The optimisation routine, on the other hand, will feel no compunction about using such values if they are consistent with the logic of the algorithm. Indeed, even if positivity constaints are explicitly imposed, some algorithms may violate them during the course of the search for a solution although not, of course, at such a solution. Other algorithms will not. So although it is correct in principle to use an analysis routine in this way, it is important to realise that doing so will possibly place a much heavier strain on the quality of the programming, and even of the underlying analysis, than normal use of the routine would do. Failure to appreciate this can lead to unpredictable and discouraging results. Another example which springs to mind is that of nonlinear parameter estimation. If 'outlying' data points are discarded as in a linear analysis – for example, by using an existing routine – the result may be that some data points will fluctuate in status during the search, thus inducing a form of noise in the objective function which might well prevent convergence. The effects can be quite subtle.

Of course, it is quite possible that the objective or constraint functions are genuinely discontinuous or otherwise ill-behaved. Discontinuous first derivatives are of course a feature of minimax problems; but they also occur because of approximations used in the evaluation of functions which are thought to be smooth. Table-lookups and their associated interpolations are rich sources of such difficulties; but they can arise in so many ways that not all eventualities can be foreseen and catered for within a closed program. Instead, users need to have their attention focussed on problems as they arise during a run, and to be presented with diagnostic information and suggestions for corrective action. Such corrections cannot be made automatically by the program, since they will often involve some degree of reformulation of the problem; such reformulation must be made as easy as possible. All this, incidentally, has not even touched upon the more fundamental question of whether the user sees his problem in clear-cut terms, with a unique objective function and constraints. Often in practice the formulation will not be so obvious or natural, and will involve a choice of priorities. For this reason, too, it must be made as simple as possible for the formulation to changed and for alternatives to be explored. It is by providing such facilities that the foundation is laid for the use of the system as a heuristic tool.

3. Algorithm Selection

The next problem is that of selecting an appropriate algorithm. This is usually catered for in Subroutine Library user manuals by providing a decision tree, which is traversed according to criteria such as whether or not analytic derivatives are available, whether constraints are present and so on. This is of course a valid approach, but tends to be limited to considerations of what the algorithms can do, rather than how the user's problem can be solved. For example, most if not all of the subroutines in the library will be intended to find local, rather than global solutions. This is because the local minimisation problem is easier to pose and solve, and because algorithms to solve the global problem are extremely time-consuming. So the user is more or less assumed to be satisfied with a local solution and little more is said about it. Now in fact local solutions are often not acceptable, as such, as answers to practical problems. For example, engineers are accustomed to assess the stability of earth embankments by searching for the failure mechanism (usually the slip circle) associated with the minimum safety factor. Local minima are of no interest. Probably, indeed, this situation is typical of most applications in which optimisation is used as a method of analysis as opposed to synthesis. In practice, local methods can be used to establish global solutions, to a sufficient degree of confidence and without undue computational effort, if the user is free to explore different starting points and to carry out sensitivity tests. To do this properly requires rather more than a set of local routines and would appear to be a task well suited to an interactive system with a built-in knowledge base.

4.   Setting parameter values

Assuming that the potential user has found an appropriate subroutine, he is now faced with another set of decisions: what values should he assign to the various parameters which all optimisation subroutines require? The existence of these parameters is perhaps the clearest indication of the weaknesses inherent in the 'subroutine' approach, and of the unreality of the first scenario given above. Their values are left for the user to assign because the algorithm designer knows that no fixed values can reasonably be expected to cover all the cases which might arise. For the same reason, he can offer the user no clear-out advice on how to set the values himself. The manual will of course suggest that if necessary different values can be tried until satisfactory results are achieved; but how does a 'naive' user distinguish between difficulties caused by inappropriate values for the parameters and the host of symptoms which can arise from other sources? Only a competent numerical analyst is likely to possess the experience to know. In his absence an expert system is likely to be of considerable help.

5.   Trouble shooting

This brings us to the highest hurdle which the user must clear. The decisions which he has made up to this point will now bear fruit in the form of a successful run (if all decisions were right, and if the problem satisfies the assumptions already referred to); or difficulties will arise simply due to 'ill-behaviour' in the functions defining the problem; or interaction will take place between all these factors to create symptoms which are difficult to diagnose without both knowledge of the underlying procedures and the means to carry out systematic investigations. Both of these can be provided by a suitable expert system; without them a satisfactory conclusion is unlikely to be achieved. Among the most common causes of difficulty at this stage is poor scaling. Few routines are proof against this, yet on occasion a simple change of units can make the difference between success and failure, or between a run requiring a few seconds and one lasting many minutes. Even if the user is quite familiar with the model he is optimising, it is unlikely that he will be accustomed to thinking of it in these terms. A suitable knowledge-based system would be able to monitor progress and identify the symptoms of poor scaling with some certainty and offer options for improving it. There are, of course, many other ways in which a run might fail; the diagnosis of them is discussed in a later section.

6.   Interpretation of the solution

Let us assume that, in spite of the difficulties discussed above, the user has completed a run and has been presented with a set of numbers allegedly representing a solution to his problem. In support of this allegation he will be given the values of first derivatives of his objective function, his constraint values and his Lagrange Multiplier values. This situation is reminiscent of the computer in a recently popular novel which, after some years of deliberation announced that the answer to the fundamental problem of the universe was 42. In both cases, what the users had hoped for was an answer which was more self-evidently true. In our case, it should be possible to explore the vicinity of the given solution in various ways and to carry out systematic sensitivity analyses of the kind described in ref 1. At the present time the OPTIMA package contains a subroutine for second order sensitivity analysis which could form the basis for this; but the system itself should provide guidance as to the interpretation of the results of such analysis. The confidence-building effects of such a facility can be very great.

However great the difficulties catalogued above may be, they do not invalidate numerical optimisation as a problem-solving technique. On the contrary, these methods together constitute what is possibly the single most powerful approach to solving practical problems by mathematical means. However, what we have hoped to show is that, at the present time, optimisation ought not to be presented to users as a method which they will be able to use simply with the help of a set of subroutines and a user manual. Instead, some means must be found to make the knowledge and experience of a specialist optimiser available to every such user, and to combine this with the means to explore alternatives and thus to arrive at a solution for which he will be happy to take full responsibility. In the next section a means of satisfying these requirements is discussed.

SECTION 3 : AN "EXPERT SYSTEM" APPROACH

In this section we shall discuss an approach to overcoming the difficulties described in Section 2. We shall first define what, in this context, we mean by the term "expert system", then we shall consider what an expert system for optimisation should be expected to do and how its component parts should contribute to this.

3.1 What is an expert system?

There are certainly three factors, and possibly four, which together now make possible a practical optimisation system. These are:

1. Availability of reliable subroutines for most of the main classes of local optimisation problem;

2. Increased availability of cheap computing power;

3. Progress in computer graphics and in interactive computing;

4. Possibly, the development of new ideas in Artificial Intelligence.

The first of these factors has perhaps been sufficiently commented upon already. Suffice to say that, for smooth, continuous well-behaved functions, the problem of unconstrained local optimisation has, for most purposes, been solved. For constrained problems the situation is not quite so satisfactory or so clear, but very powerful methods exist which, used properly, can solve a very wide range of applications. The global problem, strictly defined, can only be solved in the case of small problems because of the very heavy demands which it places on computer time; but a wide range of algorithms have been published and these certainly have their place in the type of system which we envisage. The second and third factors are currently bringing about a qualititative change in the way in which people view digital computers and how they expect to use them. The trend away from seeing computers as numerical processors is accelerating and attention is focussing more and more on the interface between user and machine. The change is becoming manifest in several ways. Home computers are training a generation of children, through game-playing, to see them as picture manipulators, with the computing made subservient and hidden in the background. Much though parents may decry this, the fact is that professionals are moving in the same direction, albeit more slowly. It is well known that a grahics terminal showing a moving simulation of a complex molecule will attract more attention at an exibition or a conference than a terminal or microcomputer producing printout. Of course, good graphics demands both large memory and serious computing power. But these are becoming every more readily available, feeding on the appetite they generate. Already the effect of this is being felt in the field of operational research, where there is an increasing tendency to replace the use of closed algorithms for scheduling with heuristic techniques based on interactive manipulation of on-screen simulations. It is significant that the latter methods are used directly by the scheduler himself, on-line, while the former were batch-run and often required the services of a consultant. The effect of these trends is to demystify both the computer and its acolytes; it leads to a situation in which the user expects to be in control, and to take full responsibility for the results which he obtains. Systems which aid him in this task are what we will denote by the term 'expert system'. It will be noted that there is no desire here to replace the user; on the contrary such a system, by removing irrelevancies, is meant to amplify his role. The only expert whose role is threatened is that of the mediator between user and machine. The reader of a text on expert systems (eg Hayes-Roth, ref 2) will note that this is a slightly specialised use of the term.

3.2 What should an optimisation system do?

An optimisation system, in the author's view, ought to provide a means whereby a user can make, test and change relevant hypotheses rapidly and without confusion, leading to a solution of the real problem. A hypothesis might be a problem formulation, a choice of algorithm, a value for a parameter or for any other variable which needs to be under his control. Advice should be available instantly on any of these topics, and absurd decisions ought to be forbidden. Otherwise, full control should be in the hands of the user. The advice on offer ought to take congnisance of any experience built up on previous runs of the same problem. During the optimisation process, monitoring should be carried out and advice given on any problems which arise. Once a possible solution has been found, it should be possible to request all the information needed to evaluate it critically, and this information should be presented in a convenient and flexible way. Graphical display of information should be used wherever it can be of advantage. Where serious difficulties arise, and indeed at any point at which he sees fit, the user should be able to interrogate the system as to the basis for its actions. Such interrogation should be possible on various levels of detail, so that if necessary, the user should be able to obtain as much information about the process as if he had carried it out manually. The components of a system capable of carrying out these functions is shown in Figure 1. The functions of these components will be discussed in more detail below.

3.3 The knowledge base

The knowledge base is that part of the system which most directly enables it to play the role of the consultant numerical analyst. Its functions is to store the information which will allow the system to respond to requests for advice, using built-in rules and basing its response on both the answers to questions put to the user and the information generated by the progress monitor. The rules which it contains must be constructed in such a way as to embody the experience of as wide a range of expert optimisers as possible. Some examples of such a rule are as follows.

Example: Rules for action in the event of convergence difficulties

Such rules are of two types:

> **Type 1 rules** relate the observed behaviour of the iterative process to one or more antecendent conditions;

> **Type 2 rules** relate antecedent conditions to recommendations for action.

Possible type 1 rules:

(1)   If: convergence on several runs from various starting points is rapid and always to the same point

      Then: the problem is well-posed and the solution is a local, and possibly a global, minimum.

(2)   If: the process stops at a significantly different point on each of several runs from different starting points

      Then: either: the convergence criterion is too loose

             or: the function is badly scaled

             or: the function is discontinuous

             or: the function is not smooth

             or: the function is constant

             or: there are many local minima

(3) If: the function value decreases only slowly

     Then: either: the function is badly scaled
           or: the function is noisy

(4) If: the gradients are being estimated by divided differences and the
           components fluctuate widely in value from one point to another
           close by

     Then: the step length is too small
           or: the function is too noisy

(6) If: the process terminates without convergence, having completed a
           number of iterations

     Then: either: the convergence criteria are inconsistent with the
           accuracy of function or gradient evaluations

           or: iterations have led to a point on the edge of a region in which
                  the function or constraints cannot be evaluated.

These rules are by no means complete, either individually or as a set, in
fact, they are probably not even consistent. Alternatives should have
probabilities associated with them. But they illustrate the form in which
some of the knowledge must be presented to the system, and they give a flavour
of the effort needed to construct even a modest set of rules which would
adequately cover all eventualities in a limited context, and which would be
mutually consistent.

3.4 The progress monitor

Nonlinear optimisation algorithms are not usually demanding of computer memory, by modern standards. This is mainly because the storage space is usually proportional to n^2, where n is number of optimisation variables, or perhaps $(n+m)^2$, where m is the number of constraints. In most cases, n and m are quite small, less than 50. This contrasts with the linear programming case, where these numbers can be of the order of thousands. These modest requirements mean that quite powerful nonlinear algorithms can be run on small computers, although in practical cases the computing time may be prohibitive. On a moderately sized computer, therefore, it is now feasible to store information about past iterations, and therefore to endow an optimisation process with a useful memory over and above that exploited by the algorithm itself. Thus, information about rates of reduction in the function values, gradients and constraints can be stored for many iterations. This information can be structured, stored and made accessible to the plan module, which then calls upon the knowledge base to assess progress against expectation.

Clearly, the availability of a file of information about past actions and their effects means that quite complex strategies can be implemented. The control of such strategies is the job of the plan module.

3.5 The plan module

It is usual in expert systems to include a module which implements some kind of solution strategy. One can clearly see the need for this in the case of, say, a system for medical diagnosis, there are clearly different ways in which the various possible alternative diagnoses can be searched for and eliminated. In the present case the need for such a module may be less clear, particularly since each optimisation algorithm in fact incorporates a solution strategy. Indeed, in the early stages of development it is anticipated that the module will play a nominal role. However, much of the effort involved in obtaining a solution to a practical problem is in fact extra-algorithmic, for example, reformulation, changing parameter values, changing the choice of algorithm and other actions which cannot be covered by a closed numerical algorithm. Eventually the knowledge about how best to go about such actions will be built into the plan module, which will use it in conjunction with information from the progress monitor and input from the user to advise on the best sequence of steps to be taken, for example to find a global minimum.

3.6   The justifier module

It is striking to note, in reading the literature on expert systems, (eg Ref 2) how often the comment is made that the most useful single feature of a given expert system is its ability to explain why it has done what it has done. Indeed, a moments consideration will show how such a facility distinguishes the expert system from virtually all traditional programs. The current trend towards demystification of computers has already been mentioned; here we see it brought to its logical conclusion - the computer being made to explain and justify even the most minute of its actions. In the past, it must be said, optimisers have made little effort, on the whole, to ensure that users could know what was going on in their subroutines. This has certainly contributed, on occasion, to the disillusionment which users must have felt when difficulties have arisen. One of the principles underlying our approach is that an optimisation system is capable in principle and ought to be capable in practice, of explaining its actions in terms which are understandable by any numerate user. Achieving this in practice will be a challenging assignment; we give below a simple example of the kind of facility we have in mind. It represents the information which might be displayed in response to the question "how was the last trial value computed?" (A Gauss-Newton algorithm is assumed to be in operation).

At the start of the -----'th iteration, the values of the optimisation variables were as follows:

variable name	=	value
variable name	=	value
variable name	=	value

The rate of change of the sum of squares with respect to each variable was as follows:

rate of change wrt variable name	=	value
rate of change wrt variable name	=	value
rate of change wrt variable name	=	value

If the sum of squares had been linear, the following corrections to the values of the optimisation variables would have resulted in a solution:

Variable	Change	New value
variable name	value %	value
variable name	value %	value
variable name	value %	value

(NOTE: changes of this amount in each variable are regarded as equivalent to step length of one; changes of half this amount to a step length of 0.5 etc)

In this case, a step length of value was found to give a satisfactory reduction in the sum of squares, leading to the following new values for the variables:

variable name   =   value

variable name   =   value

variable name   =   value

If this description should turn out not to be adequate, then the set of equations whose solution generated the correction vector could be displayed; if necessary, a step-by-step breakdown of the solution process could be given. In this way the user could, if he wished, gain access to as much information as he would have had if he had carried out the iteration by hand. This hsould enable him to diagnose the most stubborn of faults.

Summing up, in this section we have tried to underline the essential differences between the expert system which we propose, and the currently available software for function minimisation. Perhaps the fundamental difference is that the optimisation system proposed will take a higher level of responsibility for finding an acceptable solution, while interacting with the user and being subservient to him. In the final section, we shall discuss work to date on this project.

SECTION 4 : CURRENT WORK AND PROPOSED DEVELOPMENTS

In view of the limitations of the 'subroutine' approach to the provision of optimisation software, it is perhaps surprising that there is so little published work on the development of expert systems for this task. The only published work known to the authors at this time is that of Arora and Baenziger[3]. They have developed a pilot program for a system very similar in principle to that advocated in this paper, although limited in its scope to engineering design. Their system, still in the early stages of development, combines heuristics with more or less standard optimisation algorithms and the authors claim that experience with the pilot program justifies the development of a full system.

The authors of this paper feels that there are two parallel lines of development which could be followed simultaneously. These two routes, on which preliminary work at Queen's University, Belfast has already begun, are based on two ways in which optimisation problems may be classified. They are:

(A)   Breaking the task down by mathematical class, for example, nonlinear least squares, constrained problems, etc.

(B)   Breaking the task down by application area, for example, engineering structural design, economic modelling, etc.

A development path based upon classification (A) would begin by producing a system to deal with a well defined subset of the possible optimisation problems. Such a subset needs to be nontrivial but sufficiently limited so that a working and indeed useful, system can be developed with a reasonable effort - say 1 man-year. This would serve as a pilot program, highlighting the difficulties to be overcome but also demonstrating the potential advantages to be gained by further development. Such development would take the form of a systematic extension of the facilities to cover an increasingly general subset of the problem classes until a system is arrived at which is capable of dealing with constrained global minimisation problems with very slack conditions on the functions involved. This is in fact an outline for a research program extending over a considerable period of time, but one which, from near the outset, will produce useful programs and insights. Development along route (B), on the other hand, would amount initially to little more than a conscious decision, at the design stage of an application program, to incorporate as far as possible the principles discussed above. As progress was made along route (A), for example as rules were developed and tested and the knowledge base extended, this decision would tend to shape more and more the form of the finished application program. Ultimately one would hope for convergence to at least a situation in which applications programs could make direct use of program modules generated by development path A. The approach is shown schematically in Figure 2, and in the following two sections we discuss some preliminary work which has been carried out.

The pilot nonlinear least squares program

A pilot program has been written by Adams[4]. This is intended to investigate the feasibility of producing a general expert system of the type advocated in this paper, and to pinpoint the areas requiring major effort. It uses as its numerical component subroutines from the NOC OPTIMA library, Hatfield Polytechnic[6].

The program can be seen as a first step along path A of the previous section. The nonlinear least squares problem was selected as the first area of interest for a number of reasons. Firstly, this is one of the most commonly occurring classes of problem; it arises in the form of nonlienar equation solving, curve fitting and the closely related problem of parameter estimation. Secondly, it is one of the best defined problem classes, being easily recognisable and without the complications which constraints can cause. In addition, numerical algorithms exist which usually work well, and for which the causes of failure are well understood. In spite of this, nonlinear least squares problems are far from trivial for non-specialists and ought to be well representative of the general optimisation problem. The program was designed, written and tested over a 5-month period, and is best described in terms of its approach to some of the difficulties described in section 2.

Problem formulation

This is of course one of the major areas in which research is required. The program incorporated two features to aid in the formulation process. Firstly, a question and answer dialogue can take place during which the user extracts the information he needs about general aspects of problem formulation. Then, when he is in a position to proceed to a solution of his problem, the program provides him with a partially prepared subroutine for function evaluation. This is essentially a standard subroutine, displayed on the screen, with standard storage allocation (based on a three-number specification of problem size provided, on prompt, by the user). The user then types in the fortran code needed to complete the specification of the problem, including any additional storage allocation. The program is then compiled and linked automatically. This approach is clearly ony useful when the function evaluation routine is small enough to be entered easily from the terminal. In other circumstances a pre-existing program file is used.

Parameter setting

The approach taken is similar to that above, with advice given on request and default values preset.

Trouble-shooting

This is the part of the program which has the most scope for an "inference" engine. The present version of the program has only a simple inference structure.

The first line of defence (following straightforward data checking) is the standard compiler error diagnostic. The user can repeat from the beginning if an error is found at this stage, with his work up to this point being saved. Once an error-free compilation has been achieved and execution has begun, a number of checking procedures come into play. The most important is the checking of the derivative calculations against numerical estimates (in the case where analytical derivatives have been provided). Where the difference exceeds a certain level in any component, a warning is given and both sets of numbers are displayed. The user can then decide whether he wishes to accept the analytical derivatives anyway, experiment with different step lengths for the approximate derivative calculation in order to confirm the existence of an error, or check his coding of derivatives. This should eliminate a notorious source of blunders, even among experienced optimisers.

Once the program is under way, the user can exercise various levels of control over its progress. This is intended to help in those cases in which the program stops at a point which is unacceptable to the user, for whatever reason. It gives him the opportunity to reset convergence criteria, introduce or modify scaling and make various other changes. He also obtains help from the system in assessing whether or not the point at which the program stops is likely to be a local minimum (see Sensitivity Analysis below). This is based on gradient criteria and on the conditioning of the hessian matrix or an estimate of it. In addition, an analysis of the stopping point is carried out to check the validity of the Gauss-Newton assumption. The result of this can indicate whether a general minimisation algorithm rather than a specialised sum-of-squares method should be tried.

Sensitivity analysis

When a solution has been found, a range of sensitivity analyses can be carried out, based on the properties of the first and second derivatives. Eigenvalues of the hessian are computed and the user can step along the corresponding eigenvectors. He can thus interactively explore the neighbourhood of the solution and indeed cause it to be graphically displayed (so far, only in the case of two-dimensional functions). Based on this analysis, he can then decide whether to accept the point as a solution; alter parameters and start again either from this point or from another; or reformulate the problem.

A functional diagram of the program is shown in figure 3.

The program has been used on real (and difficult) problems, but it has yet to be used by a non-specialist. It will provide both a test-bed for the ideas discussed in this paper, and a framework around which a more complete nonlinear least squares program can be developed.

Nonlinear circuit analysis and design

The first step along path B is intended to be the development of a package for nonlinear microwave circuit identification, analysis and design.

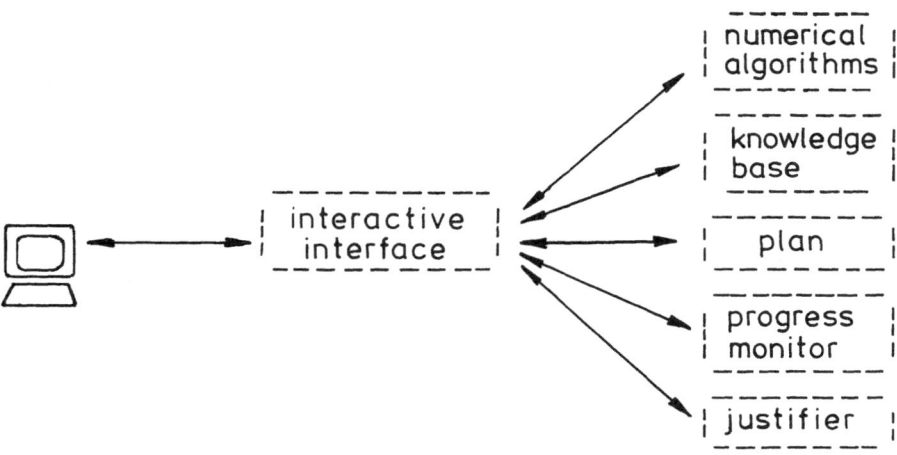

Fig. 1. GENERAL FORM OF THE EXPERT SYSTEM

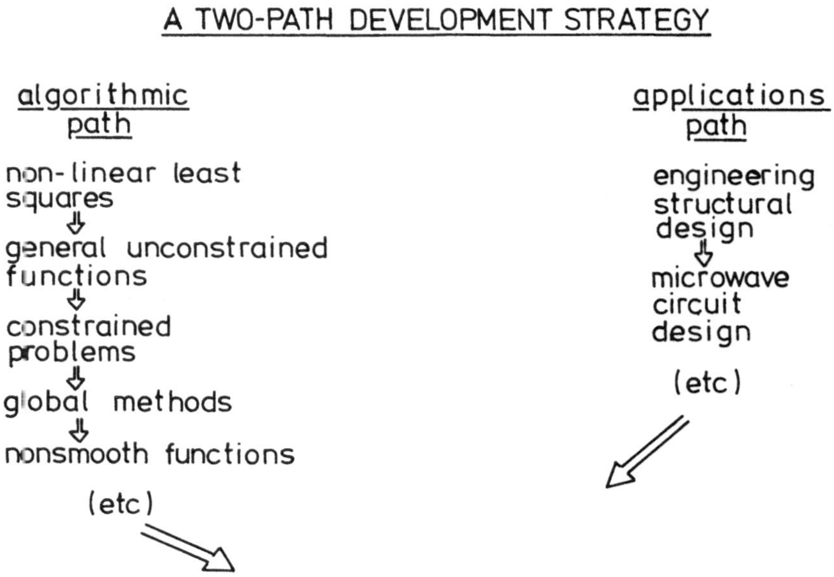

A TWO-PATH DEVELOPMENT STRATEGY

<u>algorithmic
path</u>

non-linear least
squares
⇓
general unconstrained
functions
⇓
constrained
problems
⇓
global methods
⇓
nonsmooth functions

(etc)

<u>applications
path</u>

engineering
structural
design
⇓
microwave
circuit
design

(etc)

Fig. 2. A GENERAL EXPERT OPTIMISATION SYSTEM

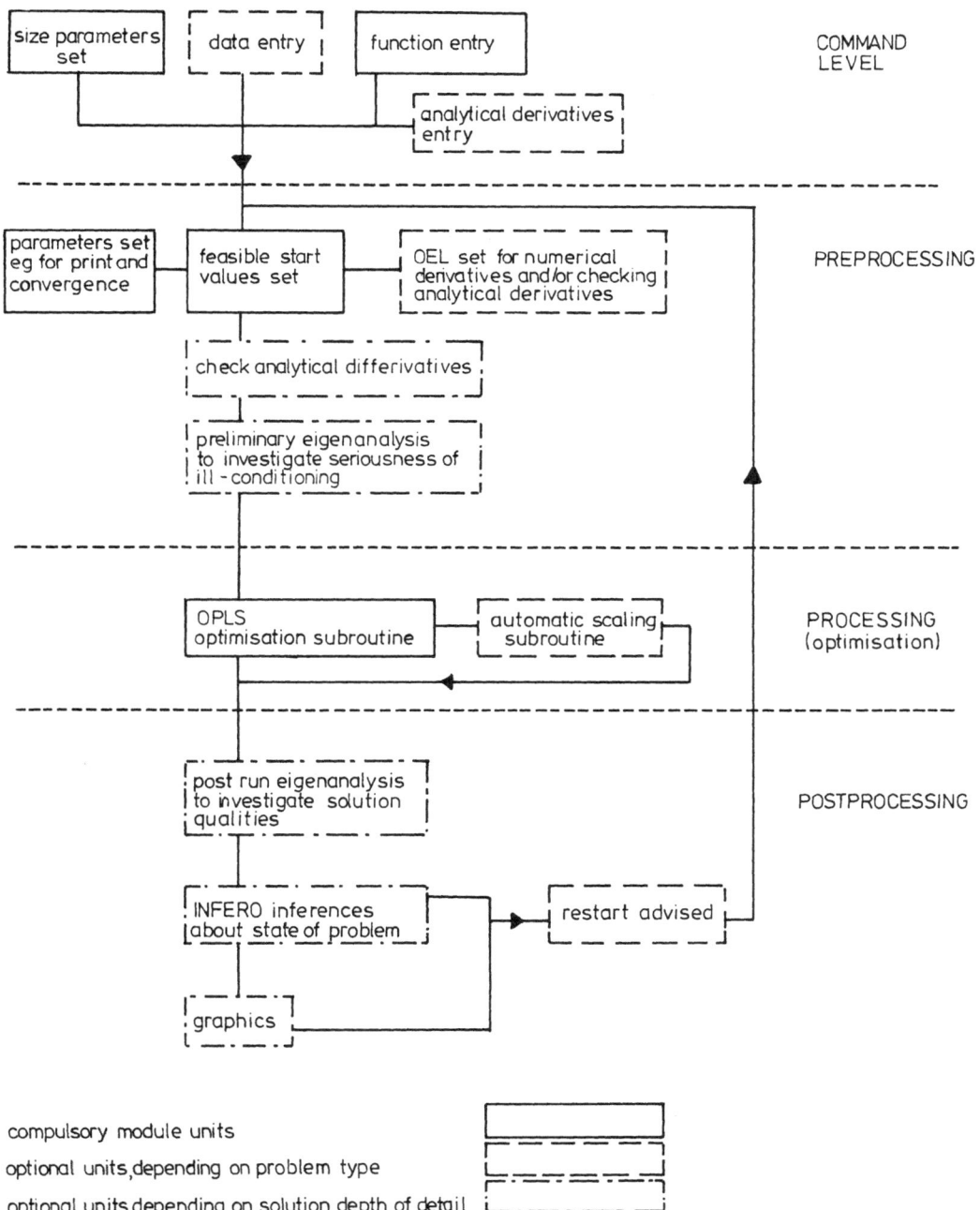

size parameters set	
data entry	
function entry	COMMAND LEVEL
analytical derivatives entry	

parameters set eg for print and convergence

feasible start values set

OEL set for numerical derivatives and/or checking analytical derivatives

PREPROCESSING

check analytical differivatives

preliminary eigenanalysis to investigate seriousness of ill-conditioning

OPLS optimisation subroutine

automatic scaling subroutine

PROCESSING (optimisation)

post run eigenanalysis to investigate solution qualities

POSTPROCESSING

INFERO inferences about state of problem

restart advised

graphics

compulsory module units

optional units, depending on problem type

optional units, depending on solution depth of detail

Fig. 3.

REFERENCES

(1) J.J.McKeown   Sensitivity Analysis with
                                   respect to independent
                                   variables. In:Nonlinear
                                   optimization LCW Dixon,
                                   E Spedicato,GP Szego,Eds.
                                   Birkhauser(1980)

(2) Frederick Hayes-Roth,
     Donald Waterman,
     Douglas Lenat   Building Expert Systems
                                        Addison-Wesley(1983)

(3) J.S.Arora,G.Baenziger   Uses of artificial
                                   intelligence in design
                                   optimisation.
                                   Comp.meth.app.math.eng.
                                   54(1986) 303-323

(4) W.Adams   Development of a user
                                   interface for nonlinear
                                   least-squares problems
                                   M.Sc. Dissertation.Faculty of
                                   Engineering,Queens University
                                   of Belfast(1986)

(5) T.Brazil,E.Choo,   Analysis and optimization of
     S.El-Rabaie,J.A.C.   the harmonic output power of
     Stewart,V.Fusco,J.J.   an FET amplifier.
     McKeown   16th European Microwave
                                   Conf.Dublin September 1986
(6) LCW Dixon,MC Bartholomew-
     Biggs et al   OPTIMA manual,Numerical
                                   Optimisation Centre,Hatfield
                                   Polytechnic

Throughput analysis of a flow-controlled communication network with buffer space limitations

J.L. van den Berg, O.J. Boxma
Centre for Mathematics and Computer Science
P.O. Box 4079, 1009 AB Amsterdam, The Netherlands

This paper studies the traffic flow in a virtual circuit of a computer communication network with window flow control. Due to finite buffer capacity, overflow of data packets is possible. Lost packets have to be retransmitted. To maintain the original order of the packets, subsequently sent packets also have to be retransmitted, causing a deterioriation of throughput.
An approximation method, based on a queueing network model, is developed to analyse the throughput behaviour. This method leads to exact results for the single-hop network. For the multi-hop circuit, it yields very accurate approximations, as is illustrated by simulation.

KEY WORDS & PHRASES: Computer communication network; flow control; virtual circuit; overflow; (negative) acknowledgment; closed queueing network; throughput.

1. INTRODUCTION

A computer communication network consists of a number of facilities (processors, links between processors, etc.) shared by competing users (messages or packets). Senders generate messages, which are subsequently transmitted over the network to receivers. The acceptance of a message by a receiver usually consists of (i) storing the message in a buffer, and (ii) sending an acknowledgment (ACK) back to the sender. As soon as the sender receives this ACK, he knows that there is no longer any need to store a copy of that message; a buffer space can be emptied. However, when the sender has not received an ACK within a certain period of time after transmission (the time-out), or when he receives a negative acknowledgment (NACK), he will retransmit that message in order to prevent endlessly long waiting times.

The finite capacities of the network facilities cause conflicts between the messages. Queues of messages are formed in front of certain facilities, and finite buffers are filled to completion so that messages are lost and have to be retransmitted. This may lead to a drastic reduction of the performance of the system, what is reflected in ever longer waiting times and a decrease of the effective throughput of the network.

If too many messages lay a claim on the available resources in the network, flow control procedures are needed to prevent the system from becoming overloaded. Generally a distinction is made between *local* and *global* flow control. Local flow control is the collection of procedures that regulate the traffic *within* the network, for example between two nodes; global flow control is the collection of procedures that regulate the externally offered traffic (Gerla and Kleinrock [8]). The principle of global flow control is to shift congestion from the interior of the network to the points of traffic admittance (Reiser [12]).

In the following our attention will be devoted to a form of global flow control, viz. end-to-end flow control, that is being exercised on so-called virtual circuits of the network (a virtual circuit is a fixed route, along which messages are transmitted between a particular sender and a particular receiver, see Fig. 1).

NATO ASI Series, Vol. F38
Flow Control of Congested Networks
Edited by A. R. Odoni et al.
© Springer-Verlag Berlin Heidelberg 1987

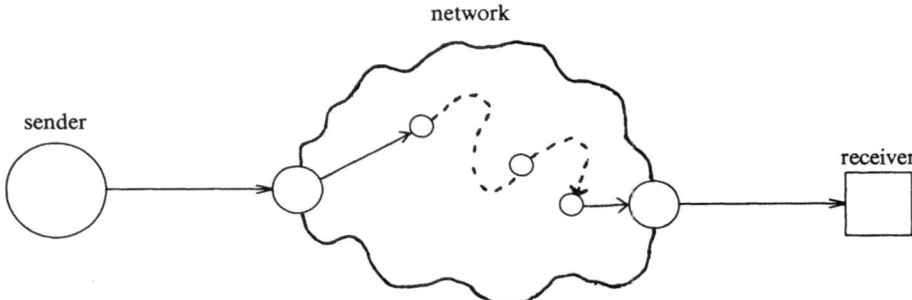

Fig. 1 A virtual circuit.

The most common end-to-end protocol is the *window flow control protocol* (see Cerf and Kahn [5] and Reiser [12,13]). The principle behind this protocol is that an upper limit is imposed on the number of not yet acknowledged messages that can simultaneously be present in the virtual circuit. This maximum is called the *window size*.

Several variants of the window flow control protocol are in existence. One of them, the *sliding window protocol*, plays an important role in this study, and will therefore now be discussed in more detail (see Reiser [12,13]).

The sliding window mechanism operates as follows (see Fig. 2).

1. For each message (or packet) that is being transmitted over the virtual circuit, a counter (which is initially set to N, the window size) is decreased by one.

2. As soon as the counter reaches the value zero, no more new messages are admitted to the virtual circuit (the sender stops transmission).

3. Each message is individually acknowledged by the receiver upon reception. As soon as the sender receives an ACK, the counter is increased by one. The sender is reactivated when the counter goes up from zero to one.

It is often assumed that ACK's are received after a negligibly short network delay (see Fig. 2).

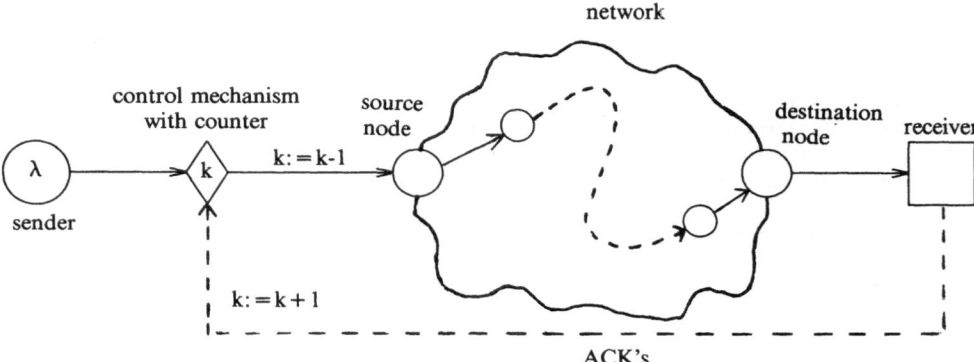

Fig. 2 Virtual circuit operating a sliding window.

A virtual circuit can easily be represented by a *queueing network*. In that case the sender is a source which, with a certain intensity, generates customers (messages); these customers travel along the successive links of the virtual circuit. Each link can be modelled as a service system with one server, with service times that equal the transmission times of messages over this link. Thus the service time of a message at a service system is determined by the length of that message.

For a further analysis it is important to remark that the sliding window mechanism keeps the sum of the total number of customers and of the ACK's present on the virtual circuit, and of the counter, constant and equal to N. Consequently, a virtual circuit with a sliding window mechanism can be modelled as a closed cyclic queueing network (see Fig. 3, and cf. Reiser [12]). It is thereby of course assumed that each node in the original system has a buffer capacity of at least N, so that retransmission is never necessary. The queue length at the service system with service intensity λ now represents the counter in the original system. Of course, the server with service intensity $1/\alpha_i$ represents the link between the i-th and (i+1)-th node of the virtual circuit (with average transmission time α_i), $i = 1, ..., M$.

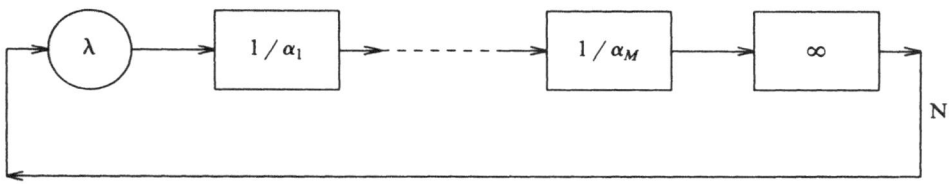

Fig. 3 Queueing model of the virtual circuit pictured in Fig. 2.
The queue with infinite service intensity represents the receiver.

A complete network, with sliding window mechanisms on the virtual circuits, can hence be modelled as a queueing network consisting of several closed chains. If we suppose that the transmission times of one and the same message at successive links are independent (Kleinrock's "Independence Assumption", see Kleinrock [10]) and negative exponentially distributed, then an exact analysis of the joint queue-length distribution in the model is possible (Baskett et al. [1]). However, the computational complexity rapidly grows with the number of closed chains so that, actually, exact solutions are only possible in rather simple cases (see Reiser and Kobayashi [14]). In [12], Reiser presents an efficient heuristic method to handle this problem.

As remarked above, the buffer capacities of the nodes - in particular in a local area network - cannot always be assumed to be infinite. Therefore it is possible that a message, upon arrival at a node, cannot be accepted because the buffer space is full. In that case this message, K say, has to be retransmitted by the sender. There are various procedures to let the sender know that K has been lost. First of all, a time-out mechanism may be used: a clock starts running at the time a message transmission starts, and when no ACK has been received for this message within a certain period of time, the time-out, the sender assumes that the message has been lost. Another procedure is the following. When the receiver receives a message with a higher sequence number than K without having received K, he sends a negative acknowledgment (NACK) - which contains K's sequence number - back to the sender. Usually the order of the messages has to be maintained, in order to prevent the need for sorting messages at the receiver. Consequently, not only K has to be retransmitted, but also those messages that were subsequently sent.

These blocking and retransmission phenomena generally make an exact mathematical analysis prohibitive. There is a strong need for mathematical models that reasonably accurately reflect the behaviour of real networks with overflow and flow control, and that are still amenable to an exact or approximate analysis. The goal of the present paper is to develop and analyse such a model. It is a queueing model of a virtual circuit with a sliding window mechanism, in which only the destination node has a finite buffer capacity. The model is an extension of the cyclic queueing model of Reiser [12], which was described above. An approximation method for the analysis of this queueing model is developed. Attention is restricted to the throughput of the circuit, and in particular to the maximum attainable throughput (in the case of an infinitely fast source). Throughput is a very important performance measure in these flow-controlled networks. Another useful performance measure, the end-to-end delay of messages over the network, will be the subject of a future study.

The rest of the paper is organized in the following way. Below we describe the queueing network model which represents a virtual circuit with limited buffer capacity at the destination node. Section 2 is devoted to an approximate analysis of the model. This analysis is shown to be exact in the case of a single-hop circuit (i.e., no intermediate nodes between source node and destination node). In Section 3 we compare the approximation results with results obtained by simulation. Several tables and graphs are presented, which illustrate the accuracy of our approximation and which expose the influence of retransmissions on the (maximum) throughput in a virtual circuit with overflow and flow control. In Section 4 we summarize our findings and we mention some points which are considered for future research.

Queueing model description

We have already seen that a virtual circuit which operates a sliding window protocol can be modeled as a closed queueing network. We now add the assumption that the last node of the circuit has finite buffer space of size L $(<N)$. In Fig. 4 the corresponding queueing network has been drawn.

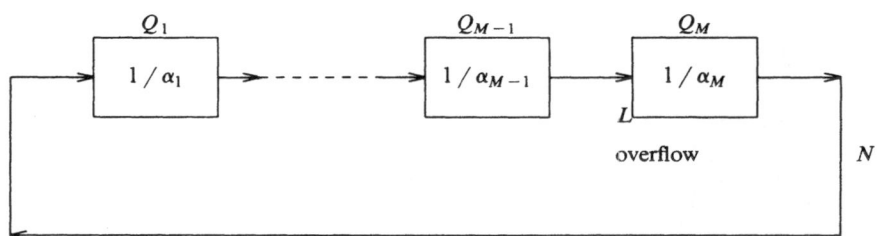

Fig. 4 Queueing model of a virtual circuit with buffer overflow.

The service systems $Q_1, \ldots, Q_M$ represent the M nodes of the virtual circuit with their outgoing links. Q_i is a single-server system, $i = 1,2,...,M$. Service times at Q_i are independent, negative exponentially distributed stochastic variables with mean α_i, $i = 1,2,...,M$; service times at different queues are also independent. The service times represent transmission times of messages between the nodes of the virtual circuit plus, possibly, an additional nodal delay. The receiver is not modelled because we assume that as soon as a message has obtained its service at Q_M, it has reached the receiver; the receiver immediately sends an acknowledgment back to the sender, and this ACK has infinitely short transmission time and delay. In reality ACK's are either piggybacked on other messages which are sent along the reverse route, or are stand-alone. Stand-alone ACK's are often assigned a higher priority. In that case the assumption of infinitely short transmission time and delay for ACK's (which are much shorter than ordinary dataflow messages) is not unrealistic; see also Remark 2.4.

We have not modelled the sender, because we assume that its message generation intensity is infinite: as soon as the sender has received an ACK, a new message is admitted to Q_1. In this way we can study the maximum attainable throughput of the virtual circuit (to study throughput per se we would have to include another queue, with service rate equal to the intensity of the sender). We have thus arrived at a closed cyclic queueing model. The number of customers in the system equals N, the window size.

$Q_1, \ldots, Q_{M-1}$ have "infinite" waiting rooms (waiting rooms of size at least $N-1$). Q_M has a finite waiting room of size $L-1$: at most L customers can be in service or waiting at Q_M. We assume that $N > L$. When a customer, say K, leaves Q_{M-1} and finds L customers in Q_M upon arrival, then overflow occurs. Earlier, several procedures were mentioned to let the sender know that a message has been lost (time-out, NACK). In the present study we shall model the NACK mechanism but not the time-out mechanism. Our model could be adapted to incorporate the time-out mechanism (cf. Remark 2.3), but the minor complications arising from this inclusion would obscure the essence of the approximation method.

Suppose that message K is the first one that is lost. The receiver only notices that K has been lost, when he receives a message with a higher sequence number than K. The receiver now sends a NACK for K back to the sender. As in the case of ACK's, it is assumed that NACK's experience no transmission time and delay. The implications for the queueing model are the following. Suppose that customer K is the first one to cause overflow, and suppose that customer $\hat{K}$ is the first one, sent after K, to be admitted to Q_M. As soon as the service of $\hat{K}$ in Q_M is completed, K and all customers who have left Q_1 after K instantaneously return to Q_1 (to fit the overflow phenomenon in the context of our closed queueing network, we might introduce a "wait" queue WQ; overflown customers immediately enter WQ, and stay there until a NACK has reached the sender). Consequently, immediately after the service completion of $\hat{K}$ in Q_M, all customers are in Q_1. Because the original order of the customers has to be maintained, K is the first one to be taken into service in Q_1, preempting another service if necessary. Apart from this, the service discipline in all queues is first-come first-served.

REMARK 1.1
In the present study it is assumed that, upon the arrival of a NACK at the source node, old messages are instantaneously removed from the nodes of the virtual circuit. An interesting question is whether a protocol which effectively accomplishes this can be implemented efficiently. Such a protocol should work as follows. Suppose A and B are communicating over a virtual circuit. At a certain epoch B sends a NACK to A (because B received a message with a wrong sequence number). The source and destination addresses of this NACK control message (B and A, respectively) are used to recognize messages, sent from A to B, in the buffers of the intermediate nodes of the virtual circuit. At each node visited by the NACK, messages with destination address B and source address A can be removed because copies of these messages will be retransmitted upon arrival of the NACK at the source node.

2. ANALYSIS OF THE QUEUEING MODEL
In this section we study the maximum attainable throughput, T, of the queueing model described above, with

$$T := E[\text{number of customers leaving } Q_M \text{ per unit of time who do not have to be retransmitted}], \quad (2.1)$$

or alternatively,

$$T := E[\text{number of customers entering } Q_M \text{ per unit of time who do not have to be retransmitted}]. \quad (2.2)$$

Let t_j denote the j-th epoch, after time 0, at which a customer completes service in Q_M, having been the first admitted customer in this queue after the occurrence of overflow; t_j corresponds to the epoch of the j-th transmission of a NACK. A basic observation is that the joint queue length process at

$Q_1, \ldots, Q_M$ is a *regenerative process* (cf. Cohen [7]), with regeneration epochs $t_1, t_2, \ldots$. So the queue length process, and in particular also the departure process from Q_{M-1}, repeats itself probabilistically speaking after each t_j epoch. Moreover, the lengths $r_j := t_j - t_{j-1}$ of the successive regeneration intervals are independent. Let r denote a stochastic variable with the same distribution as $r_1, r_2, \ldots$. Because of the regenerative character of the queue length and departure processes,

$$T = \frac{E[\text{number of customers entering } Q_M \text{ in } [0, r] \text{ who do not have to be retransmitted}]}{E[r]}. \tag{2.3}$$

So we can restrict ourselves to one arbitrary regeneration interval $[0, r]$. It appears to be natural to divide this interval into three consecutive subintervals I_1, B, I_2. Here
I_1 is the period from 0 until the first arrival of a customer at Q_M (note that all N customers are at Q_1 at time 0);
B is the period from the end of I_1 until the first occurrence of overflow at Q_M;
I_2 is the period from the end of B until the end of the regeneration interval.
Denote the lengths of these periods by i_1, b, i_2, respectively. Then

$$r = i_1 + b + i_2.$$

During the i_1-period no contribution to the throughput is made. During the I_2-period at least one customer is admitted to Q_M, but no contribution to the throughput is made because all those admitted during this period must be retransmitted. Hence only during the B-period a contribution to the throughput can be made. Let α denote the mean interdeparture time from Q_{M-1} during a B-period. Then, cf. (2.3),

$$T = \frac{1}{\alpha} \frac{E[b]}{E[r]} = \frac{1}{\alpha} \frac{E[b]}{E[i_1] + E[b] + E[i_2]}. \tag{2.4}$$

Trivially,

$$E[i_1] = \sum_{j=1}^{M-1} \alpha_j. \tag{2.5}$$

Below we introduce one single approximation assumption which immediately yields a simple approximation for $E[b]$ and $E[i_2]$ and hence for T.

Approximation assumption
During B- and I_2-periods, the arrival process at Q_M is a Poisson process with intensity $1/\alpha$; $1/\alpha$ equals the throughput in the closed cyclic queueing system with N customers, obtained from the one under consideration by removing Q_M.

Before motivating this approximation assumption, we first demonstrate how it yields simple approximations for $E[b]$ and $E[i_2]$, and hence for T.

Approximation for $E[b]$
According to the assumption, Q_M behaves during a B-period like an M/M/1 queue, say Q, with arrival intensity $1/\alpha$ and mean service time α_M. At the beginning of a B-period, the queue length at Q (Q_M) increases from zero to one. Let $\mu_{j,k}$, $j,k \geqslant 0$, denote the first entrance time in state k starting from state j for the queue length process in Q. Then, *under the approximation assumption*,

$$E[b] = E[\mu_{1,L+1}]. \tag{2.6}$$

Introducing

$$a := \alpha_M / \alpha,$$

we can prove the following lemma:

LEMMA 2.1

$$E[\mu_{1,L+1}] = \alpha_M \frac{(L+1)(a-1) + (1/a)^{L+1} - 1}{(1-a)^2} - \alpha, \quad a \neq 1,$$ (2.7)

$$= \alpha(\frac{1}{2}L^2 + \frac{3}{2}L), \qquad\qquad a = 1.$$

PROOF

First observe that

$$E\{\mu_{i,j+1}\} = E\{\mu_{i,j}\} + E\{\mu_{j,j+1}\}, \quad i < j,$$ (2.8)

and

$$E\{\mu_{0,1}\} = \alpha.$$ (2.9)

Based on the assumption that there are j customers in Q, $E\{\mu_{j,j+1}\}$ can be calculated by just looking at the next event (the departure or the arrival of a customer). Denoting by τ_M a service time at Q_M, by τ a negative exponentially distributed random variable with $E\{\tau\} = \alpha$, and by (A) the indicator function of event A, it holds that

$$E\{\tau_{j,j+1}\} = E\{\tau(\tau_M \geq \tau)\} + E\{\tau_M(\tau_M < \tau)\} + E\{\mu_{j-1,j+1}(\tau_M < \tau)\}, \quad j = 1, 2, \dots.$$ (2.10)

From (2.8), because of the mutual independence of $\mu_{j-1,j+1}$, τ_M and τ in the right hand side of (2.10), it follows that

$$E\{\mu_{j,j+1}\} = E\{\tau(\tau_M \geq \tau)\} + E\{\tau_M(\tau_M < \tau)\} + Pr\{\tau_M < \tau\}[E\{\mu_{j-1,j}\} + E\{\mu_{j,j+1}\}].$$

From this $E\{\mu_{j,j+1}\}$ can easily be solved. Using

$$E\{\tau(\tau_M \geq \tau)\} + E\{\tau_M(\tau_M < \tau)\} = \frac{\alpha_M}{1+a},$$

we find

$$E\{\mu_{j,j+1}\} = \frac{\alpha_M + E\{\mu_{j-1,j}\}}{a}.$$ (2.11)

It is now easy to derive (2.7) from (2.8), (2.9) and (2.11).

Approximation of $E[i_2]$
Under the approximation assumption,

$$E[i_2] = (L+1)\alpha_M + \alpha(\frac{1}{1+a})^{L-1}.$$ (2.12)

Indeed, i_2 consists of $L+1$ services in Q (Q_M) and, possibly, an idle period. The probability that Q (and hence Q_M) becomes empty in an I_2-period before a customer is admitted, equals $(1/(1+a))^{L-1}$ (remember that admission is not possible before the queue length has decreased to $L-1$, and that $Pr\{\tau > \tau_M\} = 1/(1+a)$).
In fact it is possible to obtain the *distributions* of b and i_2 under the approximation assumption, but they are not needed in the present study.

From (2.4)-(2.7) and (2.12) the following approximation for the maximum attainable throughput is obtained:
For $a \neq 1$,

$$T \approx \frac{a\dfrac{(L+1)(a-1) + (1/a)^{L+1} - 1}{(1-a)^2} - 1}{(L+1)\alpha_M + \alpha(\dfrac{1}{1+a})^{L-1} + \alpha_M \dfrac{(L+1)(a-1) + (1/a)^{L+1} - 1}{(1-a)^2} - \alpha + \sum_{j=1}^{M-1} \alpha_j},$$ (2.13)

For $a = 1$,

$$T \approx \frac{\frac{1}{2}L^2 + \frac{3}{2}L}{(L+1)\alpha + \alpha(\frac{1}{2})^{L-1} + \alpha(\frac{1}{2}L^2 + \frac{3}{2}L) + \sum_{j=1}^{M-1} \alpha_j}.$$

We would like to emphasize that (2.13) has been derived after the introduction of one single approximation assumption.

Motivation of the approximation assumption

We draw upon results from [3] concerning a closed cyclic queueing model that is identical to the one of Fig. 4, with one exception: all waiting rooms are infinite. In [3] exact results from [4] are used to analyse the influence of the "slowest server" (the server with largest mean service time) on cycle time- and sojourn time distributions. Of particular interest to us is the result for the mean cycle time, $E[C]$, because the throughput T_0 equals $N / E[C]$. It is shown in [3] that, when α_i and α_j are the largest and one-but-largest mean service time:

$$E[C] = N\alpha_i[1 + O((\frac{\alpha_j}{\alpha_i})^N)], \quad N \to \infty, \tag{2.14}$$

so

$$T_0 = \frac{1}{\alpha_i}[1 - O((\frac{\alpha_j}{\alpha_i})^N)], \quad N \to \infty. \tag{2.15}$$

As an example, for $M = 2$,

$$T_0 = \frac{\alpha_2^N - \alpha_1^N}{\alpha_2^{N+1} - \alpha_1^{N+1}} = \frac{1}{\alpha_2}\left[1 - \frac{\frac{\alpha_2}{\alpha_1} - 1}{(\frac{\alpha_2}{\alpha_1})^{N+1} - 1}\right], \quad \alpha_2 \neq \alpha_1.$$

Approximation of $E[C]$ by $N\alpha_i$ and of T_0 by $1/\alpha_i$ yields remarkably accurate results, even when N is rather small and α_j is close to α_i. The worst case is when all mean service times are equal; in that case $T_0 = N / ((N+M-1)\alpha_1)$. The explanation for the high accuracy is that in a closed cyclic system the queue with the slowest server is seldom empty - so the departure process of this queue is closely approximated by a Poisson process, and the throughput in the system is closely approximated by the service rate at this queue.

Now consider our closed cyclic model, with finite waiting room in Q_M, during a B-period. We distinguish two cases.

Case (i): α_M is not the largest mean service time.

The output rate from Q_{M-1} during a B-period is in this case hardly influenced by the presence of Q_M. Hence this output rate may be put equal to the throughput in the cyclic queueing model consisting of $Q_1, \ldots, Q_{M-1}$. The output process is closely approximated by a Poisson process, corresponding to the departure process from the queue with the slowest server.

The above reasoning still holds if there is not one unique slowest server in the set $\{Q_1, \ldots, Q_{M-1}\}$. The assumption of a Poisson process is less justified, but the departure rate from Q_{M-1} is still closely approximated by the throughput in the $(M-1)$-queue cyclic model.

Case (ii): α_M is the largest mean service time.

If L would have been infinite, the throughput would have been dominated by Q_M. But for finite L, the other queues still contain at least $N-L$ customers; the queue with the slowest server out of these $M-1$ queues will hardly ever be empty, and the output process from Q_{M-1} is still determined by that queue.

A clear illustration of this fact is provided by the model with $M = 2$ queues. During a B-period Q_1 contains at least $N - L > 0$ customers, and the departure process from Q_1 is a Poisson process with rate $1 / \alpha_1$, regardless of Q_2. Indeed, our approximation procedure (which involves only one approximation assumption) yields exact results for the case of $M = 2$ queues.

The above reasoning can also be applied to the I_2-period. The approximation will be slightly less accurate, because Q_M may contain close to L customers during the larger part of this period, and because there may be less than $N - L$ customers in $\{Q_1, \ldots, Q_{M-1}\}$ due to overflow.

In the next section numerical results will be presented, which illustrate the accuracy of our approximation and which reveal the influence of window flow control and retransmissions on the throughput in a virtual circuit. We close the present section with some remarks.

REMARK 2.1

Additional motivation for the approximation assumption is provided by referring to the Chandy-Herzog-Woo theorem (Norton's theorem in the context of electrical circuits; cf. Chandy et al. [6] and Lavenberg [11]). This theorem applies to the entire class of product-form networks, but for our purpose it is sufficient to restrict attention to the cyclic product-form network of Fig. 4 with $L = \infty$. The principle is as follows. Study one queue, say Q_M, by replacing the network by a two-queue closed cyclic system consisting of Q_M and one service center, C. C should be "flow-equivalent" to $\{Q_1, \ldots, Q_{M-1}\}$, i.e., the service rate at C, when it contains j customers, should equal the arrival rate at Q_M when $Q_1, \ldots, Q_{M-1}$ together contain j customers. According to the Chandy-Herzog-Woo theorem, this replacement procedure yields *exact* results for Q_M when C's service rate with j customers is chosen to be the throughput in the $(M-1)$-queue closed cyclic system $\{Q_1, \ldots, Q_{M-1}\}$ with j customers.

If this theorem would still hold in our model, then we could study Q_M during B- and I_2-periods by introducing an arrival rate $\mu(j)$, equal to the throughput in the $(M-1)$-queue model $\{Q_1, \ldots, Q_{M-1}\}$ with j customers. But this is just the idea we have used, apart from the fact that we have taken $\mu(j)$ equal to $\mu(N) = 1 / \alpha$ for all j. Note that $j \geq N - L$ in the B-period, while the throughput hardly varies with the number of customers. We could have calculated $\mu(j)$ for all j-values, but this would have complicated the approximation without a clear improvement of accuracy.

REMARK 2.2

In [2] the throughput in a virtual circuit is studied for the (idealized) case in which *overflow is observed as soon as it occurs*. Again the regenerative approach is applicable. In this case the I_2-period has length zero, and at the beginning of an I_1-period there are $N - L$ customers in Q_1 and L customers in Q_M. Using M/M/1-theory, it is easy to calculate the probability p_j of there being j customers in Q_M at the end of the I_1-period. The mean length of the B-period is subsequently estimated by calculating mean entrance times $E[\mu_{j, L+1}]$, multiplying by p_j and summing over j.

In Section 3 we numerically compare the maximum attainable throughput for the case with NACK's and for this idealized situation in which overflow immediately leads to retransmission. For the latter case, too, the approximation procedure yields exact results when $M = 2$.

REMARK 2.3

The approximation procedure can also be modified to incorporate a time-out mechanism. It is reasonable to assume that the time-out is chosen so large, that the probability of a message time-out excess due to very long sojourn times (instead of overflow) is negligibly small. Now a similar regeneration approach as before can be used, with similar choices of the I_1- and B-periods. The I_2-period should be replaced by a period between overflow and time-out excess. The length of this period can be estimated.

REMARK 2.4

In the present model the network delay of ACK's and NACK's is neglected. Such delays can be modelled by the introduction of an infinite server queue between Q_M and Q_1. The approximation method can easily be adapted to this new situation.

REMARK 2.5

A performance measure which has not been discussed until now is the overflow intensity (O) defined as

$$O = \frac{1}{E[\text{time between two subsequent arrivals of a NACK at the source node}]}.$$

Obviously

$$O = \frac{1}{E[i_1] + E[b] + E[i_2]}.$$

Hence, the approximation method for the throughput can also be used to approximate the overflow intensity.

3. NUMERICAL RESULTS

In this section we present tables and graphs to show the accuracy of our throughput approximation (2.13), and to reveal the influence of window flow control and overflow on the throughput. From the large variety of possible parameter combinations, some representative examples have been chosen. For these examples the approximation results are compared to results obtained by simulation.

Table I and Figure 5 expose the dependence of the throughput T on the buffer size L; Table II and Figures 6a, 6b show the influence of the number of customers N (window size) on the throughput, and Table III and Figure 7 consider models with a large number of queues.

In all tables three possible situations are distinguished:

(i) $\max\{\alpha_i \ i = 1, ..., M - 1\} < \alpha_M$,
(ii) $\max\{\alpha_i \ i = 1, ..., M - 1\} = \alpha_M$,
(iii) $\max\{\alpha_i \ i = 1, ..., M - 1\} > \alpha_M$.

The tables show that the approximation yields very good results, even when the overflow intensity is large (as in case (i) above). The relative approximation errors indicated in the tables are defined as:

$$\frac{\text{approximation result} - \text{simulation result}}{\text{simulation result}} \times 100\% .$$

We now discuss the results in some detail.

In Table I the approximation is tested for different values of the buffer size (L) at Q_M. $T_{sim.}$ and $T_{approx.}$ denote the simulation result and the approximation result for the throughput, respectively. In all cases $M = 4$, $N = 6$ and $\alpha_4 = 1$. Of course the throughput grows when L becomes larger.

The last four cases in Table I (with $\alpha_1 = 2 >> 0.2 = \alpha_2 = \alpha_3$) provide additional justification for the idea behind our approximation assumption: if there is a unique queue in $\{Q_1, ..., Q_{M-1}\}$ with largest mean service time, then the arrival process at Q_M during a B-period is very well approximated by a Poisson process with intensity $1/\max\{\alpha_i, i = 1, ..., M - 1\}$ (cf. Section 2). Indeed, the approximation yields extremely accurate results for these cases.

A typical curve of the throughput as function of the buffer size is shown in Figure 5. The results are compared with the throughput in the corresponding model with infinite buffer space at all queues (see Figure 3). Of course the throughput remains constant as soon as L becomes larger than the number of customers (N) in the system. We also display the throughput for the idealized case in which overflow is observed as soon as it occurs (cf. Remark 2.2). As expected, the throughput for that case is somewhat higher.

In Table II it is shown that the approximation method accurately reflects the influence of the window size (N) on the maximum throughput. Again the results are presented for a model consisting of $M = 4$ queues. It can be concluded that the dependence of the throughput on N increases when the overflow intensity becomes smaller. In the first cases of Table II (with $\alpha_4 = 2 > 1 = \max\{\alpha_1, \alpha_2, \alpha_3\}$, hence with a high overflow intensity) the throughput hardly depends on N.

In Figure 6a the throughput, as a function of N, is depicted for a model consisting of four queues of which Q_4 has two buffer places ($M = 4$, $L = 2$). To show the influence of the finite buffer space, the throughput results for the corresponding model with infinite buffer space are indicated in the same figure. It is interesting to see how, due to the finite buffer space (which makes retransmissions necessary in case of overflow), the throughput decreases if N is enlarged from two to three, whereas the throughput increases for the model with infinite buffer space. From Figure 6b it follows that this is not a general phenomenon. The shape of the throughput curve depends on the relation between α_m and $\max\{\alpha_i, i = 1, ..., M - 1\}$. Indeed, for the case of Figure 6b, $\alpha_M \ll \max\{\alpha_i, i = 1, ..., M - 1\}$ (small overflow intensity) whereas for the case depicted in Figure 6a, $\alpha_M = \max\{\alpha_i, i = 1, ..., M - 1\}$ (large overflow intensity).

Table III contains some results for larger values of M, the number of queues. In all cases $M = 11$ and $\alpha_i = 1$, $i = 1, ..., 10$. The approximations for these larger networks are still very accurate.

Figure 7 shows the influence of M on the throughput. It is seen that the difference with the throughput in the model with infinite buffer space becomes smaller when the number of queues grows. Obviously this is due to the fact that the overflow intensity in the model with finite buffer space decreases when M becomes larger (the customers are spread over more queues).

L	α_1	α_2	α_3	$T_{sim.}$	$T_{approx.}$	% error
1	0.5	0.2	0.2	0.359	0.361	+0.6
2	0.5	0.2	0.2	0.565	0.571	+1.1
3	0.5	0.2	0.2	0.672	0.682	+1.5
4	0.5	0.2	0.2	0.742	0.748	+0.8
1	1	1	1	0.248	0.247	-0.4
2	1	1	1	0.415	0.420	+1.2
3	1	1	1	0.519	0.524	+1.0
4	1	1	1	0.587	0.589	+0.3
1	2	1	1	0.217	0.213	-1.8
2	2	1	1	0.348	0.349	+0.3
3	2	1	1	0.415	0.417	+0.5
4	2	1	1	0.449	0.451	+0.4
1	2	0.2	0.2	0.240	0.242	+0.8
2	2	0.2	0.2	0.372	0.374	+0.5
3	2	0.2	0.2	0.438	0.436	-0.5
4	2	0.2	0.2	0.467	0.467	+0.0

Table I. Comparison of approximation results with simulation results for various values of the buffer size (L). In all cases $M = 4$, $N = 6$, $\alpha_4 = 1$.

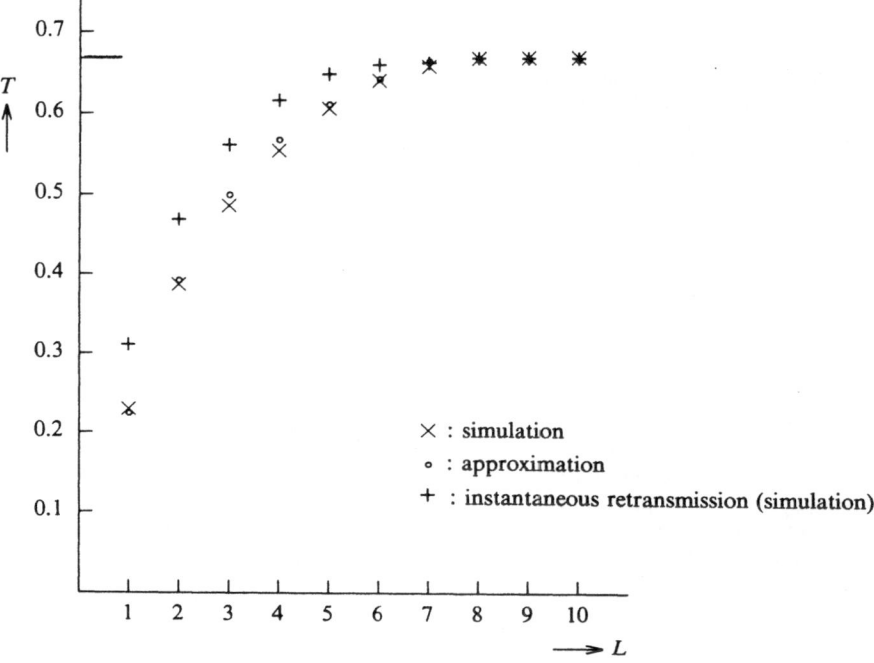

Fig. 5 The influence of L on the throughput. $M = 5$, $N = 8$, $\alpha_i = 1$, $i = 1, ..., 5$.

N	L	α_1	α_2	α_3	α_4	$T_{sim.}$	$T_{approx.}$	% error
4	3	1	0.5	0.5	2	0.336	0.335	-0.3
6	3	1	0.5	0.5	2	0.336	0.335	-0.3
8	3	1	0.5	0.5	2	0.338	0.336	-0.6
10	3	1	0.5	0.5	2	0.338	0.336	-0.6
4	5	2	1.5	1.5	2	0.323	0.323	0.0
6	5	2	1.5	1.5	2	0.339	0.339	-0.0
8	5	2	1.5	1.5	2	0.341	0.344	+0.9
10	5	2	1.5	1.5	2	0.342	0.347	+1.5
4	2	0.2	1	1	0.1	0.770	0.778	+1.0
6	2	0.2	1	1	0.1	0.825	0.836	+1.3
8	2	0.2	1	1	0.1	0.858	0.868	+1.2
10	2	0.2	1	1	0.1	0.875	0.888	+1.5

Table II. Comparison of approximation results with simulation results for various values of the window size (N) In all cases $M = 4$.

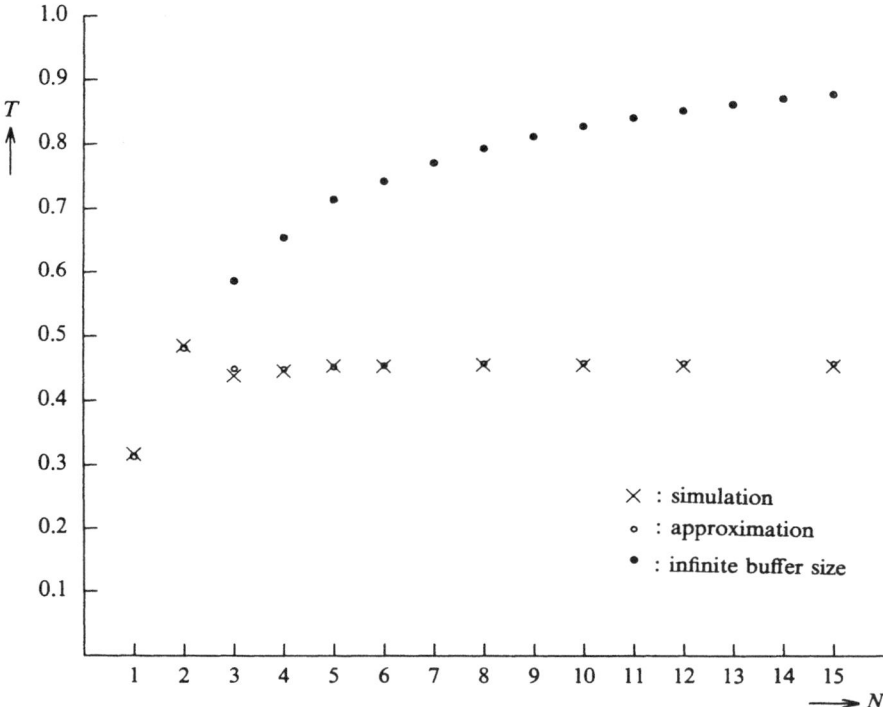

Fig. 6a The influence of N on the throughput. $M=4$, $L=2$, $\alpha_1=0.2$, $\alpha_i=1$, $i=2,3,4$.

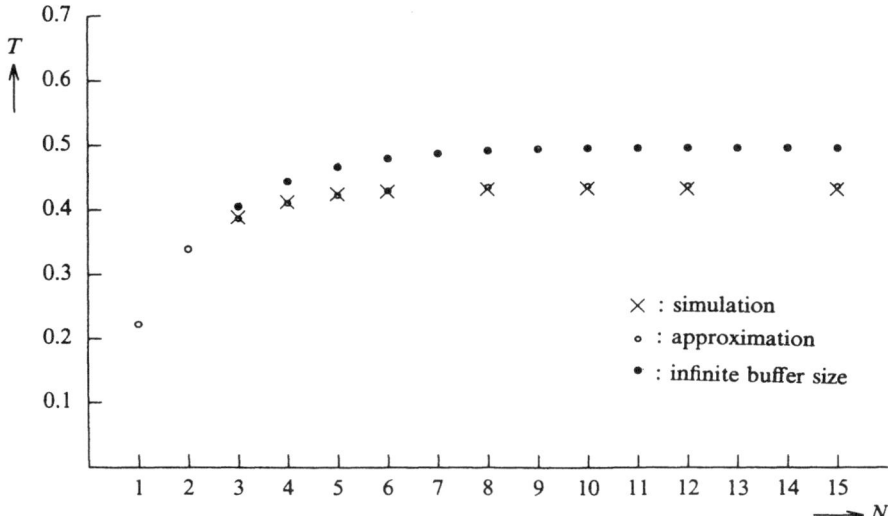

Fig. 6b The influence of N on the throughput. $M=4$, $L=2$, $\alpha_1=2$, $\alpha_2=\alpha_3=1$, $\alpha_4=0.5$.

N	L	α_{11}	$T_{sim.}$	$T_{approx.}$	% error
30	5	5	0.138	0.136	-1.4
30	5	1	0.562	0.565	+0.5
30	5	0.2	0.767	0.769	+0.3
30	10	5	0.164	0.162	-1.2
30	10	1	0.695	0.713	+2.6
30	10	0.2	0.767	0.769	+0.3

Table III. Comparison of approximation results with simulation results for the case that the system consists of $M = 11$ queues. In all cases $\alpha_i = 1$, $i = 1,..,10$.

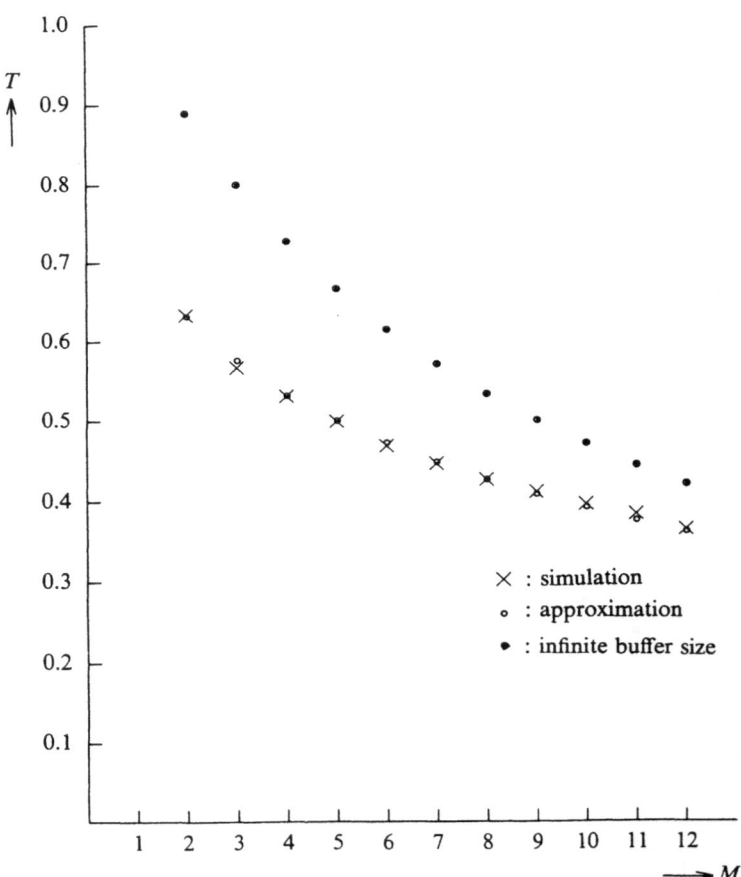

Fig. 7 The influence of the number of queues (M) on the throughput. $N = 8$, $L = 3$, $\alpha_i = 1$, $i = 1,...,M$.

4. CONCLUSIONS

A simple throughput approximation for virtual circuits in computer communication networks with window flow control and overflow has been derived and investigated. The results can be summarized as follows.

- The approximation is exact for the case of a single-hop circuit (no intermediate nodes between the source node and the destination node).
- In general, the relative error of the approximation is at most a few percent.
- The approximation formula (2.13) and its construction give much insight into the qualitative and quantitative behaviour of the throughput as a function of several system parameters (window size, buffer size of the destination node, number of nodes, transmission speeds).
3 For example, the throughput is seen to be almost independent of the window size when the overflow intensity is high.

Finally, we mention some points which are considered for future research.

(i) The end-to-end delay of messages over a virtual circuit of the network is an important performance measure. The approach of the present paper may be useful for obtaining end-to-end delay approximations. Preliminary results are contained in [2].

(ii) The ultimate goal of studies like the present one, is to obtain insight into the behaviour of a network consisting of several virtual circuits. When studying one circuit of such a network in isolation, the presence of other circuits can be represented by external traffic entering and leaving the nodes of the isolated circuit. In the case of infinite buffer sizes and negative exponentially distributed service times (cf. Fig.3), the assumption that the external arrival streams are Poisson leads to an easily analyzable queueing network of BCMP type (cf. Hayes [9], Section 11.3.2). It appears that one can eliminate the external streams in this case by adjusting the service rates. If node i has mean service time α_i and external arrival rate λ_i, then $1 / \alpha_i^{adjusted} = 1 / \alpha_i - \lambda_i$ (and no external traffic) leads to exact throughput results for the virtual circuit. This justifies the study of virtual circuits *without* interfering traffic, in the case without buffer size restrictions. It remains to be investigated whether a similar approach can be used when not all buffer sizes are infinite.

(iii) In relation to Remark 1.1 concerning the instantaneous removal of messages from the nodes of the virtual circuit, we mention the investigation of a model in which old packets are neither removed nor ignored. Generally speaking, the presence of old packets in the network is the source of throughput degradation in a situation with an increasing offered load. The method described in the present paper can easily be modified to incorporate the presence of old packets; introduction of one additional approximation assumption is needed. Preliminary numerical results are encouraging.

REFERENCES

1. F. BASKETT, K.M. CHANDY, R.R. MUNTZ AND F.G. PALACIOS (1975). *Open, closed and mixed networks of queues with different classes of customers,* J. ACM 22, 248-260.

2. J.L. VAN DEN BERG (1986). *Queueing analysis of a virtual circuit in a computer communication network with window flow control,* Report OS-N8602, Centre for Mathematics and Computer Science.

3. O.J. BOXMA (1985). *Response times in cyclic queues - the influence of the slowest server,* preprint 380, Mathematical Institute, University of Utrecht.

4. O.J. BOXMA, F.P. KELLY AND A.G. KONHEIM (1984). *The product form for sojourn time distributions in cyclic exponential queues,* J. ACM 31, 128-133.

5. V.G. CERF AND R. KAHN (1974). *A protocol for packet network intercommunication,* IEEE Trans. Commun. 22, 637-648.

6. K.M. CHANDY, U. HERZOG AND L.S. WOO (1975). *Parametric analysis of queueing networks,* IBM J. Res. Develop. 19, 43-49.

7. J.W. COHEN (1976). *On Regenerative Processes in Queueing Theory,* Springer-Verlag, Berlin.

8. M. GERLA AND L. KLEINROCK (1980). *Flow control: a comparative survey,* IEEE Trans. Commun. 28, 553-574.

9. J.F. HAYES (1984). *Modeling and Analysis of Computer Communication Networks,* Plenum Press, New York.

10. L. KLEINROCK (1964). *Communication Nets - Stochastic Message Flow and Delay,* McGraw-Hill, New York.

11. S.S. LAVENBERG (1983). *Computer Performance Modeling Handbook,* Academic Press, New York.

12. M. REISER (1979). *A queueing network analysis of computer communication networks with window flow control,* IEEE Trans. Commun. 27, 1199-1209.

13. M. REISER (1982). *Performance evaluation of data communication systems,* Proc. IEEE 70, 171-196.

14. M. REISER AND H. KOBAYASHI (1975). *Queueing networks with multiple closed chains: theory and computational algorithms,* IBM J. Res. Develop. 19, 282-294.

Flow Control in Local-Area Networks of Interconnected Token Rings[1,2]

Werner Bux and Davide Grillo*
IBM Research Division
Zurich Research Laboratory,
8803 Rüschlikon, Switzerland

Abstract - We investigate flow-control issues in local-area networks consisting of multiple token rings interconnected through bridges. To achieve high throughput, bridges perform only a very simple routing and store-and-forward function, but are not involved in error- or flow-control. In case of congestion, bridges discard arriving frames, which will be recovered through an appropriate end-to-end protocol between the communicating stations. The end-to-end protocol considered is the IEEE 802.2 type-2 Logical-Link-Control (LLC) protocol. Extensive simulations show that performance can be severely degraded if, in such a network, the LLC protocol is employed as defined today. Therefore, we suggest an enhancement to this protocol in the form of a dynamic flow-control algorithm. As our results demonstrate, this enhancement guarantees close-to-optimal network performance under both normal traffic load and overload conditions.

1. INTRODUCTION

Local-area networks must be capable of interconnecting a large number of stations over distances of several kilometers. Whenever the limitations of a single ring or bus subnetwork are reached with respect to the maximum number of attachments or maximum distance, means to interconnect subnetworks become necessary.

In this paper, we consider a network of interconnected token rings. Rings operate as specified in the ECMA and IEEE Standards [1],[2]. Ring interconnection is provided by nodes called bridges [3]. To meet the high throughput requirements for interconnecting high-speed rings at reasonable costs, the

_____

*Present address: Fondazione Ugo Bordoni, Rome, Italy.
[1]Copyright © 1985 IEEE. Reprinted, with permission, from IEEE TRANSACTIONS ON COMMUNICATIONS, Vol. COM-33, No. 10, pp. 1058-1066, October 1985.
[2]This paper was presented in part at INFOCOM' 84, San Francisco, CA, April 1984.

bridge functions have to be simple which excludes their performing any complex flow or error control. In case of congestion, bridges simply discard frames they cannot handle momentarily. Discarded frames will be recovered through the end-to-end protocol between the communicating stations. We consider a network architecture in which type 2 of the IEEE 802.2 Logical-Link-Control (LLC) standard [4] is employed as end-to-end protocol.

A potential problem in such a network is that recovery of frames lost in congested bridges may worsen congestion, but, without a detailed analysis, it is very difficult to predict how severe this problem will be. Therefore, we developed a detailed model of a multi-ring network including all the relevant medium-access-control and logical-link-control functions. Since the complexity of such a model is far beyond the one of queueing models that can be analytically treated today, we employed simulation. Specifically, we used the RESQ2 simulation tool [5] which, for our modeling purposes, turned out to be a very flexible and powerful instrument. Our simulations show that network performance can be severely degraded in case of congestion. To overcome this problem, we suggest providing a dynamic flow-control mechanism in the LLC protocol which, in case of congestion, throttles the input traffic to the network.

In the next section, the elements of the multi-ring network are described. Section 3 shows how this network performs when the IEEE 802.2 type-2 LLC protocol as defined today is employed. In Section 4, we introduce a new flow-control mechanism in this protocol, and show the improvement attained. Section 5 summarizes our findings.

2. ELEMENTS OF THE MULTI-RING NETWORK

The network topology underlying our study is depicted in Fig. 1. Several token rings are interconnected through bridges and a "backbone" token ring, the latter serving to interconnect bridges.

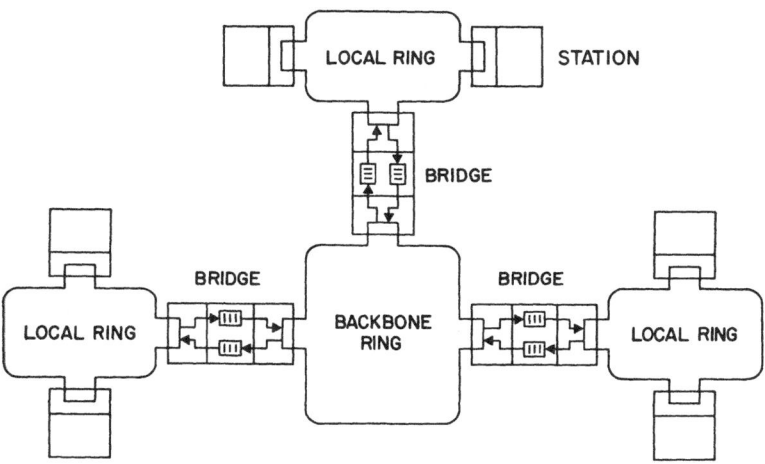

Fig. 1. Network topology.

2.1. Token Rings

The token-ring standards [1], [2] define a priority mechanism by which eight levels of priorities can be provided. We study two different kinds of ring operation. The so-called "non-priority mode" assumes that all stations and the bridges operate at the same ring-access priority level and are only allowed to transmit a single frame per token. In "priority mode", bridges operate at a higher ring-access priority level than normal stations. Whenever a bridge has a frame ready for transmission on a local ring, it will either seize the low-priority token, if available, or make a reservation for high-priority access. This forces the currently transmitting station to issue a priority token which will then be used by the bridge. After the bridge has completed transmission, circulation of the low-priority token is resumed at the next station downstream from the one which last transmitted. In priority mode, bridges are allowed to transmit continuously until their transmit buffer has been emptied, i.e., we assume a very long token-holding timeout. Stations transmit only one frame per token. On the backbone, priority access is not employed and each bridge transmits a single frame per token.

For simplicity, we assumed that both token-passing overhead and transmission errors can be neglected.

2.2. Bridges

Bridges provide a basic routing and store-and-forward function [3]. Frames are buffered in a bridge until they can be transmitted on the local or backbone ring. We assume that the bridge memory space is partitioned into two separate buffer pools, one for each data-flow direction, see Fig. 2. The buffer pools are structured in segments of a fixed size. Frames which do not find a sufficient number of free segments upon their arrival at a bridge are lost.

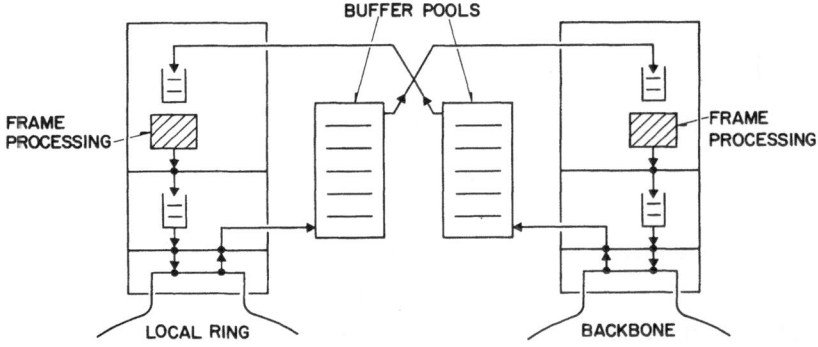

Fig. 2. Bridge.

A bridge is controlled by two processors, each of which handles one direction of data flow. These processors are modeled by two independent single servers which process all frames on a first-come, first-served basis.

2.3. Stations

A conceptual representation of a user station is shown in Fig. 3.

Application and Higher Layers: An application is represented by a traffic source in the sending station, which generates messages with given length and interarrival-time distributions, and a corresponding sink at the receiving station. The following two functions pertaining to the "higher layers", i.e., layers above LLC are modeled in detail: 1) segmentation/reassembly of messages when the maximum frame length specified for LLC is exceeded, and 2) the "higher-layer interface" on top of LLC, which works as follows. Data units

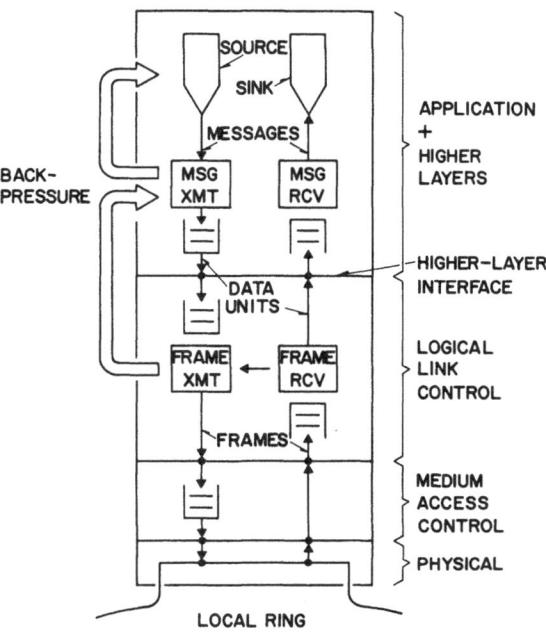

Fig. 3. Station.

provided by users of the LLC service are transmitted by LLC in the form of information (I-) frames. If LLC cannot transmit I-frames at the rate data units are supplied, it will apply backpressure on the higher layers. Specifically, a data unit supplied by a higher-layer entity is not accepted by LLC, when the total number of I-frames and data units being currently handled by the LLC entity has reached a threshold value B. This number encompasses: 1) the number of I-frames already transmitted but not yet acknowledged; 2) the number of I-frames waiting to be transmitted, and 3) the number of accepted data units not yet processed, i.e., formatted as I-frames. An unaccepted data unit together with all subsequent data units pertaining to the same message are queued by the higher-layer entity. At the same time, the generation of further messages is halted. When LLC service again becomes available, queued data units, if any, are handled first and the application resumes generation of new messages.

Logical Link Control: Particular emphasis was placed on a complete and detailed representation of all LLC functions. We subsequently give a brief outline of these functions; for a detailed specification, the reader is referred to [4].

Flow Control: Flow control is realized by a window mechanism, i.e., a sender is permitted to transmit up to a fixed number W (the window size) I-frames without having to wait for an acknowledgment. In our LLC implementation, for each I-frame successfully received, one Receive-Ready (RR) frame is transmitted back carrying the acknowledgment. We do not make use of the Receive-Not-Ready (RNR) function provided in [4].

Error Recovery: If the send sequence number of a received I-frame is not equal to the one expected, the receiver will return a Reject (REJ-) frame. It then discards all I-frames until the expected one has been correctly received. The sender, upon receiving a REJ-frame, retransmits I-frames starting with the sequence number received within the REJ-frame.

In addition to reject recovery, a timeout mechanism is provided. At the instant of transmission of an I-frame, a timer is started when not already running. When the sender receives an acknowledgment, it restarts the timer when there are still unacknowledged I-frames outstanding. When the timer expires, the station performs a "checkpointing" function by transmitting an RR-frame with a dedicated bit (the "P-bit") set to one. The receiver, upon receiving this frame, returns an RR-frame with the "F-bit" set to one. When this RR-frame is received by the sender, it either proceeds with transmitting new I-frames or retransmits previous I-frames, depending on the sequence number contained in the RR-frame received. The checkpointing function itself is protected by timeout. The timer is started upon transmission of the P-bit, and stopped when the F-bit is received. When the timer expires, transmission of an RR-frame with the P-bit set is repeated.

Medium-Access Control: User systems are attached to the rings through ring adapters which implement the medium-access control functions described in Section 2.1. We assume that ring adapters do not cause any noticeable increase in delay or decrease in throughput.

3. NETWORK PERFORMANCE FOR IEEE 802.2 TYPE-2 LLC

In this section, we first define the performance measures and list the assumptions underlying our examples. We then discuss simulation results pertaining to a network in which the IEEE 802.2 type-2 LLC as defined in [4] is employed.

3.1. Performance Measures

We shall restrict the discussion to two basic performance measures, throughput and end-to-end-delay, both measured at the higher-layer interface, because there, data units are delivered without errors and in the proper sequence.

Throughput of a connection is defined as the mean number of bits received at both ends across the higher-layer interface per unit time. In the subsequent examples, we shall usually show the total throughput of all connections in the network.

End-to-end delay is defined as the time elapsed between supplying a data unit to LLC at the higher-layer interface in the source node until receiving it across this interface at the sink node.

For the subsequent discussion, we need to specify a further quantity called "offered data rate". This is defined as the mean number of bits generated by an application for transmission per unit time under the condition that the application is not halted because of backpressure (c.f. Section 2.3).

3.2. Assumptions

The choice of values for the simulation parameters given below is based on the parameters specified in [1], [2], [4], and on experience gained in experimental implementations [6].

The transmission rates investigated are 4 Mbps for the local rings and 4 or 16 Mbps for the backbone. Bridges are assumed to need 300 μs to process one frame, unless otherwise specified. The default size of each of the two bridge buffer pools is assumed to be 4 kbytes. The maximum I-field length is 0.5 kbyte; the framing overhead of I-frames and the length of S-frames is 24 bytes. The time intervals between generation of messages are assumed to be exponentially distributed. The mean message length is 1 kbyte; the coefficient of variation 1.5. This message-length distribution, together with the effect of message segmentation and the superposition of S-frames, results in an overall frame-length distribution that resembles the bimodal distribution observed in measurements [7] with a mean of about 250 bytes.

Execution of the LLC protocol is assumed to require the following processing times: a) transmit I-frame (first time): 2 ms; b) receive in-sequence I-frame and transmit RR-frame: 2.5 ms; c) receive RR-frame and delete I-frame(s): 0.75 ms; d) receive out-of-sequence I-frame and transmit REJ-frame: 1 ms; e) receive REJ- or RR-frame with F-bit and retransmit I-frame(s): 1 ms; f) receive out-of-sequence I-frame and transmit nothing: 0.5 ms; g) handle timer interrupt and transmit RR-frame: 1 ms; h) receive RR-frame with P-bit and transmit RR-frame with F-bit: 1 ms.

The LLC timeout value chosen is 250 ms. The backpressure threshold value B defined n Section 2.3. is set to the window size plus four. This gives a source node sufficient flexibility to prepare I-frames for later transmission also during times when the LLC window is closed.

3.3. Results

Fig. 4 shows the total throughput as a function of the total offered data rate. Each of the 12 stations attached to a ring is assumed to generate the same amount of traffic and to have a logical link set up to a station on a different

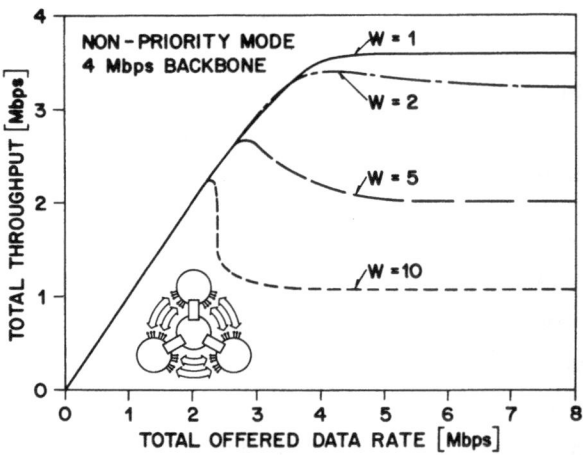

Fig. 4. Total throughput vs total offered data rate for different window sizes W.

ring. I-frame transmission on each logical link is two-way. The backbone transmission speed is 4 Mbps. Bridges operate in non-priority mode.

When the offered data rate is increased from zero, throughput initially follows linearly. As the backbone becomes noticeably loaded, queues of frames waiting to enter the backbone build up in the bridges, and eventually buffer overflow occurs. Loss of an I-frame leads to the retransmission of one or more I-frames, depending on the number of I-frames a station has outstanding when it receives a REJ or has performed checkpointing. Since the window size sets an upper limit to the number of frames to be retransmitted per lost frame, the additional traffic created by retransmissions decreases with smaller window sizes. This explains the significant differences between the throughput values pertaining to different window sizes.

For the same example, Fig. 5 shows the mean end-to-end delay as a function of the total throughput. At small throughput values, the message delay is higher for small window sizes. This is caused by the time periods during which stations cannot transmit pending I-frames because the window is closed.

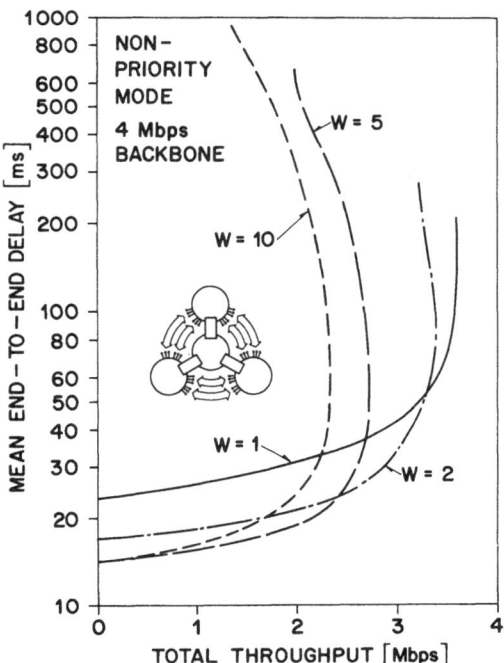

Fig. 5. Mean end-to-end delay vs total throughput for different window sizes W.

As throughput increases, delay also increases, owing to contention at the various network resources. When the throughput approaches the maximum value attainable, delays increase very steeply. Further increase of the offered data rate leads to both a decrease in throughput and an increase in delays. In the traffic-load range beyond the maximum throughput, delays grow to unacceptably high values. Taking into account possible fluctuations in the traffic, it is advisable to configure the network such that the offered data rate is sufficiently smaller than the maximum throughput. However, given the peakedness of data traffic, it is an open question whether this can always be guaranteed in a real installation.

We next consider the network performance under the same assumptions as above, except for a significantly faster backbone, i.e., 16 Mbps, see Fig. 6. In contrast to the previous example, here the system bottleneck is not the backbone but the local rings. At higher ring utilizations, queues of frames waiting to enter the local rings build up in the bridges, and eventually overflow of the buffer pools in direction backbone to local rings occurs. Frame losses have the same effect as described in the context of Fig. 4; hence, we obtain a similar throughput characteristic.

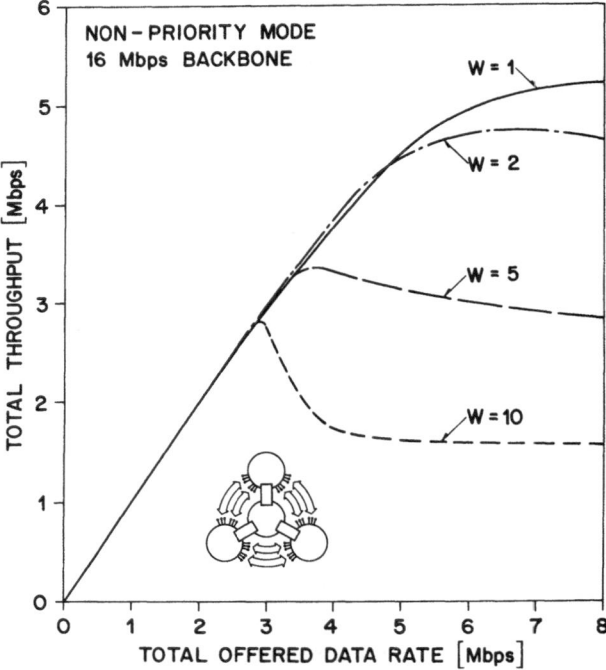

Fig. 6. Total throughput vs total offered data rate for different window sizes W.

Fig. 7 shows the delay-throughput characteristic for the same scenario. Except for smaller absolute delay values, we observe the same tendencies as for the lower backbone speed.

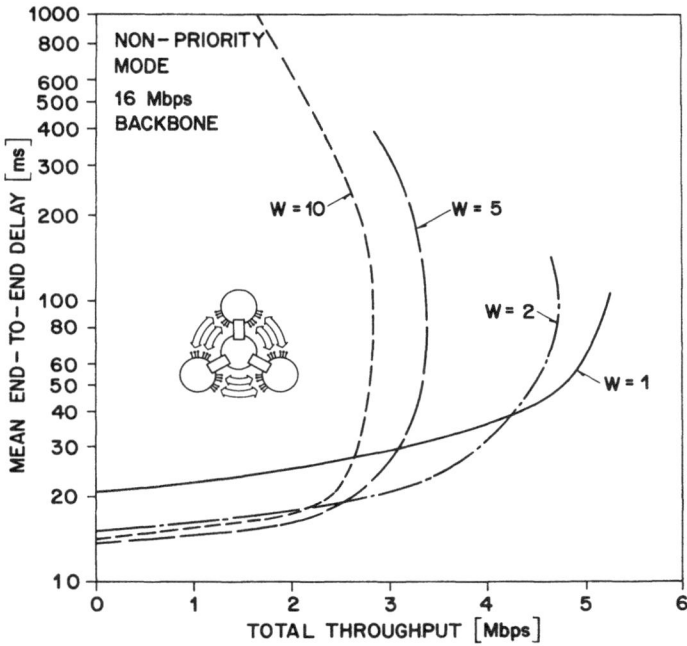

Fig. 7. Mean end-to-end delay vs total throughput for different window sizes W.

In both of the above examples, bridges do not make use of priority access to the local rings. In Fig. 8, we consider the same scenario as in the previous example, however, bridges employ priority access to the local rings. We observe substantial improvement compared to non-priority mode (Fig. 6). In fact, for all the window sizes investigated, no or negligibly few frame losses were observed. The explanation of this remarkable effect is as follows. When priority access for bridges is employed, access of stations to their local ring is delayed, slowing down both the injection of new I-frames into the network and the returning of acknowledgments. Delayed acknowledgments, in turn, further throttle the transmission of I-frames because of the LLC window flow-control mechanism. The overall effect is similar to that of flow-control schemes suggested for wide-area packet-switching networks, in which packets are handled with higher priority the closer they are to their destination [8]-[10].

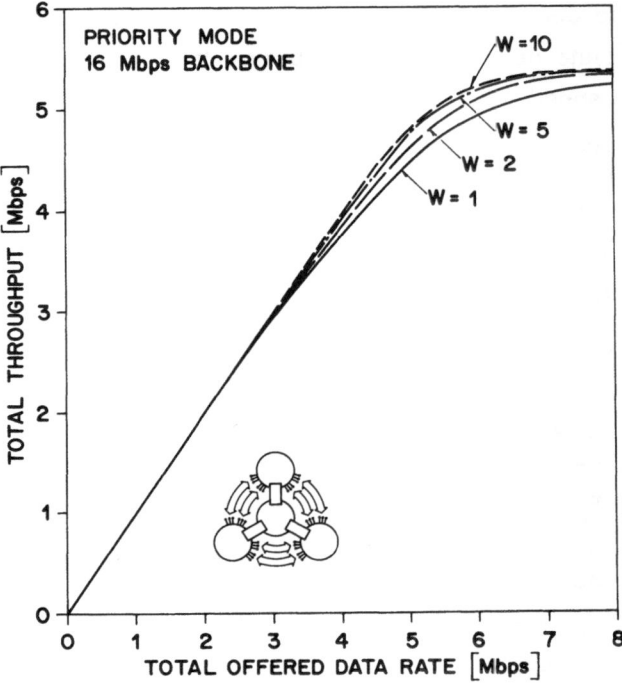

Fig. 8. Total throughput vs total offered data rate for different window sizes W.

Fig. 9 shows the corresponding delay-throughput characteristic. It should be pointed out that [as will become clear from a subsequent example (Fig. 13)] the effectiveness of the priority-mode operation is due to a great extent to the symmetry of the traffic pattern assumed in this example.

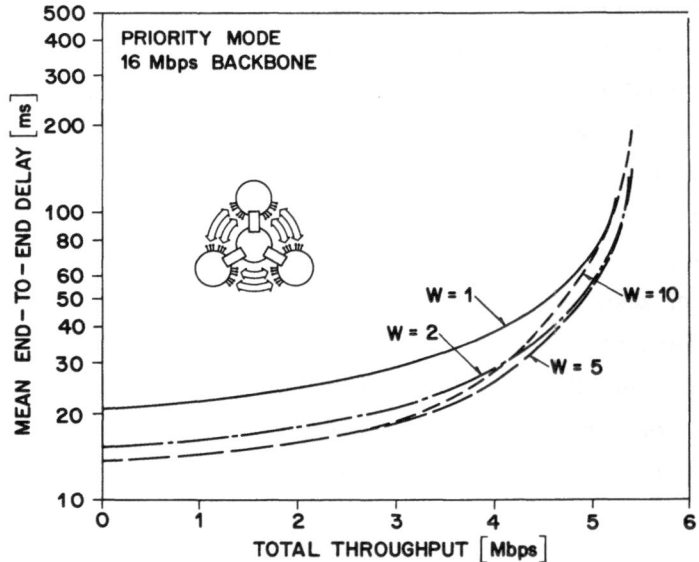

Fig. 9. Mean end-to-end delay vs total throughput for different window sizes W.

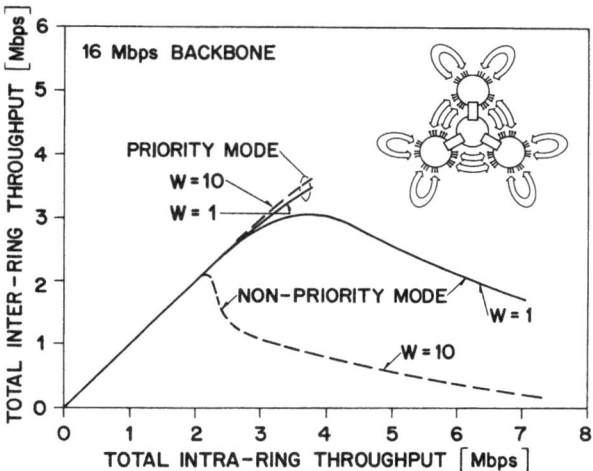

Fig. 10. Total inter-ring vs total intra-ring throughput for different window sizes W.

A discussion of priorities for bridges would be incomplete without considering their impact on how fair local-ring bandwidth is shared between intra-ring and inter-ring connections. Fig. 10 addresses this question for the following scenario: 24 stations are attached to each ring; 12 of them communicate with 12 stations attached to two other rings; each of the other 12 stations has a logical-link set up with a station on the same ring. I-frame flow is two-way on each connection. All stations generate the same amount of traffic. The figure shows the throughput of all inter-ring and all intra-ring connections as the offered data rate is varied. When priority mode is employed, we observe a fair sharing of the bandwidth between the two connection types in the sense that the total throughput of both is equal, almost completely independent of the traffic load and the LLC window size. In non-priority mode, throughput becomes very unbalanced when the local rings are heavily utilized: high throughput for intra-ring connections, low throughput for inter-ring connections. The explanation for this effect is as follows. In non-priority mode, the intra-ring connections as a whole obtain much better service than the inter-ring connections, because the latter have to share the single-access points of the bridges to the local rings. Furthermore, at larger window sizes, a significant portion of the bandwidth available for inter-ring traffic is lost owing to retransmissions; therefore, unfairness between inter- and intra-ring connections is even more pronounced.

Figs. 11 and 12 present further results for a network carrying both inter- and intra-ring traffic. The situation considered here is that 12 stations attached to

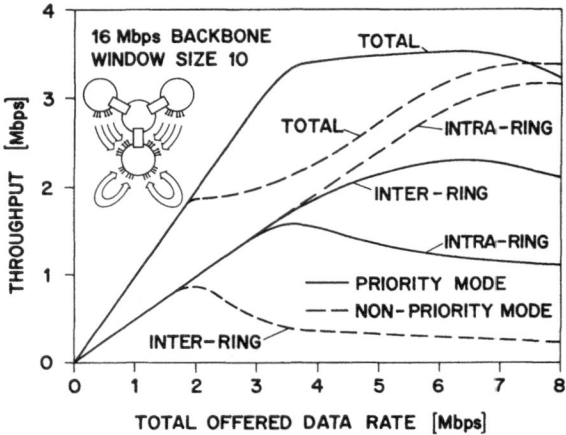

Fig. 11. Inter-ring, intra-ring, and total network throughput vs total offered data rate.

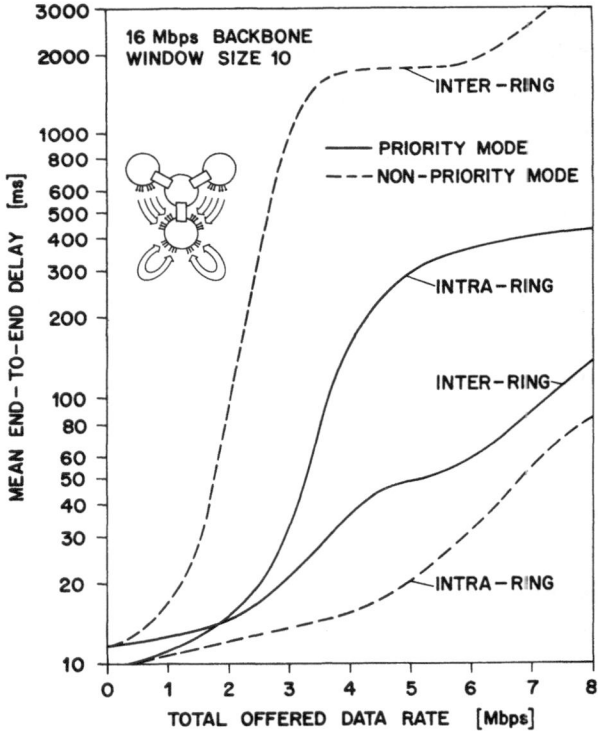

Fig. 12. Mean inter-ring and intra-ring end-to-end delay vs total offered data rate.

two rings transmit information frames to 12 stations on the third ring. Further connections are set up among 12 additional stations on the third ring. We consider both priority and non-priority modes for a window size of ten. Fig. 11 shows the throughput on the inter- and intra-ring connections and the total network throughput as a function of the total offered data rate; Fig. 12 gives the corresponding end-to-end delay results. In non-priority mode, we again observe unfair sharing of the ring bandwidth between intra- and inter-ring connections. It is interesting that this unfairness already exists at relatively small offered data rates and becomes very pronounced at high loads. In priority mode, fair ring bandwidth sharing is achieved up to rather high offered data rates. At very high traffic loads, priority mode tends to favor the inter-ring connections and then delays on both connection types deviate distinctly. However, this unfairness effect is by far less severe than the one in non-priority mode. Furthermore, Fig. 11 reveals that also from a total network throughput point of view, priority mode is superior to non-priority mode.

The conclusion from the examples shown so far is that use of priority ring access for bridges has two distinct advantages. It yields better overall efficiency and fairer sharing of the ring bandwidth.

It is important, however, to understand that priority mode does not avoid congestion problems under all circumstances, as the following two examples demonstrate.

1) When in the scenario of Figs. 4 and 5, priority mode is employed, we observe basically the same throughput and delay characteristics as for non-priority mode. This is not surprising, since the network bottleneck is the backbone, and hence, providing bridges with priority access to their local rings cannot change performance significantly.

2) Even the combination of a fast backbone with ring-access priority for bridges is not always sufficient to avoid congestion problems, as our next example in Fig. 13 demonstrates. Six stations attached to one ring and the same number of stations attached to a second ring transmit I-frames to 12 stations on the third ring. For larger window sizes, we observe the typical throughput characteristic, indicating congestion in the bridge connecting the backbone with the third ring when the offered data rate exceeds a certain value.

A further important question is to what extent the bridge-buffer size affects performance. Fig. 14 shows the throughput characteristic for the same scenario

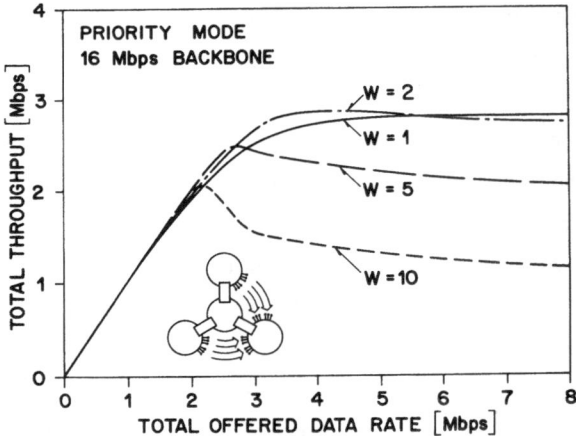

Fig. 13. Total throughput vs total offered data rate for different window
sizes W. 2 × 4 kbyte bridge buffer.

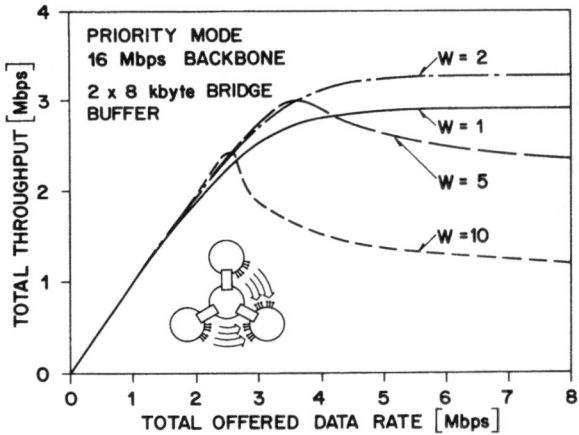

Fig. 14. Total throughput vs total offered data rate for different window
sizes W. 2 × 8 kbyte bridge buffer.

as in Fig. 13, except that bridges now have twice as much buffer space as
before. Comparison of Figs. 13 and 14 indicates that there is some gain in
throughput by enlarging the bridge-buffer size from two times 4 kbytes to two
times 8 kbytes. However, the gain is modest, especially in an overload situation, for the following reason. When the input traffic to a bridge approaches
its output capabilities, its buffers tend to fill up completely, irrespective of the
absolute buffer size, and therefore, frame losses cannot be reduced by
increasing the bridge-buffering capacity.

All the results shown so far have indicated that a small window size is an effective means of minimizing congestion. However, a small window size can also be a disadvantage, e.g., when only a small number of stations is simultaneously active. In Fig. 15, we show the throughput characteristic of a network in which only two stations on each ring communicate with two stations attached to a different ring. We observe that the total throughput can be substantially improved by increasing the window size. This indicates that at small window sizes, a small number of stations is not able to make full use of the available bandwidth, because acknowledgments do not return sufficiently fast.

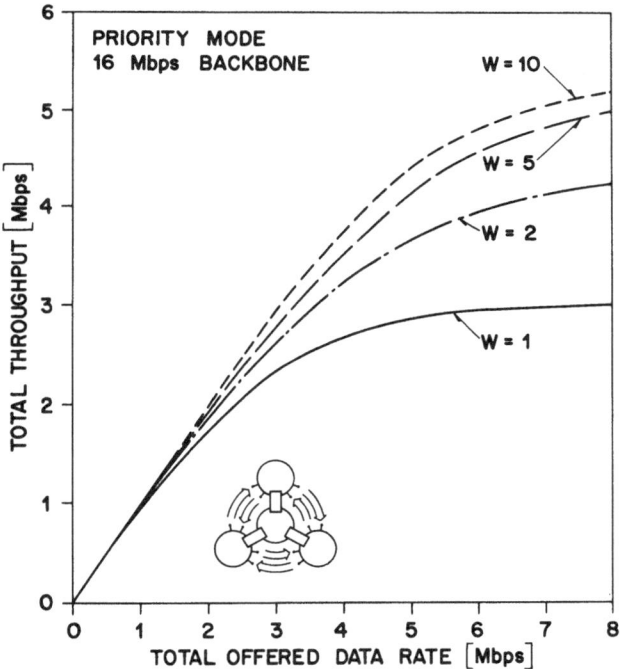

Fig. 15. Total throughput vs total offered data rate for different window sizes W.

4. DYNAMIC FLOW CONTROL

4.1. Concept

The results shown in the previous section demonstrate that in case of congested bridges, network performance can be severely degraded. This suggests that the architecture should be enhanced by providing an effective flow-control mechanism to ensure efficient operation under both normal traffic load and overload.

Our specific proposal is to introduce the following dynamic flow-control algorithm into the IEEE 802.2 type-2 protocol. Initially, stations use the window size as defined during the set-up of the logical link. Whenever a station needs to retransmit an I-frame (either because of a received reject frame or after check-pointing), it sets its window size to one. Afterwards, the window size is increased by one (up to the initial value) for every n-th successfully transmitted (i.e., acknowledged) I-frame.

The rationale behind this algorithm is as follows. Under normal conditions, i.e., no congestion, the actual window size used is the one initially chosen by the communicating stations. By reducing the window size to one, whenever there is an indication of a possible congestion, we achieve a high responsiveness of the flow-control mechanism in the sense that an immediate and very effective throttling of the network input traffic is performed. Subsequent to reduction, stations again attempt to increase their window sizes. This process is tightly coupled to the reception of acknowledgments. Hereby, we achieve control of the speed by which the window size is increased, by the momentary ability of the network to transport frames successfully.

We subsequently show how the multi-ring network performs when this enhanced LLC protocol is employed.

4.2. Results

We first address the question of selecting an appropriate value for the parameter n in the dynamic window-size algorithm. The value of n specifies how many I-frames a station needs to transmit successfully (following a window-size reduction) before it increases its window size by one.

For the scenario previously studied in Figs. 4 and 5, in Fig. 16 we show the total network throughput as a function of the total offered data rate for different values of n. The initial window size is ten. We observe substantial improvement in throughput when the window size is dynamically adjusted; the gain is higher for larger n. The incremental gain in throughput decreases, however, as n increases; see, e.g., the small difference between n = 8 and n = 16. For small values of n, the window is opened too rapidly, i.e., the throttling effect does not last long enough.

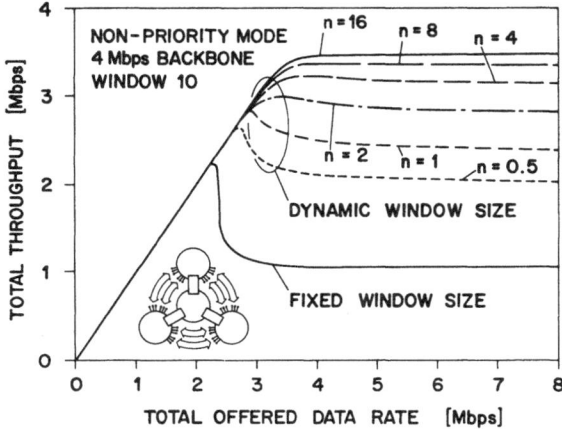

Fig. 16. Total throughput vs total offered data rate for dynamic window-size
LLC with initial window size ten. (Note: n = 0.5 means that the
window size is increased by two per acknowledgment.)

For the same example, Fig. 17 shows the end-to-end delay as a function of the
total throughput. It can be seen that the delay characteristic is better for

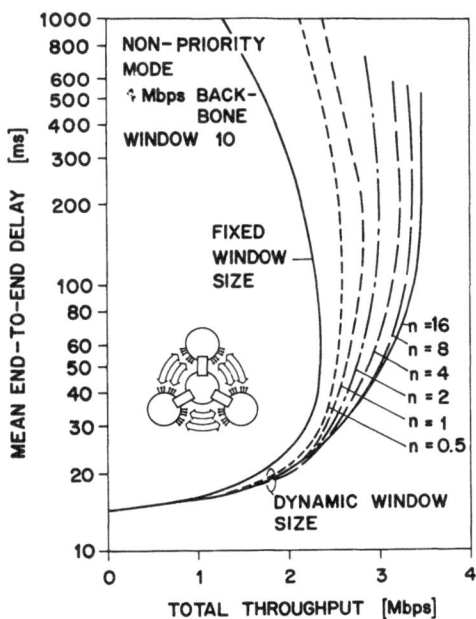

Fig. 17. Mean end-to-end delay vs total throughput for dynamic window-size
LLC with initial window size ten. (Note: n = 0.5 means that the
window size is increased by two per acknowledgment.)

larger parameter n; in particular for n = 8 or greater, the behavior is almost ideal.

For the scenario previously considered in the context of Fig. 13 where the entire I-frame traffic is flowing to one ring, Fig. 18 shows the improvement attained through the dynamic window algorithm. Again, a value of n = 8 already yields excellent performance. A general observation from numerous simulations has been that, even under extreme overload, the frame-loss frequency n bridges is never substantially greater than one percent, when the dynamic window-size algorithm with $n \geq 8$ is employed. Under this condition, only a very small fraction of the bandwidth is lost for retransmissions, and hence, performance is excellent. For this reason, we chose to use n = 8 in the subsequent examples, intended to show further interesting aspects of the dynamic flow-control algorithm.

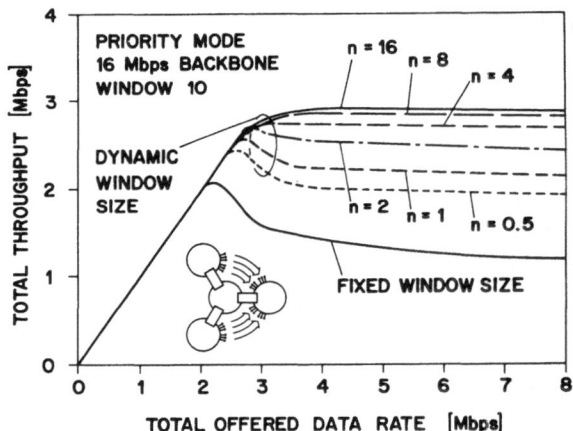

Fig. 18. Total throughput vs total offered data rate for dynamic window-size LLC with initial window size ten. (Note: n = 0.5 means that the window size is increased by two per acknowledgment.)

The first question we address is whether the dynamic window-size algorithm has an impact on fairness. We consider the same example as in Fig. 11. For the dynamic window-size algorithm with an initial window of 10 and n = 8, Fig. 19 shows the throughput of the inter- and intra-ring connections and the total network throughput for both priority and non-priority modes. The figure demonstrates that the overall efficiency in terms of the total throughput is excellent for both modes when the dynamic window-size algorithm is employed. In the case of non-priority mode, throughput of the inter-ring con-

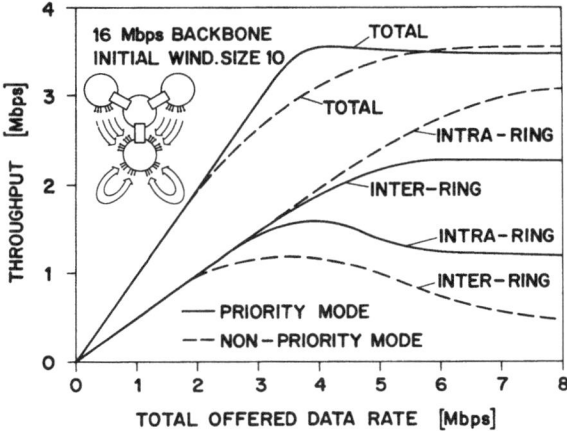

Fig. 19. Inter-ring, intra-ring, and total network throughput vs total offered data rate.

nections is significantly improved compared to the fixed-window LLC; at medium and high traffic loads, however, there still exists a distinct unfairness between the two connection types in non-priority mode. In priority mode, fairness is good for offered data rates up to the ring transmission rate of 4 Mbps. At high overload, priority mode tends to favor the inter-ring connections.

For the scenario underlying Figs. 13, 14, and 18, we compare in Fig. 20 the throughput characteristics of the dynamic window-size LLC and the fixed window-size LLC for three different bridge-buffer sizes. We have already observed

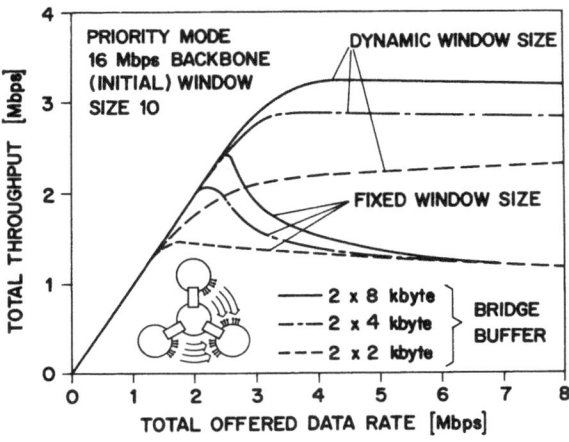

Fig. 20. Total throughput vs total offered data rate for different bridge-buffer sizes.

(see Figs. 13 and 14) that for larger fixed window sizes, the overload behavior is not improved through bigger bridge buffers. On the other hand, the dynamic window-size algorithm yields a stable throughput behavior even for relatively small bridge-buffer sizes, in the sense that throughput never decreases with increasing offered data rate. With the dynamic window, throughput is improved by bigger bridge buffers, but increasing the buffer size beyond the two times 8 kbytes shown in the figure does not yield any noticeable further improvement.

For both fixed and dynamic window-size LLC, Fig. 21 shows the total throughput of the network with a 16-Mbps backbone ring and priority ring access for bridges. For the case of 0.3 ms mean bridge-processing time per frame, we have already seen from Fig. 8 that performance is excellent even with the fixed window-size LLC. Consequently, the dynamic window-size algorithm yields equally good results in this case. If the frame-processing times in the bridges are substantially longer, i.e., 1.2 ms on the average, the bridge processors become the system bottlenecks at larger values of the offered data

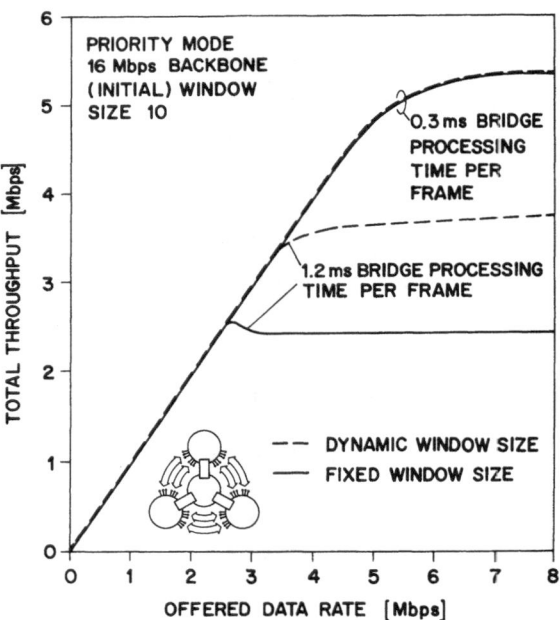

Fig. 21. Total throughput vs total offered data rate for different bridge-processing times.

rate; hence, we again observe the negative effects of frame losses and retransmissions. The dynamic window helps to improve performance also in this case. The algorithm works in such a way that the bridge processors are very highly utilized, but at the same time, the frame-loss frequency remains small, typically on the order of one percent.

In our last example, we reconsider the scenario of a 4-Mbps backbone ring and completely symmetrical traffic, however, the number of active stations is varied. For the fixed-window protocol, we observe from Fig. 22 that congestion generally becomes worse when the number of stations increases. In contrast, network performance is no longer sensitive to the number of stations, when the dynamic window-size algorithm is employed. These observations shed additional light onto the necessity of a dynamic flow-control scheme in multiring networks. We conclude the discussion of the dynamic flow-control scheme with two final remarks:

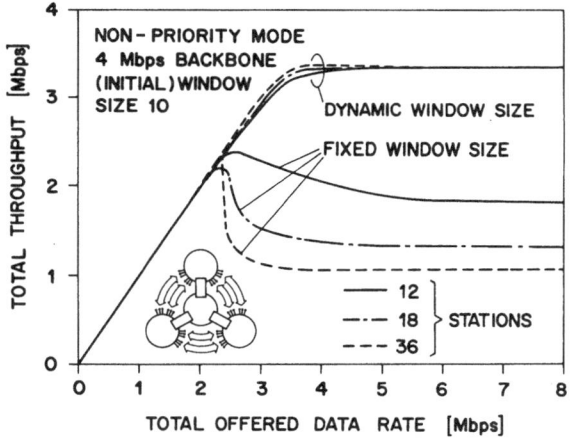

Fig. 22. Total throughput vs total offered data rate for different number of active stations.

1) In all cases where the fixed window-size protocol works without congestion problems, the dynamic flow-control algorithm yields equally good performance since the window size is never or only very rarely reduced. A typical example is the one of Fig. 15, for which the throughput characteristic of the dynamic-window LLC with an initial window size of ten is identical to the result for the fixed window of ten, namely, almost ideal.

2) If I-frames are lost owing to transmission errors and not congestion, the window will be unnecessarily reduced. However, for normally functioning rings with bit error rates of, e.g., 10^{-9} or less, this will not cause any noticeable performance degradation. On the other hand, in failure situations, e.g., transient periods with high error rates or short ring interruptions, our flow-control method may be very helpful. After a failure, heavy overload can occur owing to back-logged traffic and increased recovery activity. Initial experiments suggest that this overload can be very effectively controlled by the dynamic window mechanism. This is the subject of an ongoing study.

5. CONCLUSION

In this paper, we have investigated the performance of local networks consisting of interconnected token rings. We first considered networks in which the IEEE 802.2 type-2 LLC protocol as defined today is employed end-to-end. The key observations of this part of the study are: 1) in a congested network, large fixed window sizes can lead to severe performance degradation. On the other hand, small windows are unnecessarily restrictive and can lead to poor performance if the network is not congested. 2) If bridges are provided with priority to access the local rings, significant improvements both with respect to overall efficiency and fair sharing of bandwidth can be achieved. 3) To a limited extent, performance can be improved by providing larger buffers in bridges. However, large bridge buffers are not sufficient to overcome congestion problems in general. 4) Generally, the congestion problem becomes more severe for larger numbers of active stations in the network.

From these observations, we concluded that the architecture should be enhanced by a suitable flow-control mechanism. Our proposed solution is to add a dynamic window-size algorithm to the LLC protocol. Simulations demonstrate that the algorithm suggested yields close-to-optimal network performance under both normal traffic load and overload conditions, different traffic patterns, different number of stations, and even relatively small bridge-buffer sizes. Besides its effectiveness in minimizing congestion, the new flow control method has the following attractive properties: i) it is simple to implement; ii) it is local to the station sending information frames, hence no additional information exchange is required; iii) it is compatible with the fixed window-size protocol; iv) bridges need not be involved in flow control; and v) the medium-access control protocol is not affected.

ACKNOWLEDGEMENT

The authors would like to thank D. W. Andrews, N. A. Bouroudjian, K. Kümmerle, D. A. Pitt and K. K. Sy for many helpful discussions.

REFERENCES

[1] Standard ECMA-89: Local Area Networks Token Ring Technique.
[2] ANSI/IEEE Standard 802.5 − 1985, Token Ring Access Method and Physical Layer Specifications.
[3] Sy K. K. and Pitt D. A., "An architecture for interconnected token rings," Proposal to IEEE 802.5 Committee, February 1984.
[4] ANSI/IEEE Standard 802.2 − 1985, Logical Link Control.
[5] Sauer C. H. and MacNair E. A., "Simulation of computer communication systems". Prentice Hall, Englewood Cliffs, N.J., 1983.
[6] Bux W., Closs F., Kümmerle K., Keller H., and Müller H. R., "A reliable token ring for local communications," IEEE J. Select. Areas Commun., vol. SAC-1, pp. 756-765, 1983.
[7] Shoch J. F. and Hupp J. A., "Measured performance of an Ethernet local network," Commun. ACM, vol. 23, pp. 711-721, 1980.
[8] Gerla M. and Kleinrock L., "Flow control: A comparative survey," IEEE Trans. Commun., vol. COM-28, pp. 553-574, 1980.
[9] Giessler A., Jaegemann A., Maeser E., and Haenle J. O., "Flow control based on buffer classes," IEEE Trans. Commun., vol. COM-29, pp. 436-443, 1981.
[10] Lam S. S. and Reiser M., "Congestion control of store-and-forward networks by input buffer limits - An analysis," IEEE Trans. Commun., vol. COM-27, pp. 127-134, 1979.

DISTRIBUTED ASYNCHRONOUS OPTIMAL ROUTING IN DATA NETWORKS

by

John N. Tsitsiklis and Dimitri P. Bertsekas

Massachusetts Institute of Technology

Cambridge, Mass. 02139

U. S. A.

Abstract

In this paper we summarize the results regarding a class of distributed optimal routing algorithms of the gradient projection type that operate under weaker and more realistic assumptions than those considered thus far [8]. In particular, convergence can be shown to an optimal routing without assuming synchronization of computation at all nodes and measurement of link lengths at all links, while taking into account the possibility of link flow transients caused by routing updates. This demonstrates the robustness of these algorithms in a realistic distributed operating environment.

I. Introduction

Routing is an important packet network function that affects both the throughput of the network and the average packet delay. The most common approach is shortest path routing which, however, may cause low throughput, poor response to traffic congestion, and oscillatory behavior, depending on its implementation [14]. A more sophisticated alternative is optimal distributed routing based on multicommodity flow models, whereby congestion on each network link is measured in terms of the average traffic carried by the link.

The most popular formulation of the optimal distributed routing problem is based on a multicommodity flow optimization whereby a separable objective function of the form

NATO ASI Series, Vol. F38
Flow Control of Congested Networks
Edited by A. R. Odoni et al.
© Springer-Verlag Berlin Heidelberg 1987

$$\Sigma_{(i,j)} \ D^{ij}(F^{ij})$$

is minimized with respect to the flow variables F^{ij} subject to multicommodity flow constraints [1]-[3], [9], [14]. Here (i,j) denotes a generic directed network link, and D^{ij} is a strictly convex differentiable, increasing function of F^{ij} which represents in turn the total traffic arrival rate on link (i,j) measured, for example, in packets or bits per second.

We want to find a routing that minimizes this objective. By a routing we mean a set of active paths for each origin-destination (OD) pair (set of paths carrying some traffic of that OD pair), together with the fraction of total traffic of the OD pair routed along each active path.

A typical example of adaptive distributed routing, patterned after the ARPANET algorithm [4], operates roughly as follows.

The total link arrival rates F^{ij} are measured by time averaging over a period of time, and are communicated to all network nodes. Upon reception of these measured rates each node updates the part of the routing dealing with traffic originating at that node. The updating method is based on some rule, e.g., a shortest path method [2], [4], or an iterative optimization algorithm [1], [5], [6].

There are a number of variations of this idea; for example, some relevant function of F^{ij} may be measured in place of F^{ij} [such as average delay per packet crossing link (i,j)], or a somewhat different type of routing policy may be used, but these will not concern us for the time being.

Most of the existing analysis of distributed routing algorithms of the type above is predicated on several assumptions that are to some extent violated in practice. These are as follows.

1) The quasi-static assumption, i.e., the external traffic arrival process for each OD pair is stationary over time. This assumption is approximately valid when there is a large number of user-pair conversations associated with each OD pair, and each of these conversations has an arrival rate that is small relative to the total arrival rate for the OD

pair (i.e., a "many small users" assumption). An asymptotic analysis of the effect of violation of this assumption on the stationary character of the external traffic arrival rates is given in [7].

2) The fast settling time assumption, i.e., transients in the flows F^{ij} due to changes in routing are negligible. In other words, once the routing is updated, the flows F^{ij} settle to their new values within time which is very small relative to the time between routing updates. This assumption is typically valid in datagram networks but less so in virtual circuit networks where, existing virtual circuits may not be rerouted after a routing update. When this assumption is violated, link flow measurements F^{ij} reflect a dependence not just on the current routing but also on possibly several past routings. A seemingly good model is to represent each F^{ij} as a convex combination of the rates of arrival at (i,j) corresponding to two or more past routing updates.

3) The synchronous update assumption, i.e. all link rates F^{ij} are measured simultaneously, and are received simultaneously at all network nodes who in turn simultaneously carry out a routing update. However, there may be technical reasons (such as software complexity) that argue against enforcing a synchronous update protocol. For example, the distributed routing algorithm of the ARPANET [4] is not operated synchronously. Furthermore, in an asynchronous updating environment, the rates F^{ij} are typically measured as time averages that reflect dependence on more than one update.

In [8] we study gradient projection methods, which are one of the most interesting classes of algorithms for distributed optimal routing. A typical iteration in a gradient method consists of making a small update in a direction which improves the value of the cost function, e.g., opposite to the direction of the gradient. A gradient projection method is a modification of this idea, so that constrained optimization problems (such as the multicommodity flow problem of this paper) may be handled as well; namely, whenever an update leads to a point outside the feasible set (which is determined by the constants of the problem), feasibility is enforced by projecting that point back into the feasible set. In the context of the optimal routing problem, a gradient projection method finds at each iteration for each origin-destination pair a path that is shortest with respect to link lengths that are first derivatives of link costs, and shifts some flow from other paths to that path. The first application of this type of method in data communication routing is due to Gallager [1] as explained later in [10]. Gallager's method operates in a space of link flow fractions. Related gradient projection methods which operate in the space of path flows are given in [3], [5], [11], and [12]. This latter class of methods is the starting point for the analysis of the present paper. We conjecture, however, that qualitatively similar results hold for Gallager's method as well as for its second derivative version [6].

Our main result states that gradient projection methods for optimal routing are valid even if the settling time and synchronous update assumption are violated to a considerable extent. Even though we retain the quasi-static assumption in our analysis, we conjecture that the result of this paper can be generalized along the lines of another related study [7] where it is shown that a routing algorithm based on a shortest path rule converges to a neighborhood of the optimum. The size of this neighborhood depends on the extent of violation of the quasi-static assumption. A similar deviation from optimality can be caused by errors in the measurement of F^{ij}. In our analysis, these errors are neglected.

A practical routing algorithm that nearly falls within the framework of the present paper is the one implemented in the CODEX network [13], [14]. There destination nodes of OD pairs asynchronously assign and reroute virtual circuits to paths that are shortest with respect to link lengths that relate to first derivatives of link costs. Only one virtual circuit can be rerouted at a time, but several virtual circuits can be rerouted before new measurements are received. More precisely, a destination node assigns (or reroutes) a virtual circuit to a path for which the assignment (rerouting) results in minimum cost. This is equivalent to assignment (rerouting) on a shortest path with respect to link lengths which are first derivatives of link costs evaluated at a flow that lies between the current flow and the flow resulting once the assignment (rerouting) is effected. Another difference is that, in the CODEX network, each virtual circuit may carry flow that is a substantial portion of a link's capacity. This may place a lower bound on the amount of flow that can be diverted to a shortest path at each iteration.

Our analysis is based on models for asynchronous distributed algorithms, and a general theory for showing convergence of such algorithms (a survey is given in [15]; the subject has a long history and some representative papers are [16]-[20]). We provide here a brief discussion of this theory. In a typical algorithm (aimed at solving an optimization problem) each processor i has in its memory a vector x^i which may be interpreted as an estimate of an optimal solution. Each processor obtains measurements, performs computations, and updates some of the components of its vector. Concerning the other components, it relies entirely on messages received from other processors. We are mainly interested in the case where minimal assumptions are placed on the orderliness of message exchanges.

There are two distinct approaches for analyzing algorithmic convergence. The first approach is essentially a generalization of the Lyapunov function method for proving a convergence of centralized iterative processes. The idea here is that, no matter what the precise sequence of message exchange is, each update by any processor brings its vector x^i closer to the optimum in some sense. This approach applies primarily to problems involving monotone or contraction mappings with respect to a sup-norm (e.g., a distributed shortest path algorithm); it is only required that each processor communicates to every other processor an infinite number of times.

The second approach is based on the idea that if the processors communicate fast enough relative to the speed of convergence of the computation, then the evolution of their solution estimates x^i may be (up to first order in the step-size used) the same as if all processors were communicating to each other at each time instance. The latter case is, however, mathematiacally equivalent to a centralized (synchronous) algorithm for which there is an abundance of techniques and results. Notice that in this approach, slightly stronger assumptions are placed on the nature of the communication process than in the first one. This is compensated by the fact that the corresponding method of analysis applies to broader classes of algorithm.

Our analysis of [8] is close in spirit to the second approach outlined above. Unfortunately, however, the results available cannot be directly applied to the routing problem studied in this paper and a new proof is required. One reason is that earlier results concern algorithms for unconstrained optimization. In the routing problem, the nonnegativity and the conservation of flow introduce inequality and equality constraints. While equality constraints could be taken care of by eliminating some of the variables, inequality constraints must be explicitly taken into account . Another difference arises because, in the routing algorithm, optimization is carried out with respect to path flow variables, whereas the messages being broadcast contain estimates of the link flows. In earlier results the variables being communicated were assumed to be the same as the variables being optimized. Finally, the transient behavior of the network (which results from the fact that we do not make the fast settling time assumption) adds a few more particularities to the model and the analysis.

REFERENCES

[1] R. G. Gallager, "A Minimum Delay Routing Algorithm Using Distributed Computation", IEEE Trans. Commun., Vol. COM-25, pp. 73-85, 1977.

[2] M. Schwartz and T. E. Stern, "Routing Techniques Used in Computer Communication Networks", IEEE Trans. Commun., Vol. COM-28, pp. 539-559, 1980.

[3] D. P. Bertsekas, "Optimal Routing and Flow Control Methods for Communication Networks", in Analysis and Optimization of Systems, A. Bensoussan and J. L. Lions, Eds. New York: Springer-Verlag, 1982, pp. 615-643.

[4] J. M McQuillan, I. Richer, and E. C. Rosen, "The New Routing Algorithm for the ARPANET", IEEE Trans. Commun., Vol. COM-28, pp. 711-719, 1980.

[5] D. P. Bertsekas and E. M. Gafni, "Projected Newton Methods and Optimization of Multicommodity Flows", IEEE Trans. Automat. Contr., Vol. AC-28, pp. 1090-1096, 1983.

[6] D. P. Bertsekas, E. M. Gafni, and R. G. Gallager, "Second Derivative Algorithms for Minimum Delay Distributed Routing in Networks", IEEE Trans. Commun., Vol. COM-32-, pp. 911-919, 1984.

[7] E. M. Gafni and D. P. Bertsekas, "Asymptotic Optimality of Shortest Path Routing", Mass. Inst. Technol., Cambridge, MA, LIDS Report P-1307, July 1983; also in IEEE Trans. Inform. Theory, Jan. 1987.

[8] J. N. Tsitsiklis and D. P. Bertsekas, "Distributed Asynchornous Optimal Routing in Data Networks", IEEE Trans. on Aut. Control, Vol. AC-31, 1986, pp. 325-332.

[9] L. Kleinrock, Communication Nets: Stochastic Message Flow and Delay. New York: McGraw-Hill, 1964.

[10] D. P. Bertsekas, "Algorithms for Nonlinear Multicommodity Network Flow Problems", in Proc. International Symposium on Systems Optimization and Analysis, A. Bensoussan and J. L. Lions, Eds. New York: Springer-Verlag, 1979, pp. 210-224.

[11] D. B. Bertsekas, "A Class of Optimal Routing Algorithms for Communication Networks", in Proc. 5th Int. Conf. Comput. Commun., Atlanta, GA, Oct. 1980, pp. 71-76.

[12] D. P. Bertsekas and E. Gafni, "Projection Methods for Variational Inequalities with Application to the Traffic Assignment Problem", Math. Progr. Study, Vol. I7, D. C. Sorensen and R. J.-B. Wets, Eds. Amsterdam, The Netherlands: North-Holland, 1982, pp. 139-159.

[13] P. Humblet, Private communication, 1985.

[14] D. B. Bertsekas and R. G. Gallager, Data Networks, Prentice-Hall, N.J., 1986.

[15] D. P. Bertsekas, J. N. Tsitsiklis, and M. Athans, "Convergence Theories of Distributed Iterative Processes: A Survey", Mass. Inst. Technol., LIDS Report P-1412, Oct. 1984.

[16] D. Chazan and W. Miranker, "Chaotic Relaxation", Linear Algebra and Its Applications, Vol. 2, 1969, pp. 199-222.

[17] J. C. Miellou, "Iterations Chaotiques a retards, etudes de la convergence dans le case d'espaces partiellment ordonnes", Computes rendus de l'Academie des Sciences, Paris, Serie A 278, 1974, pp. 957-960.

[18] G. M. Baudet, "Asynchronous Iterative Methods for Multiprocessors", Journal of the ACM, Vol. 2, 1978, pp. 226-244.

[19] D. P. Bertsekas, "Distributed Dynamic Programming", IEEE Trans. on Automatic Control, Vol. AC-27, 1982, pp. 610-616.

[20] D. P. Bertsekas, "Distributed Asynchronous Computation of Fixed Points", Math. Programming, Vol. 27, 1983, pp. 107-120.

ANALYTIC MODELS FOR
TREE COMMUNICATION PROTOCOLS

Philippe Flajolet and *Philippe Jacquet*

INRIA, Rocquencourt

78150 Le Chesnay (France)

Abstract: The tree protocol for local area networks, together with a number of its variants, can be exactly analysed under a Poisson arrival model. This note surveys some of the evaluations that have been obtained for characteristic parameters including delay, session length or probability of immediate access to the channel. The mathematical techniques involved are: functional equations and Mellin transforms.

1. Introduction

Protocols for regulating access to a channel shared by several stations were first designed in the sixties, and started with the ALOHA concept: Each station transmits as soon as it has a message to send; the message is broadcast, and picked up by its intended receiver unless two (or more) stations collide; in that case, every station detects the collision, and senders schedule a later retransmission. The rule for retransmitting is precisely the *communication protocol*.

Stations' "feedback" from the channel is thus limited to a ternary information: ACK (acknowledgement, i.e. successful transmission); LACK (lack of transmission, i.e. silence); NACK (no ACK, i.e. collision). Since stations are not distinguishable from each other in general, a key idea to resolve collisions is to let a *random* component enter their retransmission policy. If the protocol is suitably designed, one may hope for the best, namely expect the channel to succesfully transmit messages with reasonable delays as long as the traffic load is not too high.

In this paper, we consider the case of a *slotted time channel* where transmissions start at discrete instants $0, 1, 2, \ldots$ and messages are calibrated so that their duration does not exceed one slot. Basically, the ALOHA protocol is the following simple rule:

A. In case of a collision, wait a random amount of time (i.e. slots) uniformly distributed over the interval $[1..\delta_0]$ before attempting a retransmission. Parameter δ_0 is a design parameter whose value is fixed and common to all stations, its value being chosen based on the network configuration.

It was soon realised (Fayolle et al.; Kleinrock et al.) that the ALOHA protocol is unstable: If messages arrive according to a Poisson process with intensity $\lambda > 0$, then with probability 1, the "backlog" (messages awaiting retransmission because of previous collisions) tends to infinity. Intuitively, the protocol maintains a virtual time window of a fixed size that, sooner or later, is doomed to become saturated.

NATO ASI Series, Vol. F38
Flow Control of Congested Networks
Edited by A. R. Odoni et al.
© Springer-Verlag Berlin Heidelberg 1987

The next idea, which gave rise to the Ethernet protocol, was to use a "sliding" parameter δ whose value changes dynamically with stations and time, and whose control is meant to have the protocol adjust to traffic variations. The simple idea is for a station to retransmit randomly in the interval $[1..\delta]$ where:

E. Initially, upon a message arrival, δ is 1. After each collision experienced by its message, the station doubles it own value of δ.

It took some time and effort (Aldous) to realise that Ethernet is itself unstable: in practice, this may mean fairly suboptimal channel utilisation, delay inefficiencies or poor response to bursts of traffic. Meanwhile, the *tree protocol* was invented circa 1977 by Capetanakis, based on the following elegant idea:

T. If a group G of stations collide ($|G| \geq 2$), that group is split by coin flippings into two subgroups G_0 and G_1. The stations in G_0 *first* recursively resolve their collisions. *Then* the group G_1 resolve their collisions independently.

The interest of this protocol is to use a dichotomy to separate colliders, and an execution is simply described by a tree. It is not immediately clear however that it can be implemented without having the stations communicate some extraneous informations about their coin flippings. A decentralised formulation (providing a practical implementation) of that protocol was arrived at independently by Tsybakov et al. and there, each station manages a *stack*.

Partial stability regions were characterised in the initial papers, and we now know that, under the Poisson model, the protocol is stable until an arrival rate of about 35%. Actually, going to details, there are two ways of implementing the protocol:

- *Free Access:* The immediate access rule (in the style of ALOHA) is enforced. Thus the resolution of collisions by a group may involve newly arrived messages; the system operates in a somewhat last-in first-out fashion.

- *Blocked Access:* There collisions are resolved in sessions. Newly arrived stations wait until a collision resolution "session" is over (in case any is taking place) before they are allowed to enter the competition, and start a new session.

The purpose of this paper is to present the analytic methods involved in the evaluations, putting into perspective some of our own results [FFHJ 1985], [FFH 1986], [MF1985], [Jacquet 1987] and [JR 1986]. It is quite interesting that there is a fairly rich mathematical structure behind the tree protocol, and many relevant parameters can be exactly analysed. We shall try to illustrate the mathematical methods at stake here.

Note: The reader can refer to the survey paper by Massey [Massey 1981] and to a special issue of the IEEE Transactions [Massey 1985] for detailed references on the subject that are not duplicated here.

2. Blocked Access: A Basic Analysis

Strangely enough, the basic tree protocol with blocked access had been analysed *before* it was invented, by Knuth (1973). The reason is the generality of the recursive splitting process based on random choices that turns out to be the exact model for the trie data structure, and for a variety of searching methods in computer science. Below is a revised presentation of Knuth's result.

Let L_N be the random variable (RV) denoting the time taken to resolve N collisions using the tree method, assuming no further arrivals. If the group of size N is split into subgroups of size K and $N - K$ ($N \geq 2$) then:

$$L_N = 1 + L_K + L_{N-K} \qquad \text{with} \qquad L_0 = L_1 = 1. \tag{1}$$

When a fair coin is used, the "splitting" probabilities that the RV K has value k is:

$$\pi_{N,k} = \frac{1}{2^N} \binom{N}{k} \tag{2}$$

so that, taking expectations of (1):

$$l_N \equiv E[L_N] = 1 + \sum_{k=0}^{N} \pi_{N,k}(l_k + l_{N-k}), \qquad (N \geq 2) \tag{3a}$$

a relation that permits to compute inductively each of the l_N. The form of (3a) suggests using exponential generating functions (egf's). If $l(z) = \sum_{N \geq 0} l_N \frac{z^N}{N!}$, then (3a) becomes:

$$l(z) = e^z - 2 - 2z + 2e^{z/2} l(\frac{z}{2}), \tag{3b}$$

a *difference equation*. From there two routes are possible:

- Direct solution: Set $\Lambda(z) = e^{-z} l(z)$ and determine the relation satisfied by it. It is of the form: $\Lambda(z) = 2\Lambda(\frac{z}{2}) + $ a known function. Thus the coefficients of Λ can be exactly recovered, whence by convolution with those of e^z, the coefficients of $l(z)$. This provides a finite sum for l_N, which involves exponential cancellations and does not yield easily to asymptotic analysis (though the so-called "Rice integrals" method from the calculus of finite differences can be used).

- Iterative solution: A functional equation like (3b) is of the form:

$$\phi(z) = \alpha(z) + \beta(z)\phi(\gamma(z))$$

with ϕ the unknown function, and it can be solved by iteration. If this is done here, and the solution is expanded, one finds a well-conditioned sum:

$$l_N = 1 + 2\sum_{k=0}^{\infty} 2^k \left[1 - (1 - \frac{1}{2^k})^N - \frac{N}{2^k}(1 - \frac{1}{2^k})^{N-1}\right] \tag{4}$$

an expansion that however reserves some surprises!

It is interesting *per se* to consider approximations of l_N, and it is not difficult to conjecture that $l_N \approx \frac{2N}{\log 2}$. A good physical reason for interest in this asymptotic problem is that, if $l_N \sim cN$, then the constant c is (asymptotically) an average service time. Thus the protocol should be stable for arrival rates λ such that $c\lambda < 1$, which suggests a maximum admissible throughput for the tree protocol of $\lambda = \log 2/2 = 0.34657$.

That conjecture is almost true, but not as simple as it looks. First, using in (4) the approximation $(1-a)^N \approx e^{-aN}$ —which is easy to justify— we find:

$$l_N = 2F(N) + O(\sqrt{N}) \qquad \text{with} \qquad F(x) = \sum_{k\geq 0} 2^k[1 - (1 + \frac{x}{2^k})e^{-x/2^k}]. \qquad (5)$$

That sum is a so-called harmonic sum and the best way to treat it is to determine its Mellin transform defined as:

$$F^*(s) = \int_0^\infty F(x)x^{s-1}\,dx. \qquad (6)$$

Here, we find:

$$F^*(s) = \frac{(s+1)\Gamma(s)}{1 - 2^{s+1}} \qquad \text{for } s \text{ in the strip } -2 < \Re(s) < -1. \qquad (7)$$

From there, using the Mellin inversion formula, and computing the integral by residues, we get:

$$F(x) = \frac{1}{2i\pi} \int_{-3/2-i\infty}^{-3/2+i\infty} F^*(s)x^{-s}\,ds \sim \sum_{\Re(s)\geq -1} \text{Res}[F^*(s)x^{-s}], \qquad (8)$$

where the last sum has the character of an asymptotic expansion in non-increasing powers of x. However (There is the rub!), $F^*(s)$ has poles with a non-zero imaginary part, and since $x^{it} = e^{it\log x}$, these correspond to *periodic fluctuations*. Thus l_N is not a "smooth" function of N, though the periodic fluctuations are of limited importance since their amplitude is $< 10^{-5}$. We have arrived at:

THEOREM 1: (i). [Knuth 1973] *Quantity l_N satisfies asymptotically:*

$$l_N = \frac{2}{\log 2}N + NP(\log_2 N) + O(\sqrt{N})$$

where $P(.)$ is a periodic function with amplitude $< 10^{-5}$.

(ii). *The supremum $\lambda_{\max}$ of arrival rates λ for which the tree protocol with blocked access is stable satisfies:*

$$|\lambda_{\max} - \frac{\log 2}{2}| < 10^{-5}.$$

Therefore $\lambda_{\max} \approx 0.34657$.

3. Free Access: A Functional Equation...

It is natural to try and extend the previous analysis to the case of the free access version of the protocol. As we shall see, the functional equations that appear are of a different form, and they lead to some analytic difficulties. However, from the analysis, the maximum admissible throughput can be exactly determined.

Consider thus the free access tree protocol with a Poisson rate of arrival λ. The starting point is a recursive relation on RV's that extends Eq (1). Since arrivals keep coming in, the basic equation is:

$$L_N = 1 + L_{K+X} + L_{N-K+Y} \qquad \text{with} \qquad L_0 = L_1 = 1 \qquad (9)$$

where X, Y are Poisson RV's. At this stage, also consider the possibility for the coin flippings to be biased, with probabilities p and $q = 1 - p$ for head and tail. Introduce the egf $l(z)$ of the expectations l_N and the modified egf $\Lambda(z) = e^{-z}l(z)$. These quantities now depend on λ and $l_N = l_N(\lambda)$ etc., thus the l_N for blocked access coincide with $l_N(0)$. It is not too difficult to see that $\Lambda(z)$ satisfies an equation of the form:

$$\Lambda(z) = \Lambda(\lambda + pz) + \Lambda(\lambda + qz) + \alpha(z) \qquad (10)$$

where $\alpha(z)$ is a known function. That functional equation ceases to be "local": it relates the values of Λ around 0 to the values around λ, and this reflects the fact that, due to arrivals, each l_N is related by an infinite recursion to all the other l_j.

A first step is thus to consider functional equations of the general form (where ϕ is the unknown function):

$$\phi(z) = \alpha(z) + u\phi(\sigma_1(z)) + v\phi(\sigma_2(z)) \qquad \text{with} \quad \sigma_1(z) = \lambda + pz \text{ and } \sigma_2(z) = \lambda + qz. \qquad (11a)$$

Solutions by iteration of (11a) involve a sum over the "iteration semigroup" H defined as the set of all compositions of σ_1 and σ_2 (observe that H is non commutative in general):

$$\phi(z) = \mathbf{S}[\alpha(.); z] = \sum_{\tau \in H} (u; v)^\tau \alpha(\tau(z)) \qquad \text{where} \quad (u; v)^\tau \equiv u^{|\tau|_{\sigma_1}} v^{|\tau|_{\sigma_2}}. \qquad (11b)$$

We have thus at our disposal a summation operator $\mathbf{S}$ to solve non local difference equations of the form (11a). Applying it to the equation giving Λ yields explicit expressions, again involving sums indexed by H that are easy to evaluate numerically. More important, we see that $\Lambda(z)$, as well as the l_N, become infinite when $\lambda \to \lambda'_{\max}$ where $\lambda'_{\max}$ is determined as the solution of a certain transcental equation. Thus the maximum admissible throughput of the tree protocol with free access is precisely determined. We state here the easier case where $p = q = \frac{1}{2}$:

THEOREM 2: [FFH 1986] *The supremum* $\lambda'_{\max}$ *of arrival rates for which the tree protocol with free access using fair coins is stable, is the smallest positive root of the equation:*

$$1 + \frac{2\exp(-2\lambda)}{1-2\lambda} \sum_{i\geq 0} 2^i \exp(2\lambda/2^i)\left[\exp(-\lambda/2^i)(1-\lambda/2^i)-1-2(\lambda/2^i)^2+2(\lambda/2^i)\right] = 0. \tag{12}$$

Numerically, $\lambda'_{\max} = 0.360177147$.

Thus, free access accepts a slightly er traffic rate than blocked access before destabilising. Viewed from the stations, free access is also easier to implement since there is no need of a continuous monitoring of the channel during inactivity periods. Therefore it seems to be the method of choice in this class of methods, especially after further optimisations to be discussed later are applied.

The asymptotic analysis is trickier than before. The Mellin transform of a harmonic sum involves a Dirichlet series related to amplitudes and frequencies:

$$\int_0^\infty \left(\sum_k a_k f(b_k x)\right) x^{s-1}\, dx = \omega(s) \int_0^\infty f(x) x^{s-1}\, dx \qquad \text{with } \omega(s) = \sum_k a_k b_k^{-s}. \tag{13}$$

In the case examined in Section 2, we just had $\omega(s) = (1-2^{s+1})^{-1}$, by summation of a geometric progression. Now, there appears sums indexed by semi-group H, of the form:

$$\omega(s) = \sum_{\tau \in H} r(\tau(0))(p^s; q^s)^\tau \tag{14}$$

where $r(u)$ is a standard C^∞ function (a combination of exponentials). To carry out the asymptotic analysis requires determining the singularities of such an $\omega(s)$.

There is an interesting topological property of semi-group H: the images of a given point z_0 under H are *dense* over the real interval $[\frac{\lambda}{p}, \frac{\lambda}{q}]$ determined by the fixed points of σ_1, σ_2. Furthermore, these images are in a sense asymptotically *uniformly distributed*, and this fact yields the poles of $\omega(s)$, from which the asymptotic analysis can be completed.

THEOREM 3: *[FFH 1986] For the tree protocol with free access under a Poisson flow of arrivals of parameter λ with $\lambda < \lambda'_{\max}$, we have, neglecting small fluctuations:*

$$l_N \approx c(\lambda)N + o(N) \qquad \text{as } N \to \infty. \tag{15}$$

Thus, the burst response of the protocol is also fully characterised.

4. Free Access: Delay and Other Parameters

The previous section has demonstrated how to analyse the session (or collision resolution interval, CRI) lengths when there are N initial colliders –both exactly and asymptotically– characterising in passing the stability region. In particular, we introduced in (11b) a rather powerful summation operator **S** over semi-groups.

The problem now is to determine the steady state behaviour of important parameters of the protocol like delay etc. The basic approach is in two steps:

1. For parameters that are inductively defined on the tree structure, generating functions of the form (11a) can be set up. Thus, using the solution method (11b), we could also obtain expressions for their expectations over a session conditioned to be with N initial colliders.

2. The *values* of the generating functions, like $\Lambda(z)$, at $z = \lambda$ have a probabilistic interpretation, and they yield unconditional expectations of the parameters.

The new point is 2 above. To see it in the case of session length, observe that, by definition:

$$\Lambda(\lambda) = \sum_{N \geq 0} l_N e^{-\lambda} \frac{\lambda^N}{N!}. \tag{16}$$

Now, the coefficient of l_N in the above is nothing but the Poisson probability. But sessions start at "random" times, where the number of colliders also obey the Poisson distribution. Thus, the weighting in (16) yields the unconditional expectation of a CRI (or interval between two returns to the empty state, with no station backlogged).

That argument can be adapted to a delay analysis:

(i). The expectation of the cumulated delay experienced by all stations in a session, conditioned upon the number of initial colliders N, has a modified generating function $D(z)$ which satisfies an equation of the form (11a).

(ii). Over a large number s of sessions, by the "law of large numbers", the total delay will tend to $sD(\lambda)$. The total session length will tend to $s\Lambda(\lambda)$, with asymptotically $s\lambda\Lambda(\lambda)$ arrivals. Therefore, the unconditional (steady state) mean delay per message is equal to the quotient $D(\lambda)/(\lambda\Lambda(\lambda))$.

We shall summarise (see [FFHJ 1985] for detailed expressions) the previous discussion by:

THEOREM 4: [FFHJ 1985] *The steady state expectation of the delay experienced by a station under the free access tree protocol has an explicit expression in terms of the summation operator* S *of (11b) applied to standard functions.*

Similar results hold true for the variance of delay, the top-of-the-stack occupancy distribution etc. and the expressions obtained lend themselves to easy numerical evaluation. When $p = q = \frac{1}{2}$ they further simplify and somewhat resemble expressions found in Theorem 2. As an example of extensive numerical estimates, when $\lambda = 0.25$ and fair coins are used, the expected delay is only 4.79180 with a standard deviation of 11.2; there is probability 0.619 that a newly arriving message will be delivered immediately. Also, a detailed "low traffic" analysis can be conducted: We find that using a baised coin with $p = 2 - \sqrt{2} = 0.586$ slightly optimises delay for low traffic.

Finally, this type of analysis applies *mutatis mutandis* to other versions of the tree protocol. An unexpected result [MF 1985] is that using ternary instead of binary splittings in the basic protocol improves some of the characteristics by about 10%, at no extra cost of implementation: In the stack formulation, simply go down by two levels in the stack when a collision is encountered.

THEOREM 5: [MF 1985] *The tree protocol with free access and ternary splittings has a maximum admissible throughput of*

$$\lambda''_{\max} = 0.401599.$$

5. The Deterministic Tree Protocol

So far, the tree idea has been used, in accordance with the ALOHA principles, in a probabilistic manner. However, as already noticed by Capetanakis, it can also be implemented in a non randomised way, when the number of stations is fixed in advance.

Assume a network is configured to operate with a maximum number K of stations and consider the blocked access version of the protocol. If K is taken a power of 2, $K = 2^k$, then assign to each station a binary string of length k that uniquely identifies it. In each session, a station can use its predetermined string (instead of a random sequence) to participate in the splittings and schedule its retransmissions. This is the so-called *deterministic tree protocol.*

Quite clearly, the maximum length of a session is now $2K - 1$, corresponding to a full binary tree of height k that develops when all K stations are active; the maximum delay for a message will always be less than $4K$. Thus, it seems, a little gain occurs from this worst case guarantee at the expense of a little loss in flexibility. We shall see later that there is actually more to it!

We wish to analyse the deterministic protocol under the assumption that each station has messages arriving at a Poisson rate of λ/K, resulting in a global rate that is Poisson(λ).

A not too surprising phenomenon is that, when $\lambda < \lambda_{\max}$, the deterministic protocol behaves basically like the standard randomised version. More strikingly, however, stability is retained when $\lambda > \lambda_{\max}$, provided λ remains less than $\frac{1}{2}$. We shall call such a region the *hyperstable region* and there is a sort of "phase transition" taking place at $\lambda_{\max}$.

Simulations reveal –and analysis confirms– that good approximations, in the hypersable region, (say within a few percent when $K = 256$) are obtained by letting $K \to \infty$. In particular, the queueing phenomena at the stations can be characterised in the limit.

The starting point for evaluations is relations that parametrise with k (or $K = 2^k$) the analysis in Eqs. (1-4). Let $l_N^{(k)}$ denote the expected length of a session starting with N initial colliders in a universe of $K = 2^k$ stations. Quantities $l_N^{(k)}$ are also relevant to the analysis of tries in computer algorithms and had been determined earlier by Trabb Pardo (1977):

THEOREM 6: *(i). [Trabb Pardo] The expected length of a session for the determin-istic protocol, with N initial colliders and a universe of $K = 2^k$ stations, is:*

$$l_N^{(k)} = 1 + 2^{k+1} \sum_{j=1}^{k} \left[2^{-j} \left(1 - \frac{\binom{2^k - 2^j}{N}}{\binom{2^k}{N}} \right) - \frac{\binom{2^k - 2^j}{N-1}}{\binom{2^k}{N}} \right]. \tag{17}$$

(ii). When $k \to \infty$, each expectation $l_N^{(k)}$ tends monotonically to $l_N^{(k)}$: $l_N^{(k)} \to l_N$ with $l_N^{(k)} < l_N$.

The second assertion follows by simple computations. One can actually prove [Jacquet 1987] that, in the sense of Markov chains (i.e. the transition probabilities), the deter-ministic protocol converges to the probabilistic protocol provided $\lambda < \lambda_{\max}$. Accord-ingly, the queueing phenomena at the stations are asymptotically negligible. More interesting phenomena occur when $\lambda > \lambda_{\max}$:

PROPOSITION 7: *The system formed by queues at the stations coupled via the deterministic tree protocol is stable for $\lambda < \lambda_{\max}^{\circ}$, where:*

$$\lambda_{\max}^{\circ} = \frac{K}{2K - 1}.$$

This follows from the observation that the K stations receive service in at most $2K - 1$ slots. For large K, $\lambda_{\max}^{\circ}$ tends to $\frac{1}{2}$.

To continue the analysis, we observe that parameters of interest (session length, delay) should be re-normalised by K, if we want asymptotically meaningful quantities. We start with a simplified session model whose analysis is closely related to (17):

- Assume the channel is idle and that, at a beginning of a new session, each station becomes active with probability x. Then the expected length $l^{(k)}(x)$ of that session satisfies:

$$\frac{l^{(k)}(x)}{2^k} = 2^{-k} + 2 \sum_{j=0}^{k} \left[2^{-j} (1 - (1-x)^{2^j}) - x(1-x)^{2^j - 1} \right].$$

Essentially, $l^{(k)}(x)$ is a generating function of the $l_N^{(k)}$. As $k \to \infty$, the $2^{-k} l^{(k)}(x)$ converge to $L(x)$ given by:

$$L(x) = 2 \sum_{j=0}^{\infty} \left[2^{-j} (1 - (1-x)^{2^j}) - x(1-x)^{2^j - 1} \right]. \tag{18}$$

This function $L(x)$ plays an essential role in our subsequent analysis:

THEOREM 8: *[Jacquet 1987] Asymptotically for large K, the RV describing the queue length at any station, under the deterministic tree protocol with arrival rate λ in the hyperstable region, has generating function:*

$$q(z) = (1 - \mu) \frac{1 - z}{1 - z \exp(\mu(1 - z))}, \tag{19}$$

where μ is determined from λ by the equilibrium equation:

$$\mu = \lambda L(\mu)$$

and $L(x)$ is given by Eq. (18).

The reader will have recognised in (19) the generating function of an $M/D/1$ process (Markovian, i.e. Poisson arrivals/ deterministic service time): An $M/D/1$ queuing process with rate μ is a discrete time process $Q(t)$ such that:

$$Q(t+1) = \left| Q(t) - 1 \right|^{+} + A(t)$$

where $A(t)$ is a Poisson variable with parameter μ. Thus Theorem 8 expresses the following fact: in the hyperstable region, the queuing system behaves asymptotically like a collection of independent $M/D/1$ processes.

Let us give a quick intuition about the probabilistic phenomena at stake. In the steady state, a fraction μ ($0 \le \mu \le 1$) of the population will be active, resulting in a session of length $\sim KL(\mu)$ during which there arrive in turn $K\lambda L(\mu)$ new messages. Whence the equilibrium equation: $\mu = \lambda L(\mu)$. Since the "server" (i.e. channel) is periodically available, each time a new session is started, the system, once normalised, resembles in the limit an $M/D/1$ process. It can, in effect, be proved rigourously that the state transition probabilities corresponding to finite values of K converge to the transition probabilities of the $M/D/1$ process.

6. Limit Distributions

We conclude our review of analytic results on the tree protocol with a few results obtained by Jacquet and Régnier [JR 1986]. Let X_N be a parameter (random variable) of the blocked access tree protocol, like session length (i.e. size of the associated tree), path length or height of the tree, when a session is started with N initial colliders. As N increases, those parameters X_N have complicated (exact) distributions that however tend to limiting distributions of a simple form.

THEOREM 9: [Jacquet, Regnier 1986] *Let S_N be the random variable representing the length of a session of the blocked access tree protocol in its randomised version, started with N initial colliders. As $N \to \infty$, the distribution of the random variable S_N tends to a limiting Gaussian distribution.*

Let $p_{N,k}$ be the probability that the modified session length $X_N = (S_N - 1)/2$ be equal to k, and introduce the bivariate generating function

$$P(z, u) = \sum_{N,k} p_{N,k}\, e^{-z} \frac{z^N}{N!} u^k. \tag{20}$$

Function P is also a Poisson generating function when z is fixed: It represents the probability generating function (pgf) of X_N when N is itself Poisson with parameter z. The rather difficult proof proceeds in stages:

1. First use the recursive nature of the tree process to set up a *non-linear difference equation* satisfied by $P(z, u)$, like what has been done before:

$$P(z, u) = u P^2(\frac{z}{2}, u) + (1 - u)(1 + z)e^{-z}. \qquad (21)$$

2. The problem is to obtain a good asymptotic approximation for $P(z, u)$ for fixed u and large z. Setting $L(z, u) = \log P(z, u)$, L satisfies a *quasi-linear difference equation*. From there, the growth of $P(z, u)$ as well as its moments $P_u(z, 1)$ and $P_{uu}(z, 1)$ can be determined using Mellin transform techniques.

3. The characteristic function $P(z, e^{it})$ –after normalisation using mean and variance estimates from Point 2– converges as $z \to \infty$ to the characteristic function of a normally distributed variable, namely $e^{-t^2/2}$.

4. There now remains to translate the previous limit result under a Poisson model with parameter z to the case where N is fixed but large (The latter is the so-called Bernoulli model). What is needed here is an argument with which, if a_N is a sequence of numbers and

$$A(z) = \sum_N a_N\, e^{-z} \frac{z^N}{N!},$$

then $a_N \sim A(N)$. A general "semi-Tauberian" theorem (requiring the estimates to be valid in some region of the complex plane for z values) is given in [Jacquet, Régnier 1986] to that effect. By this device, results can be transfered from the Poisson to the Bernoulli case.

Several parameters, including various notions of height, can be analysed in this fashion. Also the method is general enough to accomodate the case of biased coins.

7. A Local Area Network Realisation

Besides being theoretically analysable, the tree protocol has many practical advantages. Its simplicity is comparable to that of Ethernet, since stations only need to maintain an integer index (representing their stack level): It is the way that index is managed that differs. Implemented in an asynchronous mode (unslotted time), it is compatible with the IEEE norm 802.3 . It is also fairly resistant to misinterpretations of channel feedback by stations. Last but not least, the deterministic version offers good worst-case guarantees on message delay when real-time constraints are present, and increased throughput results from the hyperstability phenomenon discussed earlier.

For those reasons, the SCORE project at INRIA has designed a prototype realisation, called LYNX, of a real time local area network based on the deterministic tree protocol, which is currently under industrial development. At present, the network consists of 14 stations: It necessitates only a simple modification of the Ethernet coupling boards, in accordance with what has been said earlier concerning norm compatibility.

Extensive measures have been conducted on LYNX, as well as with a network software emulator for 256 stations. In LYNX (as in any Ethernet network), collision slots being of a shorter duration, the observed performances are actually better than our previous computations imply. It is not difficult to "tune" the mathematical models to take this fact into account. (We have only refrained from doing so to keep the discussion simple). Measures and simulations amply confirm the analyses given in this paper. For instance, it appears clearly on that configuration that Ethernet will destabilise when $\lambda \approx 70\%$ while this occurs only at $\lambda \approx 90\%$ for the deterministic tree protocol. The rejection rate of messages, due to real time constraints not being satisfied, is also appreciably lowered by this change of protocol.

8. References

The general references we have used for this paper are:

[Massey 1981]: J. Massey, "'Collision resolution algorithms and random access communications", in *Multi-User Communication Systems*, G. Longo Ed., CISM Courses and Lectures **235**, Springer, New-York (1981), pp. 73-137.

[Massey 1985]: J. Massey Editor, *IEEE Transactions on Information Theory*, Special Issue on Random Access Communications **31** (1985).

[Knuth 1973]: D.E. Knuth, *The Art of Computer Programming: Sorting and Searching*, Addison-Wesley, Reading (1973).

Papers whose results are surveyed here are:

[FFH 1986]: G. Fayolle, P. Flajolet, M. Hofri, On a functional equation arising in the analysis of a protocol for a multi-access broadcast channel, *Adv. Appl. Prob.* **18** (1986), pp. 441-472.

[FFHJ 1985]: G. Fayolle, P. Flajolet, P. Jacquet, M. Hofri, Analysis of a stack algorithm for random multiple access communication, *IEEE Transactions on Information Theory*, Special Issue on Random Access Communications **31** (1985), pp. 244-254.

[MF 1985]: P. Mathys, P. Flajolet, Q-ary collision resolution algorithms in random-access systems with free or blocked channel access, *IEEE Transactions on Information Theory*, Special Issue on Random Access Communications **31** (1985), pp. 217-243.

[JR 1986]: P. Jacquet, M. Régnier, Trie partitioning process: limiting distributions, in *CAAP'86*, Lecture Notes in Computer Science **214** (1986), pp. 196-210.

[Jacquet 1987]: P. Jacquet, Evaluation of the queues in stations for the deterministic tree protocol, INRIA Res. Rep., in print (1987).

References contained in the above papers are not duplicated here.

DELAY AND ROUTING IN INTERCONNECTION NETWORKS

Bruce Hajek and Rene L. Cruz
Department of Electrical Engineering and the Coordinated Science Laboratory
University of Illinois
1101 W. Springfield, Urbana IL 61801
USA

0. ABSTRACT

A very brief introduction to interconnection networks is given. It is then shown that, in a restricted sense, an ideal N-by-N crossbar switch can be simulated with slowdown factor $\log_d N$ by a fully extended omega network built from d-by-d switches. Next, lower bounds are given for the product of the *average* traffic per link and the number of network links. The bounds are valid for interconnection networks with an upperbound on the number of links coming out of each node. Finally, it is shown that in a probabalistic sense, the *peak* traffic per link can be made close to the bound by using a slightly extended omega network. Randomized routing plays a key role in reducing congestion.

1. INTRODUCTION

Interconnection networks are collections of processors and/or switches wired together and programmed to provide communication among the processors in a computer system [Wu and Feng, 1984], and they are critical components in the latest generation of computers [Hwang, 1984]. The interconnections are often made in a regular pattern--for example the processors may form a two- or three-dimensional array, an n-dimensional binary cube, or a pyramid, each with local interconnections. Key considerations in the design of interconnection networks include (1) physical layout and fabrication of switches and wires, including those for control and input/output communication; (2) network throughput (often called "bandwidth") and speed; (3) fault tolerance; and (4) capability to accommodate diverse demands (sometimes called "combinatorial power").

Unlike most transportation networks, air traffic control networks and even many communication networks, interconnection networks can often be layed out by the engineer from scratch. It is thus fruitful to consider a great variety of interconnection structures, as well as to study limitations of arbitrary structures. Some characteristics that serve to help classify interconnection networks are: (1) The degree of synchronization: some systems are packet synchronized, others are synchronized only after each "batch" of traffic is served (a flow control mechanism), while others are totally asynchronous; (2) The switching format: examples are packet switching, circuit switching and hybrids such as cut-through [Kermani and Kleinrock 1979, 1980]; (3) The use, or not, of switches other than the processors served by the networks and;

NATO ASI Series, Vol. F38
Flow Control of Congested Networks
Edited by A. R. Odoni et al.
© Springer-Verlag Berlin Heidelberg 1987

(4) The amount of buffer space at non-destination modules.

Details of the traffic demand placed on an interconnection network are typically not known in advance. In particular, unbalanced demand can lead to serious congestion in some parts of the network, even while the network as a whole is lightly loaded. For high speed performance, it may be best to force congestion to diffuse by using simple, distributed control mechanisms. Moreover, *randomization* has emerged as an important design principle [Valiant 1982, Upfal 1984, Pippenger 1984, Greenberg and Leiserson 1985, Mitra and Cieslak 1986]. Roughly speaking, since there are so many unknown variables it may be unwise or meaningless to design with worst-case (rather than average-case) performance in mind.

A fundamental task in the design of interconnection networks is the specification of performance goals for the interconnection network which are consistent with the overall system in which the interconnection network is embedded. For example, one could assume a given model of demand that will be placed on the interconnection network and, based on overall system requirements, specify delay requirements in the interconnection network for the given model of demand. An alternative philosophy is the following. The designer of the interconnection network can specify delay performance guarantees for a given model of demand, and the rest of the system can be "built around" these specifications. Some combination of these two philosophies is perhaps best, whereby the "system designer" and the "interconnection network designer" negotiate between themselves in an iterative way. The specification of a good model of demand for the interconnection network is seen to be a crucial point in the design process, and is an important area for further research.

In Section 3 we briefly discuss "simulating" an ideal fully connected network by an extended omega network, and in Section 4 we briefly discuss some network limitations imposed by limiting the out degree of switches and limiting the number of links. Part of this work (and work on other topics) will appear in [Cruz, 1987].

2. THE EXTENDED OMEGA NETWORK

Let d, r and n be integers with $0 \leqslant r \leqslant n-1$, $d \geqslant 2$ and $n \geqslant 1$. An extended (if $r > 0$) omega network is a periodic assembly of switches and links for connecting $N = d^n$ sources to N destinations. There are $n + r$ identical stages, numbered 1 to $n + r$, with stage 1 furthest to the left. The network is called *fully* extended if $r = n - 1$. An example of an extended omega network for $d = 2$, $r = 1$ and $n = 3$ is shown in Figure 1. Each stage consists of a column of N/d (d input)x(d output) switches, and the links connected to the output ports of the switches. The N input and N output ports in a column of switches are numbered from the top down, and can thus be indexed by sequences in $\{0,1,...,d-1\}^n$. Also, once a particular switch has been identified, any of its output ports can be specified by referring only to the last digit of the address of the output port.

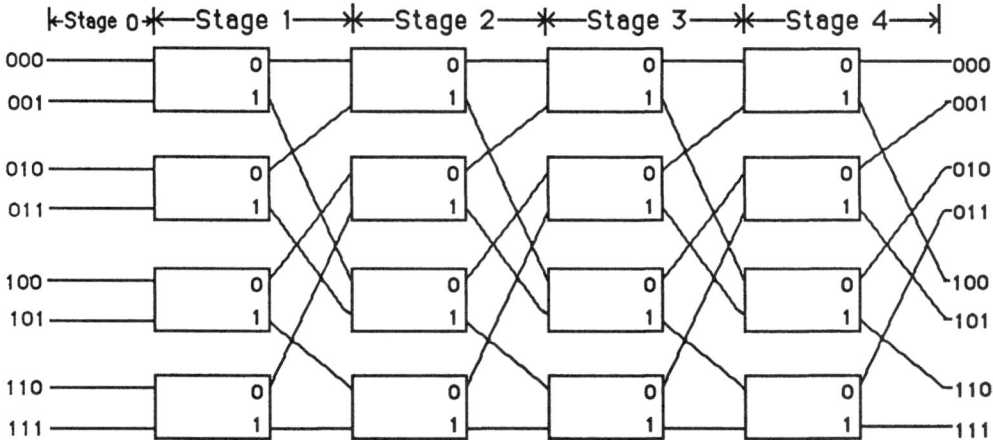

Figure 1. An example of an extended omega network

For each $a \in \{0,1,....,d-1\}^n$, source a is connected to input port a in stage 1. By convention, we say that the links connecting sources to switches in stage 1 are in stage 0. For $1 \leqslant i \leqslant n+r$, the links in stage i connect the output ports of switches in stage i to the input ports in stage i + 1, or to the destinations if i = n + r. Each link is given an address with the format $(i:a_n, ..., a_1)$ where i denotes the stage to which the link belongs and $a = (a_n, a_{n-1}, ..., a_1)$ is the address of the input port in the (i+1)th stage where the link terminates if i < n+r, or the address of the destination where the link terminates if i = n + r.

The network is configured so that output port $(a_n, ..., a_1)$ in stage i is connected to input port $(a_1, a_n, ..., a_2)$ in stage i + 1 if i < n+r, and to destination $(a_1, a_n, ..., a_2)$ if i = n + r. The link making this connection, by the previous stated rule, is $(i:a_1, a_n, ..., a_2)$. The net effect is that a packet on link $(i:a_n, ..., a_1)$ on entering a switch in the (i+1)th stage is switched to output port $b \in \{0,1,....,d-1\}$ and leaves on link $(i + 1; b, a_n, ..., a_2)$.

For any source address $s = (s_1, ..., s_n) \in \{0,1,....,d-1\}^n$ and destination address $t = (t_1, ..., t_n)$, there are d^r paths from the source to the destination. These can be conveniently indexed by a vector $c = (c_r, ..., c_1) \in \{0,1,....,d-1\}^r$. The path from s to d with index c is obtained by using output port c_i of the switch entered in stage i for $1 \leqslant i \leqslant r$ and output port d_{i-r} of the switch entered in stage i for $r < i \leqslant n + r$. Equivalently, consider a sliding window of width n which exposes n consecutive numbers from the sequence

$$t_n \cdots t_1 c_r \cdots c_1 s_n \cdots s_1$$

Then for $0 \leqslant i \leqslant n + r$, the address of the link of the path in stage i corresponds to displacing the window i units from its right-most position.

Remark. Omega networks (the case r=0) were introduced by Lawrie(1975). Chin and Hwang(1984) and Mittra and Cieslak(1986) (whose presentation we adopted above) pointed out that adding extra stages to an omega network and randomizing packet routes over the multiple routes available can reduce congestion for unbalanced loads.

3. CONGESTION AT BOUNDARIES VS. INTERNAL CONGESTION

The delay in an interconnection network cannot be smaller than the delay caused at or near the input and output ports of the network. One way to characterize the delay due to input-output restrictions is to consider the delay in an ideal full crossbar switch in which all internal links have infinite capacity and the "boundary" links have fixed capacity C--see Figure 2.

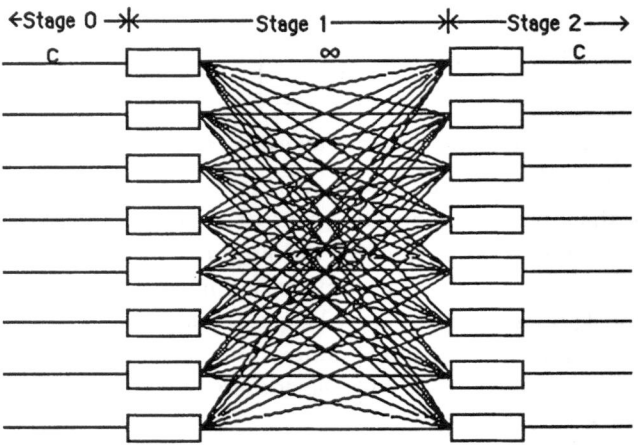

Figure 2. An ideal full crossbar switch

A reasonable goal for designing an interconnection network is to "simulate" this ideal network, possibly at a slower but guaranteed rate. As an illustration, we will compare the ideal crossbar switch to the fully extended omega network. Each is taken to have $N = d^n$ sources and N destinations. Let $\lambda = (\lambda(s,t): s,t \in \{0,\ldots,d-1\}^n)$ and assume for each (s,t) that packets input at s and destined for t (called type (s,t) traffic) arrive according to a Poisson process of rate $\lambda(s,t) \geq 0$. Assume that the processes for different (s,t) are independent. Finally, in the case of the fully extended omega network, assume that each packet is assigned to one of its d^r possible routes at random, with all d^r possibilities being equally likely.

For analytical convenience, we assume that the amount of time it takes to transmit a packet over a link in exponentially distributed with mean $1/C$ and that all transmission times are independent--this is often called "Kleinrock's independence assumption." An exception is that the links in stage 1 of the ideal network are assumed to be capable of instantaneously transmitting packets. We assume that a packet is first fully received at a switch and then instantaneously placed onto a queue on the outgoing link. The queues on each link are "served" by the links in first-come, first-serve order.

Let D_χ and D_ω denote the mean time it takes a typical packet to traverse the ideal network and the omega network, respectively.

Proposition. For any $\lambda, D_\omega \leqslant D_\chi \log_d N$.

Remark. This shows that in a certain sense, the fully extended omega network "simulates" the ideal network with slowdown factor $\log_d N$. The proposition is closely related to the results of Chin and Hwang(1984) and Mitra and Cieslak(1986), where a smaller class of demand vectors is considered. It was inspired by an extensive theory of parallel computation based on simulating idealized systems using more realistic ones (see [Alt et al. (1986)] for a good introduction). Excessive congestion at the boundaries of the network is often caused by many processors simultaneously accessing the same storage device, and can be controlled in part by hashing methods, careful task assignment, and duplicating records. We do not address the important problem of reducing congestion at the boundaries of the network in this paper. It would be nice if $D_\omega(\mathbf{s},\mathbf{t}) \leqslant D_\chi(\mathbf{s},\mathbf{t}) \log_d N$ for any $(\mathbf{s},\mathbf{t})$, where $D_\omega(\mathbf{s},\mathbf{t})$ and $D_\chi(\mathbf{s},\mathbf{t})$ denote the mean delay of type $(\mathbf{s},\mathbf{t})$ traffic in the networks. Unfortunately, this stronger statement is not true for FCFS or "round-robin" service order. An open problem is to find a simple service discipline or flow control mechanism to make that true.

Proof. Let D_i denote the mean delay a packet suffers in stage i of the omega network, and let $\nu(e)$ denote the mean rate of traffic on a link e. Then $\nu D_i = \Sigma \nu(e)/(C - \nu(e))^+$ where ν is the total traffic arrival rate, and e ranges over the addresses of all links in stage i. Fix i with $1 \leqslant i \leqslant n-1$. Then for $\mathbf{b} \in \{0,1,...,d-1\}^{n-i}$ let

$$A(\mathbf{b}) = \{(0; s_n, s_{n-1}, \cdots, s_1): (s_n, s_{n-1}, \cdots, s_{i+1}) = \mathbf{b}\}$$

and

$$B(\mathbf{b}) = \{(i; a_n, a_{n-1}, \ldots, a_1): (a_{n-i}, a_{n-i-1}, \ldots, a_1) = \mathbf{b}\}.$$

Then $\{A(\mathbf{b}): \mathbf{b} \in \{0,1,...,d-1\}^{n-i}\}$ is a partition of the links in stage 0 and $\{B(\mathbf{b}): \mathbf{b} \in \{0,1,...,d-1\}^{n-i}\}$ is a partition of the links in stage i. Now fix $\mathbf{b}$. Note that a packet passes over a link in $A(\mathbf{b})$ when it traverses the network, if and only if it also passes over a link in $B(\mathbf{b})$. Moreover, given that the packet traverses a particular link in $A(\mathbf{b})$, the link it traverses in $B(\mathbf{b})$ is uniformly distributed over the d^i links in $B(\mathbf{b})$. Thus, for $e \in B(\mathbf{b})$, $\nu(e) = d^{-i} \sum_{s \in A(\mathbf{b})} \nu(\mathbf{s})$. By the convexity of the function $\phi(y) = y/(C-y)^+$ and Jensen's inequality,

$$\nu D_0 = \sum_{\mathbf{b}} d^i \{d^{-i} \sum_{s \in A(\mathbf{b})} \phi(\nu(\mathbf{s}))\} \geqslant \sum_{\mathbf{b}} d^i \{\phi(d^{-i} \sum_{s \in A(\mathbf{b})} \nu(\mathbf{s}))\} = \sum_{\mathbf{b}} \sum_{e \in B(\mathbf{b})} \phi(\nu(e)) = \nu D_i.$$

Thus, $D_0 \geqslant D_i$ for $0 \leqslant i \leqslant n-1$. Similarly, $D_{2n+1} \geqslant D_i$ for $n \leqslant i \leqslant 2n-1$. Thus,

$$D_\omega = \sum_{i=0}^{2n-1} D_i \leqslant n (D_0 + D_{2n-1}) = nD_\chi.$$

which establishes the proposition.

4. MEAN LINK TRAFFIC VS. NUMBER OF LINKS--A LOWER BOUND

Model a network as a directed graph $G = (V,E)$ with set of nodes V and set of directed links E. Suppose there are $2N$ distinguished nodes, $s_1, s_2, \ldots, s_n, t_1, \ldots, t_n$. The nodes $s_1, \ldots, s_n$ are "source nodes" and the nodes $t_1, \ldots, t_n$ are destination nodes. We assume that each source node has only outgoing links and that each destination node has only incoming links. We let In (ν) and Out (ν), respectively, denote the set of links leading into and out of a node ν. We assume that there are at most d links leading out of each node and that the output ports of each node are labelled using $\{0,1,\ldots,d-1\}$.

Suppose that for each i, the ith source has one unit of traffic to be routed to the $\pi(i)$th destination node, where $\pi = (\pi(1),\ldots,\pi(N))$ is a permutation of $(1,2,\ldots,N)$. A *routing* for π is a vector $f = (f_{ij}(e): i,j \in \{1,\ldots,N\}, e \in E)$. We want the component $f_{ij}(e)$ to represent the flow of traffic on link e originating at source s_i and destined for destination t_j. We thus require $f_{ij}(e) \geq 0$ and the balance of flow equation

$$I_{|\nu=s_i, \pi(i)=j|} + \sum_{e \in \text{In}(\nu)} f_{ij}(e) = \sum_{e \in \text{Out}(\nu)} f_{ij}(e) + I_{|\nu=d_j, \pi(i)=j|}$$

for $\nu \in V$ and $1 \leq i, j \leq N$. We let $F(\pi)$ denote the set of routings for π, and, setting $\nu(e,f) = \sum_i f_{ij}(e)$, we define

$$\bar{m}(G,\pi) = \min_{f \in F(\pi)} \frac{1}{|E|} \sum_e \nu(e,f) \quad \text{and} \quad m^*(G,\pi) = \min_{f \in F(\pi)} \max_{e \in E} \nu(e,f).$$

It's obviously desirable to have networks G such that $\bar{m}(G;\pi)$ and m^* $(G;\pi)$ are small for all (or at least for a rich subset of all) permutations π.

We call a routing f pure if $f_{ij}(e) \in \{0,1\}$ for all i,j,e, and we define $F_p(\pi)$, $\bar{m}_p(G,\pi)$ and $m^*_p(G,\pi)$ the same way as F, $\bar{m}$ and m^*, but with the additional restriction that pure routings be used. Clearly for any G and π, $\bar{m}_p = \bar{m} \leq m^* \leq m^*_p$. We will use that fact that $\bar{m} = \bar{m}_p$ to derive a lower bound for $\bar{m}$, and hence also for m^* and m^*_p.

For any permutation π, let f^π be a pure routing for π. Then f^π is equivalent to a sequence of directed paths $p_1, p_2, \ldots, p_N$, such that path p_i is a path from source s_i to destination $t_{\pi(i)}$. We assume that f^π minimizes the mean traffic per link, which means that p_i is a *shortest length* path in G from i to $\pi(i)$ for each i. Thus $\bar{m}_p(G,\pi) = l(\pi)/|E|$ where $l(\tau)$ is the sum on the lengths of $p_1, \ldots, p_N$.

Now, assign a word to f^π, where a word is an element of $\{0,1,\ldots,d-1\}^*$, the set of all finite-length sequences with elements in the "alphabet" $\{0,1,\ldots,d-1\}$. To do this, note first that for each source s_i, a word is determined by the sequence of output ports that path p_i uses. The length of the word for source s_i is thus equal to the length of p_i. Then assign f^π the word obtained by concatenating, in order, the N words for sources $s_1, s_2, \ldots, s_N$. The word assigned to f^π together with the graph G with labelled output ports uniquely specify f^π, and the length of the word is $l(\pi)$. Furthermore, given any two distinct permutations, neither of the two associated words is a prefix of the other. Thus, by Krafts' inequality,

$$\sum_\pi d^{-l(\pi)} \leq 1 \tag{4.1}$$

where the sum is over all n! permutations. Using the fact that $l(\pi) = |E|\bar{m}(G,\pi)$. this translates into a lower bound on $\bar{m}$.

For example, if Π is a random permutation, it follows from (4.1) that

$$E[|E|\bar{m}(G,\Pi)] \geqslant H_d(\Pi) = -\sum_\pi P[\Pi=\pi] \log_d P[\Pi=\pi] \qquad (4.2)$$

(To see this, note that the left hand side is equal to $\gamma(\alpha) = -\sum_\pi P[\Pi=\pi] \log_d(\alpha_\pi)$ where α is the sub-probability distribution $\alpha_\pi = d^{-l(\pi)}$ and use the fact that γ is minimized over all probability distributions by β where $\beta_\pi = P(\Pi=\pi)$). $H_d(\Pi)$ is the entropy (base d) of Π.

In the special case that Π is uniformly distributed over all N! permutations, $H_d(\Pi) = \log_d N! \sim N\log_d N$. Also in this case, its easy to use Ineq. (4.1) to obtain the following inequality, which is perhaps more striking than Ineq. (4.2): for $0 < \delta < 1$. $P[|E|m(G,\Pi) \leqslant (1-\delta) \log_d N!] \leqslant (N!)^{-\delta}$.

Remark. We haven't found the above bounds in the literature, though our method is reminiscent of Shannon's method [Shannon, 1950] of counting network states and tree methods used extensively for related problems (see, for example, [Pippenger and Valiant, 1976]).

We close this section by showing that simple variations of the omega network nearly meet the lower bounds given.

Example 1. Suppose that $N = d^{k+c}$ where $d \geqslant 2$ and $k,c \geqslant 0$ and let $N' = d^k$. Start with a fully extended omega network with N' inputs and N' outputs (hence, with 2k-1 stages), where each switch is now represented by a node. Augment the network by replacing each source (resp. destination) of the network by the root node of a full d-ary tree of depth c directed towards (resp. away from) the root node. This yields a network G with N sources and N destinations, and for any permutation π

$$|E|m^*(G,\pi) = \left\{ \frac{2N}{d^c} \log_d \left\lfloor \frac{N}{d^c} \right\rfloor + 2Nd(1-d^{-c})/(d-1)\right\}\{d^c\}$$

$$= 2N \log_d N + 2N(d(d^c-1)/(d-1)-c)$$

$$\sim 2N \log_d N \text{ if } 0 \leqslant c \leqslant \log_d[o(\log_d N)].$$

Thus, for large N and c in the range indicated, this network has *peak* link traffic roughly twice the lower bound on average link traffic. Two extreme cases are the case $c = 0$: $|E| = 2N \log_d N$ and $m^*(G,\pi) = 1$ and the case $c = \log_d(\epsilon \log_d N)$: $|E| \leqslant (2/\epsilon + 2d/(d-1))N$ and $m^*(G,\pi) = \epsilon \log_d N$.

Example 2. Fix ϵ with $0 < \epsilon < < 1$ and replace the fully extended omega network used in the construction of Example 1 by an omega network with N' inputs and outputs (as before) and length $[k(1 + \epsilon)]$. Using an inequality of Hoeffding(1963) (also see

[Raghaven and Thompson, 1985]) we can prove that if $s > 1 + 2\epsilon$ and Π is a uniformly distributed random permutation of $(1,2,...,N)$, then

$$\lim_{N \to \infty} P[m^*(G,\Pi)|E| \geqslant sN \log_d N] = 0$$

uniformly for c in the range $[0. \log_d((\epsilon(d-1)/2d) \log_d N)]$. Thus, except for a relatively small subset of permutations, the *maximum* link traffic of the augmented, slightly-extended omega network is within a factor $1 + \epsilon$ of our lower bound on the *mean* link traffic of any network with the same number of edges.

ACKNOWLEDGEMENT

This work was supported by the National Science Foundation under contract NSF ECS 83 52030, with matching funds provided by AT&T.

REFERENCES

Alt U, Hagerup T, Mehlhorn K, Preparata FP (1986) Deterministic simulation of idealized parallel computers on more realistic ones.

Chin C-Y, Hwang K (1984) Packet switching networks for multiprocessors and dataflow computers, IEEE Transactions on Computers 33:991-1003

Cruz (1987) Maximum delay in interconnection networks. Forthcoming Ph. D. thesis

Greenberg RL, Leiserson CE (1985) Randomized routing on fat trees. Proc 26th IEEE Annual Symp on Foundations of Computer Science 26:241-249

Hoeffding W (1963) Probability inequalities for sums of bounded random variables. Journal American Statistical Society 58:13-30

Hwang K (ed)(1984) Tutorial: supercomputers--design and applications. IEEE Computer Society Press 648pp

Kermani P, Kleinrock L (1979) Virtual cut-through: a new computer communication switching technique. Computer Networks 3:267-286

Kermani P, Kleinrock L (1980) A tradeoff study of switching systems in computer communication networks. IEEE Transactions on Computers 29:1052-1060

Lawrie DH (1975) Access and alignment of data in an array processor. IEEE Transactions on Computers 24:1145-1155

Mitra D, Cieslak R (1986) Randomized parallel communications on an extension of the omega network. Proc 1986 Int Conf on Parallel Processing

Pippenger N (1984) Parallel communication with limited buffers. Proc 25th IEEE Annual Symp on Foundations of Computer Science 25:127-136.

Pippenger N, Valiant LG (1976) Shifting graphs and their applications. Journal of the Association for Computing Machinery 23:423-432

Raghaven P, Thompson CD (1985) Provably good routing in graphs: regular arrays. Proc ACM Symposium on Theory of Computing 17:79-87

Shannon CE (1950) Memory requirements in a telephone exchange. Bell System Technical Journal 29:343-349

Upfal E (1984) Efficient schemes for parallel communication. Journal of the Association for Computing Machinery 31:507-517

Valiant LG (1982) A scheme for fast parallel communication. SIAM J. Computing 11:350-361

Wu C-l, Feng T-y (eds) (1984) Tutorial: interconnection networks for parallel and distributed processing. IEEE Computer Society Press 656pp

IDENTITY AND REDUCIBILITY PROPERTIES OF SOME BLOCKING
AND NON-BLOCKING MECHANISMS IN CONGESTED NETWORKS

S. Balsamo[o], V. De Nitto Personè[o] and G. Iazeolla[oo]

Abstract: Networks of queues with blocking are used for representing resource constraints in production, communication and computer systems. Different types of blocking which have been studied in various application fields will be compared and proved to be either identical (in terms of general performance indices) or reducible one to the other. The case is also shown of identity and reducibility properties holding between some types of blocking networks and some types of non-blocking ones.

1. Introduction

Queueing networks are used as model of production, computer and communication systems and have proven to be a powerful tool for performance analysis and prediction [Klei75, Saue81, Lave83, Lazo84]. Networks of queues with blocking are used for representing resource constraints in congested networks (e.g. memory constraints or software resources constraints in computer systems, and window flow control in communication networks). In these models one or more queues have finite waiting room. A queue is said to be full when its population attains its maximum value. The blocking phenomena arises when an arriving job at a queue is hold back in the sending queue. An extensive bibliography on queueing networks models with blocking can be found in [Perr84].

When the blocking arises, different behaviors of arriving jobs and of the queue activity can be observed in the systems. We shall consider and compare some of these different behaviors by introducing so called "types" of blocking mechanisms. Systems are assumed to be deadlock-free and are analyzed in steady-state condition.

In this work three different types of blocking models, which have been studied in various application fields, are compared and an extension is obtained of the class of queueing networks with blocking which satisfy product-form. This class includes networks with more than one type of blocking. This allows efficient solution of complex models, for example, models of computer systems connected through a communication network which show different types of blocking (e.g. network links and controllers can show different types of blocking). It is also be proved that given a network

---

[o] Dip.to di Informatica, Università di Pisa, Corso Italia 40, 56100 Pisa (Italy)
[oo] Dip.to di Ingegneria Elettronica, Università di Roma II, via Raimondo, 00173 Roma (Italy)

NATO ASI Series, Vol. F38
Flow Control of Congested Networks
Edited by A. R. Odoni et al.
© Springer-Verlag Berlin Heidelberg 1987

with limited waiting rooms, it is always possible to define a conventional exponential network (i.e. an unlimited waiting room network) whose behavior is equivalent to that of the given finite waiting room network, i.e. whose steady-state probabilities are the same up to a normalizing constant and limited to the intersection of the two state spaces.

In the following section the model of three different types of blocking is introduced. Section 3 deals with identity and reducibility properties holding between distinct types of blocking. Section 4 proves that under some constraints networks with one or more of those types of blocking can also admit product-form solutions. Section 5 shows equivalences existing between some types of limited and unlimited waiting room networks.

2. The Model and Blocking Types

Consider a closed exponential network with blocking, having M nodes, N customers and routing matrix $P = |p_{ij}|$ $(i,j = 1,...,M)$. Denote by μ_i the service rate of node i and by $f_i(k)$ the service capacity of that node when there are k jobs present $(i=1,...,M; f_i(0)=0, f_i(k)>0$ for $k>0)$. Any discipline independent of service time is allowed. Let N_i denote the maximum queue length admitted at node i $(i = 1,...,M)$. System states are represented by an M-vector $n = (n_1,...,n_M)$, where n_i denotes the number of customers in node i, and the system state space $S(M,N)$ can be defined as follows:

$$S(M,N) = \{ n = (n_1,...,n_M) \mid 0 \leq n_i \leq \min\{N_i, N\}, i = 1,...,M, \sum_{i=1}^{M} n_i = N\}. \qquad (1)$$

The network model is a continuous-time ergodic Markov-chain with discrete space state $S(M,N)$. By assuming the irriducibility of routing matrix P, there exists a unique steady-state probability distribution on $S(M,N)$ [Klein75], that we shall denote by

$$\pi = \{ \pi(n), n \in S(M,N)\}$$

Vector π can be obtained by the solution of the homogeneous linear system

$$\pi = \pi Q \qquad (2)$$

subject to the normalizing condition $\sum_{n \in S(M,N)} \pi(n) = 1$ and where $\pi(n)$ is the steady-state probability of state $n \in S(M,N)$ and Q is the matrix of transition rates between states n.

In a queueing network with blocking (i.e., where $N_i \geq N$, $i = 1,...,M$) a customer which completes the service at node i, immediately goes to node j accordingly to routing probability p_{ij} $(i,j = 1,...,M)$. On the other hand, when the population in node j attains its maximum value, i.e. $n_j = N_j < N$, any other customer cannot be enqueued. In this case node j is said to be full or saturated and the blocking phenomena arises. Different blocking models or types describe different behavior both for customer arrivals at a full node and for the node activity.

We shall introduce three different blocking models used in the literature to represent different behaviors of real systems.

Type "B1" blocking model (pseudo-blocking or blocking with loop in the sending node [King69, Pitt79, Hord81]): a customer upon completion of its service at queue i attempts to enter destination queue j. If queue j at that time is full, i.e. $n_j = N_j$, the customer is looped back into the sending queue i where it receives a new independent service accordingly to the queue discipline.

Type "B2" blocking model (auto-blocking of the sending node [Hill67, Neut68, Perr81]): if upon completion of a service at queue i a customer enters a full queue then it is forced to wait in front of server i, which remains blocked until node j can accept a new job, i.e. until there is a departure from queue j. At that time the customer waiting in node i moves immediately to node j and server i is reactivated. If more than one job compete to enter a full queue j, the contention can be resolved in an arbitrary way (e.g. first blocked first enter rule). This blocking mechanism has been used to model systems such as production systems and disk I/O subsystems.

Type "B3" blocking model (conditional blocking of the sending nodes operated by the destination node [Boxm81, Gers81]): when a destination node j becomes full, it blocks the service in each of its possible sending nodes i (i.e., for each i such that $p_{ij} > 0$), provided node i is serving a customer whose destination node is j. Services will be resumed as soon as a departure occurs from queue j. This blocking mechanism has been used to model production systems and telecommunication systems.

Finally, we shall call Type "NB" (No-Blocking) any conventional network with unlimited capacity waiting rooms.

The state transition diagram of an M-node closed network with state space S(M,N) is an oriented graph in which nodes represent states $\mathbf{n} \in S(M,N)$ and arcs represent state transitions. Note that S(M,N) can be viewed as a (M–1)-dimensional subspace, by using the relationship $n_M = N - \sum_{i=1}^{M-1} n_i$.

There are some additional remarks on the state definition $\mathbf{n}$ for some blocking types. In a type B1 blocking network nodes are actually blocked (i.e., servers are never deactivated), while in a type B2 blocking network, node i can be deactivated whenever a destination node j is full. In this case, when $n_j = N_j$, the state description must specify whether the sending node i is blocked (B) or active (A). We shall denote this by two states, say $(n_1;...; n_i, B;...; N_j;...; n_M)$ and $(n_1;...; n_i, A;...; N_j;...; n_M)$, respectively. In a type B3 blocking network, on the other hand, as soon as node j is full, only sending nodes i ($p_{ij} > 0$) which actually serve customers directed to j, are deactivated. Then, for each such a sending node i, if it has more than one possible destination node (i.e., if there exists at least one node $k \neq j$ such that $p_{ik} > 0$) then the state must also specify the label of the destination node for the actually served job.

3. Equivalence properties holding between types B1, B2 and B3 blocking networks

Let X and Y be two networks with different blocking mechanisms (say, X and Y blocking types respectively). Let π_X and π_Y be the networks steady-state distributions obtained by (2). We introduce the following relations between blocking types X and Y:

- identity: blocking types X and Y are said to be identical if $\pi_X = \pi_Y$.

- reducibility: blocking type X is said to be reducible to blocking type Y if there exists an arbitrary function f such that $\pi_X = f(\pi_Y)$.

We now prove that blocking types B1 and B3 are identical on any arbitrary topology network.

Theorem 1

Type B1 and type B3 blocking models, when applied to general topologies of closed exponential networks, are identical, i.e.,

$$\pi_1 = \pi_3.$$

Proof. Consider the two state transition diagrams of two closed exponential networks with M nodes, general topology, and type B1 and type B3 blocking mechanisms, respectively. They differ only at times when the blocking phenomenon arises, i.e., when states with full nodes ($n_j = N_j$ for some $j = 1,...,M$) arise. Let us consider such a state $\mathbf{n} = (n_1,..., n_i,..., N_j,..., n_k,..., n_M)$, $i,j,k \in \{1,...,M\}$ where only nodes i and k are sending nodes for the full node j, i.e., $p_{zj} > 0$ if and only if $z = i, k$. Extension to the general case is immediate. Fig. 1 shows the portions of state transition diagrams for such a state $\mathbf{n}$ with type B1 and type B3 blocking, respectively. In the type B3 blocking diagram there appear the same transitions in and out of state $\mathbf{n}$ as in the type B1 blocking diagram, except for the loop transitions. Indeed, loop transitions in the type B1 diagram represent service completions, at node i or k, destined to the full node j. On the other hand, in the type B3 blocking, nodes i and k are blocked until node j is full.

It is easy to verify that loops on the state $\mathbf{n}$ have no effect in terms of system equations. Then the

247

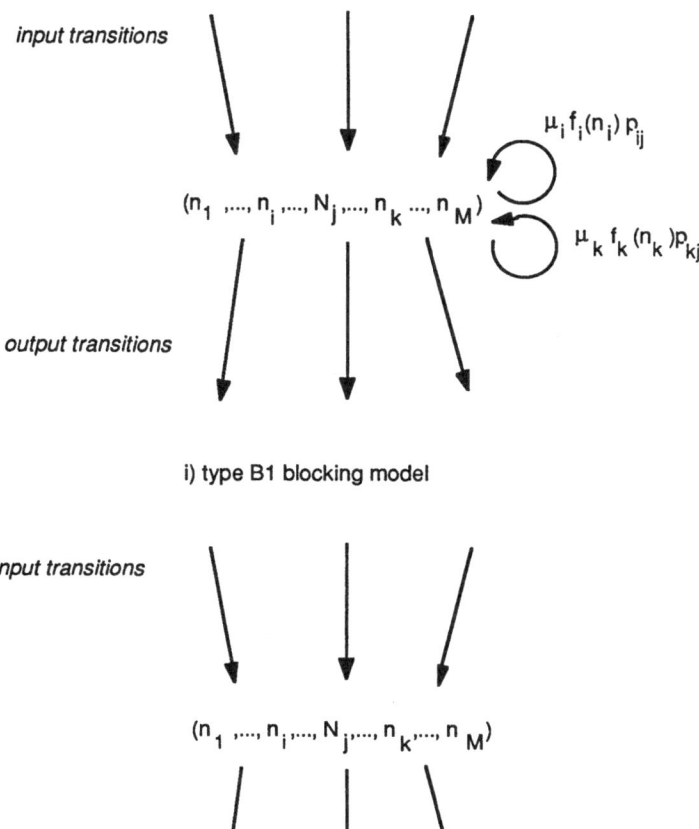

input transitions

$$\mu_i\, f_i(n_i)\, p_{ij}$$

$$(n_1 ,..., n_i ,..., N_j ,..., n_k ,..., n_M)$$

$$\mu_k\, f_k(n_k)\, p_{kj}$$

output transitions

i) type B1 blocking model

input transitions

$$(n_1 ,..., n_i ,..., N_j ,..., n_k ,..., n_M)$$

output transitions

ii) type B3 blocking model

Fig.1 - Portions of state transition diagrams of a general topology network with type B1 (i) and type B3 (ii) blocking models.

two systems are identical. Consequently, $\pi_1 = \pi_3$. [ ]

We now compare blocking types B1 and B2 on the elementary topology of Figure 2 (the two stage cyclic network), to prove that B1 is reducible to B2 and viceversa.

Theorem 2

The type B2 blocking model, when applied to the cyclic two stage network, is reducible to the type B1 blocking model and viceversa. In other words:

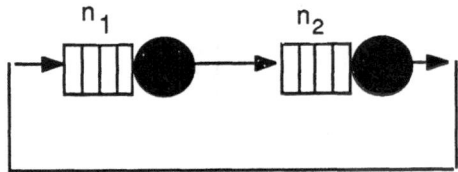

Fig. 2 - Two stage cyclic network

$$\pi_2(n_1,n_2) = C\pi_1(n_1,n_2) \qquad\qquad \forall\, (n_1,n_2) \in S(M,N),\, n_i < N_i,\, i = 1,2 \qquad (3.1)$$

$$\pi_2(N_1,n_2) = C\pi_1(N_1,n_2)\, [1 + (f_2(n_2)\mu_2) / (f_1(N_1)\mu_1)] \qquad n_2 = N - N_1 \qquad (3.2)$$

$$\pi_2(n_1,N_2) = C\pi_1(n_1,N_2)\, [1 + (f_1(n_1)\mu_1) / (f_2(N_2)\mu_2)] \qquad n_1 = N - N_2 \qquad (3.3)$$

Where C is a normalizing constant.

For the proof see [Bals86]. The proof is based on the observation of the two state transition diagrams for the network with type B1 or type B2 blocking models and on the use of the balance equations. Note that the state space of network in Fig. 2 with type B2 blocking has two states more than the space state for the type B1 blocking. In fact, in a type B2 blocking state, when a node is full, the situation of the remaining node (blocked or active) must also be considered. Indeed, if node i completes the service while node j is still full, then node i becomes blocked.

Properties shown by *Theorems 1* and *2* are synthesized in Table 1, where the reducibility entry for the general topology case is an obvious consequence of the fact that identity is just a particular form of reducibility. Similarly, the identity entry for the two stage cyclic topology is an obvious consequence of the fact that this topology is just a particular case of the general one.

TABLE 1

Equivalence Property	Closed Exponential Networks	
	General Topology	Two-stage Cyclic
Identity	B1, B3	B1, B3
Reducibility	B1, B3	B1, B2, B3

4. Cases of Product-Form Solution

A product-form solution has been proved to exist for the type B1 blocking model, under certain constraints. These are the reversible routing and the constant capacity constraints.

Introduce the blocking function of node i, $b_i(k)$, defined in the simplest case as follows:

$$b_i(k) = \begin{cases} 1 & \text{if } k < N_i \\ 0 & \text{otherwise} \end{cases}$$

with k the number of jobs present at node i. In the more general case one can define

$$\begin{cases} 0 < b_i(k) \leq 1 & \text{for } k < N_i \\ b_i(k) = 0 & \text{otherwise} \end{cases}$$

as an arbitrary load-dependent function that node i uses to regulate its input traffic. In other words, in a network with blocking the actual routing probabilities are state-dependent: quantity $p_{ij}\, b_j(k)$ gives the probability that a job leaving node i is accepted by node j when there are k jobs present at node j, and $1 - \sum_j p_{ij}\, b_j(k)$ the probability that this job is held back at node i.

For network R let $\mathbf{x} = (x_1, x_2, ..., x_M)$ denote the M-vector solution of the equation $\mathbf{x} = \mathbf{xP}$. The network routing matrix $\mathbf{P}$ is said to be *reversible* if $x_i\, p_{ij} = x_j\, p_{ji} \; \forall \; i,j$. In [Hord81, King69, Pitt78] it is proved that the steady-state distribution on S(M, N) has, in this case, the product-form

$$\pi(n) = C \prod_{i=1}^{M} \prod_{k=1}^{n_i} \frac{x_i}{\mu_i} \frac{b_j(k-1)}{f_i(k)} \qquad \text{for } \mathbf{n} \in S(M, N) \qquad (4)$$

with C a normalizing constant, and $\pi(n) = 0$ for $\mathbf{n} \in S(M, N)$.

Another circumstance in which the product-form is known to hold [Hord81] is the case of constant capacity nodes, i.e. $f_i(k) = \text{const} \overset{\Delta}{=} f_i$ regardless of the number of jobs present at node i (i=1, ..., M). In view of the assumptions $f_i(0) = 0$, $f_i(k) > 0$ for k > 0, this also implies that, for instance, none of the nodes can be empty. Closed networks in which nodes have constant service rates and no node can have an empty queue give examples of networks satisfying this property. One way of obtaining this is placing $N_i = \lfloor (N-1)/(M-1) \rfloor$ and N > M. According to [Hord81], the steady-state distribution on S(M, N) has, in this case, the product-form:

$$\pi(n) = C \prod_{i=1}^{M} \prod_{k=1}^{n_i} \gamma_i \frac{b_j(k-1)}{f_i} \qquad \text{for } \mathbf{n} \in S(M, N) \qquad (5)$$

where

$$\gamma_i = f_i \sum_{k \neq i} \mu_k f_k p_{ki} / \xi_i$$

ξ_i is the i-th component of the vector solution $\xi = (\xi_1, ..., \xi_M)$ to the equation $\xi = \xi A$ with

$\mathbf{A} = \| a_{ij} \|$,

$$a_{ij} = \mu_j f_j p_{ji} / \sum_{k \neq i} \mu_k f_k p_{ki}, \qquad i \neq j$$

$$a_{ii} = 1 - \sum_{i \neq j} a_{ij}, \text{ and}$$

C a normalizing constant.

A first result one can derive from the identity properties shown by Theorem 1 is that not only type B1 but also general topology networks with type B3 blocking can satisfy product-form, provided they are either of the reversible routing or of the constant capacity type. In other words,

Corollary 1

General topology closed exponential networks with type B3 blocking and either reversible routing or constant capacity services, have product-form solution.

A second result deals with special topology , two-stage cyclic networks. First note that such a network is itself reversible by definition. Therefore the constraints for the existence of product-form are naturally satisfied. By use of the reducibility properties shown by Theorem 2 one can then say that such a network enjoys product-form even if its blocking mechanisms is of type B2. In other words:

Corollary 2

Two stage cyclic exponential networks with type B2 blocking can be solved in product-form by use of equations (3) and (4).

A third important result is the one we can derive by putting together Theorem 1 and Corollary 1 results. In other words we can state that type B1 and type B3 general topology mixed networks can also satisfy product-form. That is

Corollary 3

General topology closed exponential networks with mixed (type B1 and B3) forms of blocking admit product-form under the reversible routing or constant capacity constraints.

A similar result can be finally derived from Theorems 1 and 2, with applications to two-stage cyclic networks. Indeed it is easy to be convinced that type B1, B2 or B3 mixed two-stage cyclic networks can also enjoy product-form. That is

Corollary 4

Two stage cyclic exponential networks with mixed (type B1, B2 or B3) forms of blocking can be solved in product-form by use of equations (3) and (4). Equations (3.2) or (3.3) will be used only if a node of type B2 is present (in position 1 or 2 respectively).

What was above said in Corollaries 1 through 4 can summarized in Table 2 where cc, rr denote "constant capacity" and "reversible routing" respectively.

TABLE 2

Combinations of blocking types and network topologies for which product form holds.		
Blocking Types	Topology	Particular Coinstraints
B1 or B3 or B1 mixed to B3	general	cc / rr
B2 or B1 mixed to B2 or B2 mixed to B3	two-stage cyclic	____

5. Equivalences holding between type B1 and NB product-form networks

In this section we shall see that, given an exponential network R with type B1 blocking and reversible routing matrix satisfying product-form (6) it is always possible to find a conventional type NB network R' whose steady-state distribution (on the common state space) is the same as that of R, up to a normalizing constant.

Theorem 3

Let R an exponential network with type B1 blocking having M nodes, N jobs, reversible routing matrix P, node functions $\{\mu_i, f_i(k), b_i(k); k=0, ..., N, i=1, ..., M\}$ and steady state-distribution $\{\pi(n), n \in S(M, N)\}$ given by (4). There exists a conventional exponential network R' with the same routing matrix P and new node functions

$$\mu_i' = \mu_i$$

$$f_i'(k) = \begin{cases} f_i(k) / b_i(k-1) & \text{for } k=1, ..., N_i \\ \text{any arbitrary positive function otherwise} \end{cases}$$

whose steady state distribution { $\pi(\mathbf{n})$ } is such that

$$\pi(\mathbf{n}) \propto \pi'(\mathbf{n}) \qquad \forall\ \mathbf{n} \in S(M, N)$$

namely,

$$\pi(\mathbf{n}) = \pi'(\mathbf{n}) / \sum_{\mathbf{n} \in S(M, N)} \pi'(\mathbf{n}).$$

For the proof see [Bals83].

It is also possible to prove that, given an exponential network R with type B1 blocking and constant-capacity nodes satisfying product-form (5), it is always possible to find a conventional exponential type NB network R' whose steady-state probabilities (limited to the common state space and up to a normalizing constant) are the same as those of R.

Theorem 4

*Let R be an exponential network with type B1 blocking, having M nodes, N jobs, routing matrix **P**, node functions { μ_i; $f_i(k) = \text{const} = f_i$; $b_i(k)$; $k=0, ..., N$, $i=1, ..., M$ } and steady-state distribution $\{\pi(\mathbf{n}),\ \mathbf{n} \in S(M, N)\}$ given by (5). There exists a conventional exponential type NB network R' with the same routing matrix **P** and new node functions*

$$\mu_i' = \mu_i\, f_i\, h_i$$

$$f_i'(k) = \begin{cases} 1 / b_i(k-1) & \text{for } k=1, ..., N_i \\ \text{any arbitrary positive function otherwise} \end{cases} \qquad (7)$$

whose steady state-distribution { $\pi(\mathbf{n})$ } is such that

$$\pi(\mathbf{n}) \propto \pi'(\mathbf{n}) \qquad \forall\ \mathbf{n} \in S(M, N)$$

$$\pi(\mathbf{n}) = \pi'(\mathbf{n}) / \sum_{\mathbf{n} \in S(M, N)} \pi'(\mathbf{n}).$$

Functions appearing in (7) are defined as follows

$$h_i \overset{\Delta}{=} e_i\, y_i$$

with e_i and y_i the i-th components of the M-vectors

$\mathbf{e} = (e_1, e_2, ..., e_M)$ and $\mathbf{y} = (y_1, y_2, ..., y_M)$ given by the solution of systems

$$\mathbf{e} = \mathbf{e} \, \mathbf{P'}, \qquad \mathbf{y} = \mathbf{y} \, \mathbf{A}$$

respectively, where

$$\mathbf{P'} = \| p_{ij}' \| \qquad i,j = 1, ..., M$$
$$p_{ij}' = f_j \, \mu_j \, p_{ji}, \qquad p_{ii}' = 1 - \sum_{j \neq i} p_{ij}'$$

$$\mathbf{A'} = \| a_{ij}' \| \qquad i,j = 1, ..., M$$
$$a_{ij}' = p_{ji}', \qquad a_{ii}' = 1 - \sum_{j \neq i} a_{ij}'.$$

Note that systems $\mathbf{e} = \mathbf{e} \, \mathbf{P'}$ and $\mathbf{y} = \mathbf{y} \, \mathbf{A'}$ both admit a solution, since in matrices $\mathbf{P'}$ and $\mathbf{A'}$ rows sum to 1 by definition. Also note that, in general, $\mathbf{A'}$ is not the transposed of $\mathbf{P'}$ except when in $\mathbf{P'}$ both its rows and its columns sum to 1. For the proof see [Bals83].

From properties proved in Theorems 1, 3 and 4, it is a matter of evidence the fact that reducibility properties existing between type B1 and type NB general topology networks (satisfying the reversible routing or the constant capacity constraint) imply the existence of similar properties holding between type B3 and NB networks. This is synthesized in Table 3, column 1. In column 2, this table synthesizes the similar situation arising for the two stage cyclic networks as by Theorems 1, 2 and 3.

A comparison of Tables 1 and 3 shows how at present status of knowledge the restriction of the network to the cc or rr case yields some information on equivalences between blocking and no-blocking networks, at the expense of restricting this information to the reducibility case only.

TABLE 3

Equivalence Property	Closed Exponential cc/rr Networks	
	General Topology	Two-stage Cyclic
Reducibility	B1, B3, NB	B1, B2, B3, NB

References

[Bals83]   Balsamo, S., Iazeolla, G., "Some equivalence properties for queueing networks with and without blocking", Proc. *Performance '83* Symposium, North Holland, (1983).

[Bals86]   Balsamo, S., De Nitto Personè, V., Iazeolla, G., "Some Equivalences of Blocking Mechanisms in Queueing Networks with Finite Capacity", Tech. Rep. R86.02, Dep. Electr. Eng., University Roma II, (1986).

[Boxm81] Boxma, O.J., Konheim, A.G., "Approximate Analysis of Exponential Queueing Systems with Blocking", *Acta Informatica*, Vol. 15, (1981), pp. 19-66.

[Gers81] Gershwin, S., Berman, U., "Analysis of Transfer Lines Consisting of two Unreliable Machines with Random Processing Times and Finite Storage Buffers", *AIIE Trans.*, 13, no. 1, (1981), pp. 2-11.

[Gord67] Gordon, W.J., Newell, G.F., "Cycling Queueing Systems with Restricted Length Queues", *Oper. Res.*, 15, (1967), pp. 266-278.

[Hill67] Hillier, F.S., Boling, R.W., "Finite Queues in Series with Exponential or Erlang Service Times. A Numerical Approach", *Oper. Res.*, 15, (1967), pp. 286-303.

[Hord81] Hordijk, A., Van Dijk, N., "Networks of Queues with Blocking", *Performance '81*, Kylstra (Ed.), North Holland, (1981), pp. 51-65.

[King69] Kingman, J.F.C., "Markovian Population Process", *Journal of Applied Probability*, 6, (1969), pp. 1-18.

[Klei75] Kleinrock, L., "Queueing Systems. Vol. 1: Theory", Wiley, NY, (1975).

[Konh76] Konheim, A.G., Reiser, M., "A Queueing Model with Finite Waiting Room and Blocking", *SIAM J. Computing*, 7, (1978), pp. 210-229.

[Lave83] Lavenberg, S.S., "Computer Performance Modeling Handbook", Academic Press, (1983).

[Laza84] Lazar, A.A., Robertazzi, T.G., "The Geometry of lattices for Markovian Queueing Networks", Columbia University, Elect. Eng. Dept., (1984).

[Lazo84] Lazowska, E.D., Zahorjan, J., Graham, G.S., Sevcik, K.C., "Quantitative System Performance", Prentice Hall, Englewood Cliffs, NJ, (1984).

[Neut68] Neuts, M.F., "Two Queues in Series with a Finite Intermediate Waiting Room", *Journal of Applied Probability*, 5, (1968), pp. 123-142.

[Perr81] Perros, H.G., "A Symmetrical Exponential Queueing Network with Blocking and Feedback", *IEEE Trans. SE*, vol. SE-7, (1981), pp. 395-402.

[Perr84] Perros, H.G., "Queueing Networks with Blocking: a Bibliography", *Perf. Eval. Rew.*, Spring 1984, pp. 8-12.

[Pitt78] Pittel, B., "Closed Exponential Networks of Queues with Saturation: the Jackson-type Stationary Distribution and its Asymptotic Analysis", *Math. Oper. Res.*, 4, (1979), pp. 367-378.

[Saue81] Sauer, C.H., Chandy, K.M., "Computer System Performance Modeling", Prentice Hall, Englewood Cliffs, NJ, (1981).

[Suri83] Suri, R., Diehl, G.W., "A variable Buffer Size Model and its use in Analytic Closed Queueing Networks with Blocking", Hardward University, Division of Applied Sciences, (1983).

[Suri84] Suri, R., Diehl, G.W., "A New 'Building Block' for Performance Evaluation of Queueing Networks with Finite Buffers", *Proc. ACM Sigmetrics*, (1984), pp. 134-142.

On Some Probabilistic Combinatorial Optimization Problems
Defined on Graphs

Patrick Jaillet
Centre de Mathematique Appliquees
Ecole Nationale des Ponts et Chaussees, B.P. 105
93194 Noisy-le-Grand
France

Abstract

A probabilistic traveling salesman problem (PTSP) is essentially a traveling salesman problem (TSP) in which the number of
nodes to be visited in each problem instance is a random
variable. One can define, as well, other probabilistic problems
in the context of network optimization. In this paper we give
an overview of some results obtained on the PTSP and on a
probabilistic version of the shortest-path problem.

I.   Introduction

Network Optimization is an important topic in Operations
Research. It covers a wide variety of models with practical
applications and its theory is one of the richest and best-
developed in Combinatorial Optimization.

Most (but not all) of the research devoted to this area
has concentrated on deterministic situations. By that, we mean
situations in which the number of nodes, their relative

NATO ASI Series, Vol. F38
Flow Control of Congested Networks
Edited by A. R. Odoni et al.
© Springer-Verlag Berlin Heidelberg 1987

positions and the size of the input/output are known with certainty before a particular optimization problem is solved. One can identify, however, a practically endless variety of problems in which one or more of these parameters are random variables, i.e. subject to uncertainty in accordance with some probability law. In general, one can recognize two types of uncertainty that have been addressed in the literature: random distances between nodes, and uncertainty with respect to the size of the input/output at nodes. In the first case, the arcs are themselves assumed to be of random length according to some probability distribution; one can cite models of this type for the traveling salesman problem, shortest-path problems, location problems, etc. (see Larson and Odoni [1981] for an introduction to the analysis of network problems under uncertainty).

In the second case (random input/output size), the approach usually taken is to formulate the problems using techniques from stochastic programming (i.e., chance-constrained optimization, or stochastic programming with recourse) that allow one to transform these problems into deterministic versions (see, for example, Stewart and Golden [1983] on vehicle routing problems). The problems can then be solved by using traditional algorithms. One of the major drawbacks of these approaches is that it becomes necessary to introduce additional parameters (penalties) whose choice in terms of form and value is at the analyst's discretion and may only vaguely be related to the original problem.

In this paper we are concerned with a somewhat different approach toward introducing and analyzing uncertainty in combinatorial problems. The approach is based on recent research by the author (Jaillet [1985], [1986a], [1986b]). Our main concern is to define and analyze probabilistic versions of well-known combinatorial optimization problems while keeping their original (combinatorial) flavor. In that context, we present an overview of some of the results obtained on probabilistic versions of the traveling salesman problem and of the shortest-path problem.

II. The Probabilistic Traveling Salesman Problem

2.1 Introduction

A "probabilistic traveling salesman problem" is
probably the most basic stochastic routing problem that can be
defined. A PTSP is essentially a traveling salesman problem
(TSP) in which the number of nodes to be visited in each problem
instance is a random variable. The generic PTSP problem can be
formally described as follows: Consider a problem of routing
through a set of n known nodes. On any given instance of
the problem only a subset consisting of k out of n nodes
($0 \le k \le n$) must be visited, with the number k determined
according to a known probability distribution (see Figure 1 for
an illustration). We wish to find a priori a tour through all
n nodes. On any given instance of the problem, the k nodes
present will then be visited in the same order as they appear in
the a priori tour. The problem of finding such an a priori tour
which is of minimum length in the expected value sense is
defined as a PTSP. What distinguishes one PTSP from another is
the probability distribution (or more generally, the probability
"law") that specifies the number k and the identity of the
nodes that need to be visited on any given instance of the
problem.

In general, PTSP's arise in practice whenever a company,
on any given day, is faced with the problem of collections
(deliveries) from (to) a random subset of its (known) global set
of customers in an area and does not wish to, or simply, cannot
redesign the tours every day. A second, very important set of
applications is in connection with strategic planning for
collection and delivery services, i.e. in connection with cases
where only a probabilistic description of demands is available
at the time of planning. In addition to applications in the
context of routing problems, the generic model, as stated, can
be of interest in any situation in which an a priori sequencing
of tasks has to be found and the relative order of tasks has

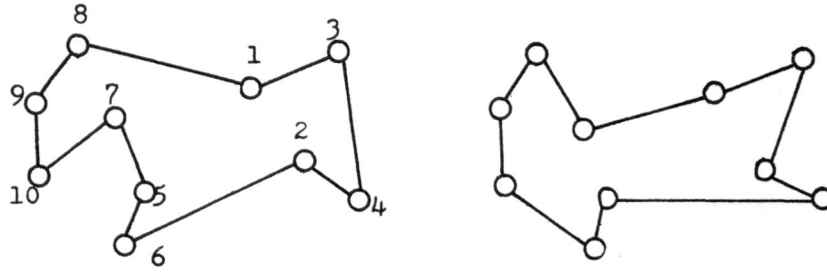

1.1: Two a-priori tours through the same set of points.

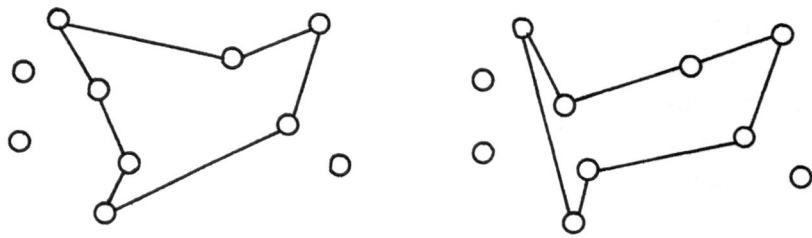

1.2: The two resulting tours when the points 4,9, and
10 need not be visited.

Figure 1: Simple graphical example of a PTSP.

to be preserved even when some of the tasks are absent. (See Jaillet [1986a] for some examples.)

The outline of this section is as follows: we first present efficient ways for computing the expected length of any given PTSP tour under very general probabilistic assumptions. We then give worst-case bounds on the absolute difference between the expected length of the optimal TSP tour and the expected length of the optimal PTSP tour. Finally, we give some results obtained in the analysis of the problem in the plane. The interested reader is referred to Jaillet [1985], [1986a], for proofs as well as additional results such as algorithmic considerations.

2.2   The expected length of a given tour   t

We consider a complete, loopless, directed weighted graph G with n+1 nodes; t represents a given hamiltonian circuit (tour) of G; by re-indexing the nodes in their order of appearance along t, we write $t = (0,1,2,...,n,0)$. "0" can be considered as the "starting city", or equivalently the "depot", where the tour starts and ends. The direct cost or distance to go from node i to node j is denoted $d(i,j)$. The set of rules defining the probabilistic assumptions will cover only the case of independently present nodes (see Jaillet [1985] for special cases of dependence). For each node i, p_i denotes its probability of presence ($p_0 = 1$, in general, but not necessarily) and is independent of all other nodes.

For a given tour t, the distance L_t covered in traversing the set of nodes actually present on each instance of the problem is a random variable. This random variable can have up to 2^n different values (each node can be either present or not, independently of the others), so that its expected value would be obtained in $O(n2^n)$ by "direct enumeration". This order of complexity is clearly not satisfactory and, if not improved, would be a strong argument for the intractability of this problem. Let us present an efficient alternative analytical approach to obtain $E[L_t]$.

Theorem 1

The expected length of an a priori tour $t = (0,1,2,\ldots, n,0)$ through $n+1$ nodes (with coverage probabilities p_j, $j \epsilon [0 \ldots n]$) is equal to:

$$E[L_t] = \sum_{i=0}^{n} \sum_{j=0}^{n} \alpha_{ij} \, d(i,j)$$

where $\alpha_{ii} = 0$ $\forall \ i \ \epsilon \ [0 \ldots n]$

$$\alpha_{i,i+1} = p_i p_{i+1} \qquad \qquad \forall \ i \ \epsilon \ [0 \ldots n-1]$$

$$\alpha_{ij} = p_i p_j \prod_{k=i+1}^{j-1} (1 - p_k) \qquad \forall \ j > i + 1$$

$$\alpha_{ij} = p_i p_j \left(\prod_{k=i+1}^{n} (1 - p_k) \right) \left(\prod_{r=0}^{j-1} (1 - p_r) \right) \text{ otherwise}$$

This result implies that $E[L_t]$ can be computed in $O(n^3)$. This worst-case bound cannot in fact be improved for general coverage probability p_j and general distances between nodes. If we assume $p_i = p \ \forall \ i \ \epsilon [0 \ldots n]$, we have $\alpha_{ij} = \alpha_{kh}$ for all quadruplets (i,j,k,h) such that $j-i = k-h$; we save $O(n)$ computations in obtaining $E[L_t]$ and we can see that by regrouping judiciously the $d(i,j)$ we can derive a special version of Theorem 1 proved in Jaillet [1985]. The same is true for the case of one (or more) black node(s) (i.e., always present).

The importance of this result is due to its providing of a closed form expression for $E[L_t]$, which becomes critical in developing procedures for solving PTSP's.

2.3 Relationship between PTSP and TSP

The TSP is a special case of a PTSP in which $p_j = 1$ for all j. It is then natural to investigate the possible links between the two problems. In other words, how well would a TSP tour through all _potential_ nodes do as a solution to a PTSP problem? We shall restrict our answers to problems in which $p_o = 1$ and $p_j = p$ for $j\varepsilon[1...n]$.

In order to obtain a non-trivial answer to our previous question, we assume the distances to be positive and to satisfy the triangular inequality (without these restrictions, one can easily construct pathological examples for which a TSP tour is arbitrarily bad for a PTSP problem, see Jaillet [1986b]).

If v and s denote respectively an optimal TSP tour and an optimal PTSP tour for a given network, we then have:

Lemma 1:

Given a graph with one depot and n nodes with coverage probability p, the worst-case performance ratio of the TSP tour is bounded by:

$$0 \leq (E[L_v] - E[L_s]) / E[L_s] \leq ((1-p)/p)(1-(1-p)^{n-1})$$

The bounds give the correct result (i.e. $E[L_v] = E[L_s]$) for p=1 and seem tight for p close to 1. When p approaches zero, the upper bound approaches (n-1), an arbitrarily large number for arbitrarily large-size problems. In fact, one can construct examples for which a TSP tour is indeed arbitrarily bad for the corresponding PTSP problem (see Jaillet [1986b]).

Finally, the cases for which one can demonstrate that the optimum TSP tour is guaranteed to solve the PTSP optimally are either trivially small problems or extremely special problems (see Jaillet [1985]).

2.4 Analysis of the PTSP in the plane

In this subsection we present a result for which set-theoretic concepts are used instead of graph-theoretic ones. We consider a set of points in 2-dimensional Euclidean space R^2, assuming the distance between points to be the ordinary Euclidean distance. More precisely, let X be an infinite sequence of points independently and uniformly distributed over $[0,1]^2$ and p be the coverage probability for each point. Let $\Pi_p(X^{(n)})$ be the expected length of an optimal PTSP tour through $X^{(n)}$, the first n points of X. The following theorem is then true:

Theorem 2:

$\forall$ p ε [0,1]   there exists a strictly positive constant c(p) such that

$$\lim_{n \to \infty} \frac{\Pi_p(X^{(n)})}{\sqrt{n}} = c(p) \qquad (a.s.)$$

This result can be seen to be a generalization of the celebrated limit law obtained for the TSP by Beardwood et al. [1959] (c(1) = β where β is the "TSP constant"). In fact, one can show that Theorem 2 can be generalized in various directions such as taking points in any bounded Lebesgue measurable set of a d-dimensional Euclidean space. One can also derive bounds for finite-size problems. The reader is referred to Jaillet [1985] for these results. A comprehensive analysis of the PTSP in the plane will also appear in a forthcoming paper in which we use various tools from probability theory such as Martingale inequalities.

Because of algorithmic applications, results like that of Theorem 2 have gained considerable practical interest (the limit law derived for the TSP was used as the main "building block" in the probabilistic analysis of heuristics for the TSP, see Karp [1977]). In addition to such algorithmic applications, Theorem 2, together with an estimate $\bar{c}(p)$ of the constant c(p), provides an approximation formula useful for predicting with high probability the expected length of an optimal PTSP tour, if the number of points is large.

III. The Probabilistic Shortest Path Problem

3.1 Introduction

The generic probabilistic shortest-path problem can be formally described as follows: consider the problem of finding a shortest path between a node source s and a node sink t in a complete network having n+2 nodes and a length associated with each arc. On any given instance of the problem only a subset consisting of k out of the n intermediate nodes $(0 \leq k \leq n)$ can be used to go from s to t, with the number k determined according to a known probability distribution. We wish to construct a priori a path to go from s to t in the complete network. On any given instance of the problem, the sequence of nodes which defines the path will be preserved but only the "permissible" nodes will be traversed (the others have to be skipped). The problem of finding such an a priori path which is of minimum length in the expected value sense is defined as a PSPP.

This problem resembles the PTSP (as defined previously) in many aspects, but includes several fundamental differences that are worthy of immediate identification. First of all, the rationale behind these two problems is different: skipping a node in the PTSP context is considered a "bonus": we skip a node having no request in order to shorten the length of a tour. In the PSPP context, we skip a node because we are not allowed to go through it (a penalty) and we will then talk of a "node failure". A second major difference is that a PSPP path does not have to include all the intermediate nodes.

One important consequence is the assumptions that must be made on the distances between nodes. In the PSPP context, the distances will not be assumed to satisfy the triangular inequality. Otherwise the optimal PSPP path would simply be the arc (s,t); a trivial but not interesting case.

The generic PSPP, as stated, can be of interest in many applications. In the context of urban networks it can be used to model congestion occurring at intersections (congestion in the "worst" sense: the traffic is blocked or not, a zero-one situation). In the context of network reliability it can describe operating strategies in case of node failures, as well as alternative measures of "cost versus reliability" for a given network. Also in the context of flying operations, nodes can represent geographical areas that can or cannot be flown over by aircraft going from one place to another (for example, because of weather conditions, unexpected military restrictions, etc.).

In the previous section we presented results obtained on the PTSP, including combinatorial properties (2.2 and 2.3) and asymptotic analysis (2.4). Similar combinatorial properties have been obtained for the PSPP. We will limit ourselves to presenting a branch-and-bound scheme for solving the problem on networks with positive distances. All the paths are assumed to be simple, i.e. without repeating nodes.

Before presenting this solution procedure, we first mention efficient ways to evaluate the expected length of any path under general probabilistic assumptions.

3.2 The expected length of a given path

We consider a complete, loopless, directed weighted graph G with two special nodes s and t and n intermediate nodes; h represents a path from s to t going through m ($0 \leq m \leq n$) intermediate nodes; by re-indexing the nodes in their order of appearance along h, we write h = (0,1,2,...,m,m+1) with $0 \equiv s$ and $m+1 \equiv t$. It is then easy to see that the expected length of this path is given by:

$$E[L_h] = \sum_{i=0}^{m} \sum_{j=i+1}^{m+1} \alpha_{ij} \, d(i,j)$$

where α_{ij} is defined as in Theorem 1.

3.3 A branch and bound scheme

A mathematical programming formulation for the PSPP can be given by:

$$\min \quad f(X)$$
$$\text{s.t.} \quad \underline{N}X = b$$
$$x_{ij} \; \varepsilon \{0,1\}$$

where $\quad X = (x_{ij})$ and $\quad x_{ij} = \begin{cases} 1 & \text{if arc } (i,j) \text{ is in the} \\ & \qquad\qquad \text{optimal path} \\ 0 & \text{otherwise} \end{cases}$

$\underline{N}$ is the node-arc incidence matrix

$$b = (b_i) \text{ and } b_i = \begin{cases} +1 & \text{if } i=s \\ -1 & \text{if } i=t \\ 0 & \text{otherwise} \end{cases}$$

and where $f(X)$ can be decomposed as the sum of a linear function $L(X)$ and a nonlinear function $NL(X)$, polynomial of order n in the x_{ij}'s. Without going into details, one observes that when the probabilities of failure are not too high, the linear function is the most important part of $f(X)$. The idea is then simple: if we have a lower bound on $NL(X)$, say ϕ, and a known feasible solution $\bar{X}$, we can reduce the feasible region by limiting our search for an optimal PSPP path among the paths X such that $L(X) + \phi \leq f(\bar{X})$; in other words, one can discard paths satisfying $L(X) + \phi > f(\bar{X})$. A branch-and-bound scheme would then be based on the minimization of $L(X)$. In the PSPP case, this problem becomes very easy; indeed it can be shown to be equivalent to the problem of finding a shortest path between s and t on the initial network with the following modified distances:

$$d'(s,t) = d(s,t)$$
$$d'(s,j) = p_j \, d(s,j) \qquad\qquad \forall \; j \; \varepsilon \, [1...n]$$
$$d'(j,t) = p_j \, d(j,t) \qquad\qquad \forall \; j \; \varepsilon \, [1...n]$$
$$d'(i,j) = p_i p_j \, d(i,j) \qquad\qquad \forall \; i,j \; \varepsilon \, [1...n]$$

Various lower bounds can be derived for NL(X). One of the methods is based on a partition of the feasible region; namely, one derives bounds which depend on the number of intermediary nodes contained in the path X. The branching part of the procedure can then be based on this partition.

This branch-and-bound procedure is currently being investigated experimentally by the author. In the case of very small probabilities of node failures, it seems to give good results for moderate-size problems.

IV. Conclusion

In this paper we have introduced probabilistic versions of two well-known combinatorial optimization problems, the traveling salesman problem and the shortest-path problem. We have surveyed some results, showing that a rigorous treatment of truly probabilistic combinatorial problems can be performed, but that these problems deserve very often special treatment due to drastic changes in some of their properties when uncertainty is introduced. We presented some algorithmic considerations for the PSPP (see Jaillet [1985] for the PTSP) but much has yet to be discovered and tested in this area.

In addition to the PTSP and PSPP, such an approach has been applied to numerous other combinatorial optimization problems defined on a graph, including: capacitated traveling salesman problems, traveling repairman problems, directed spanning tree problems, etc. There are several motivations behind this work: among them, two are particularly important. The first one is the desire to formulate and analyze models which are more appropriate for real-world problems where randomness is present. The second motivation is an attempt to analyze the "robustness" (with respect to optimality) of optimal solutions for deterministic problems when the instances of these problems are modified (in our case, we are confined to problems on graphs and the perturbation of the problem instance is simulated by the presence or not of the given nodes).

References

Beardwood J, Halton J, Hammersley J (1959) The shortest path through many points. Proc Camb Phil Soc 55:299-327

Jaillet P (1985) The probabilistic traveling salesman problems (PhD thesis) Operations Research Center Technical Report 185, MIT Cambridge MA 02139

Jaillet P (1986a) The probabilistic traveling salesman problems: a preliminary analysis. To be published in Op Res

Jaillet P (1986b) Stochastic routing problems. Course notes, to be published in Proceedings of Advanced School on Stochastics in Combinatorial Optimization, CISM Udine Italy

Karp R (1977) Probabilistic analysis of partitioning algorithms for the traveling salesman problem in the plane. Math Oper Res 2:209-224

Larson R, Odoni A (1981). Urban Operations Research. Prentice Hall

Stewart W, Golden B (1983) Stochastic vehicle routing: a comprehensive approach. Eur J Oper Res 14:371-385

The Flow Management Problem in Air Traffic Control

Amedeo R. Odoni
Operations Research Center (E40-169)
Massachusetts Institute of Technology
Cambridge, MA 02139
U.S.A.

Abstract - A system of flow management is one of the most promising short-term approaches to alleviating the severe network-wide congestion problems that air traffic in the United States and in Europe is currently experiencing. To design such a system one must address the flow management problem (FMP), a description and discussion of which is the subject of this paper. Even simplified versions of the FMP, such as the "generic FMP" which is based on a "macroscopic" model, are very challenging. The problem is inherently stochastic and dynamic and requires a discretized representation of flows. Additional complications are caused by the need to consider the distributive effects of flow management strategies as well as by certain peculiar characteristics of the capacity/demand and flow conservation relationships associated with elements of the ATC network. A brief literature review indicates that research on the FMP is still in its very early stages.

1. INTRODUCTION

In recent years, congestion has begun to affect severely air traffic in the United States and, to a lesser extent, in Europe on a network-wide basis. The phenomenon is still somewhat limited to the busiest commercial airports and to a few "sectors" of the air traffic control (ATC) network, but the distribution of traffic is such that the effects are felt by a large fraction of air travelers and by practically every North American and West European airline. For example, it is estimated that in 1986 a full 60% of the 415 million passenger enplanements in the United States took place at one of the 22 major commercial airports (out of 650 that receive some scheduled airline service) which are considered by the Federal Aviation Administration (FAA) to be operating at or close to their maximum capacity. Airlines claim that the direct cost to them of airport and ATC delays amounted to $2 billion in 1986 alone, in the United States. The size of this figure can best be appreciated if one realizes that the total profits of the entire U.S. airline industry were about $800 million for 1986. The

NATO ASI Series, Vol. F38
Flow Control of Congested Networks
Edited by A. R. Odoni et al.
© Springer-Verlag Berlin Heidelberg 1987

cost of delays to passengers -lost time, missed connections, etc.- may be of similar size.

It is likely that traffic will continue to grow in the foreseeable future. Typical 1986 forecasts call for a 4-7% annual world-wide rate of increase in the number of airline passengers and 2-3% annual increase in the number of scheduled airline operations over the next decade, with regional factors accounting for some significant variations from these averages - higher growth rates are, for example, expected in the Pacific and Southeast Asia region and lower rates in Africa and the Middle East. There exists, therefore, a very real threat of even more severe congestion in ATC and airport systems which are already operating close to their capacity. The problem is further exacerbated by the recent tendency of airlines in the United States and in Western Europe to operate "hub and spoke" systems, i.e., to use one or more airports as the focal points of their operations (hubs). This means that schedules are designed, so that a large number of flights from the "spokes" of an airline's system arrive at approximately the same time at the hub airport, passengers and their luggage are redistributed according to their ultimate destination and a large number of flights then depart from the hub, once again approximately at the same time. These spurts of scheduled activity, known as "banks", obviously tax further the capacity of the ATC system and of airports. For example, in Atlanta, the hub of Eastern Airlines and of Delta Airlines, there are at least six such "banks" during a typical weekday, each involving the arrival and then the departure of as many as 100 or more flights within roughly one-hour periods.

There are long-, medium- and short-term approaches for dealing with the air traffic congestion problem. In the long-term ("time constant" of the order of 5-10 years) an obvious approach is to try to increase the capacity of the system through improved ATC technologies (better navigation, surveillance and control resulting in more efficient utilization of airspace and runways) and/or more "concrete" (additional or better runways, new airports). Many countries have sizable and costly programs already under way along both* of these directions (see FAA (1986) for a description of the current FAA ATC improvement plan). In the United States, the federal government alone is expected to spend at least $15 billion for ATC

---

*The "more runways and airports" solution, which, in principle, may be the most effective one, may, in practice, be politically infeasible at many of the locations most critically in need of additional airport capacity.

improvements and another $10 billion for airport improvements over the next decade, while the Port Authority of New York and New Jersey has recently announced a $5 billion capital improvement program for the three commercial airports serving the New York City area.

It is also possible that, in the long run, the entire industry may slowly modify its pattern of operations in response to congestion-related costs and constraints. For example, the trend toward larger aircraft may be accelerated, so that more passengers can be carried per flight, thus reducing the rate of growth in the number of aircraft operations; or strong fare differentiation may be used to reduce time-of-day peaking of passenger demand; or short-takeoff-and-landing (STOL) aircraft may at some point play an important role in short-haul air travel, reducing demand on the existing long runways.

In the medium-term (6 months to 2 years) it is possible to attempt to restrict demand or to modify the time-of-day patterns of demand through a variety of means including (see Cohen and Odoni (1985) for a more detailed description): imposition of time-varying landing fees and user charges at airports to encourage off-peak uses; imposition of "quota" on airport use, i.e., bounds on the number of operations that can be scheduled over various intervals of time (5-minute, 15-minute, 30-minute and 1-hour intervals have been used by various administrative bodies); use of airline scheduling committees, in combination with a system of quota, for the allocation of available time "slots" among prospective users; permission to airlines to buy and sell such time slots; and resort to more "exotic" devices, such as auctioning of slots for access to quota-restricted airports or even allocation of such slots through a lottery.

Further discussion of long- and medium-term remedies lies beyond the scope of this paper, whose focus is on short-term measures. In the short-term, i.e., on a daily basis and with a planning horizon of at most 6-12 hours, the best the ATC system can do is attempt to optimize operations for given demand and capacity levels. To accomplish this, one must seek to control the flow of air traffic so as to best match the demand with available capacity over time and across the various components of the ATC and airports network. Leaving aside for now the question of what "optimizing" means in this context, we shall refer to this henceforth as the ATC Flow Management Problem (FMP).

It is the purpose of the rest of this paper to provide a non-mathematical description and discussion of the FMP. Section 2 provides the background and presents a general description of the problem, emphasizing its dynamic and stochastic nature. Section 3 then describes a macroscopic, strategically-oriented version of the problem, the generic FMP. Section 4 discusses the need to consider distributive (in addition to aggregative) criteria in solving the FMP. Section 5 points out a number of complications that arise in connection with versions of the FMP concerned with more microscopic issues than the generic version. Section 6 reviews briefly the state-of-the-art on the FMP through a survey of the existing professional literature. Finally, Section 7 summarizes the principal points of the paper. Throughout, we attempt to highlight some of the important differences between the FMP and flow control problems in other contexts, such as data communications networks and urban and highway networks.

In concluding this introduction, we should note that since, until recently, ATC congestion was a purely local problem limited to a few major airports, research on examining network-spanning problems such as the FMP is only now beginning in the ATC area. Thus, unlike other application areas, the available literature is very sparse and recent. In fact, to our knowledge, this paper represents one of the first attempts to describe in a systematic way the FMP in ATC. Our purpose here is to stimulate interest in this very challenging area of research which, because of the special requirements imposed by the ATC and air transportation context, seems to call for approaches unlike those encountered in flow control problems in other fields of application.

2. DESCRIPTION OF THE FMP

The FMP can be described in terms of an idealized network model of the ATC system, such as the one shown in Figure 1. This network consists of four types of "elements":

(i)   Airports, the "sources" and "sinks" of flows on the network.

(ii)   Airways, the "arcs" on which flows travel.

(iii)   Waypoints, the network's "nodes" at which airways intersect, merge or diverge.

(iv)   Sectors, collections of waypoints and contiguous segments of airways.

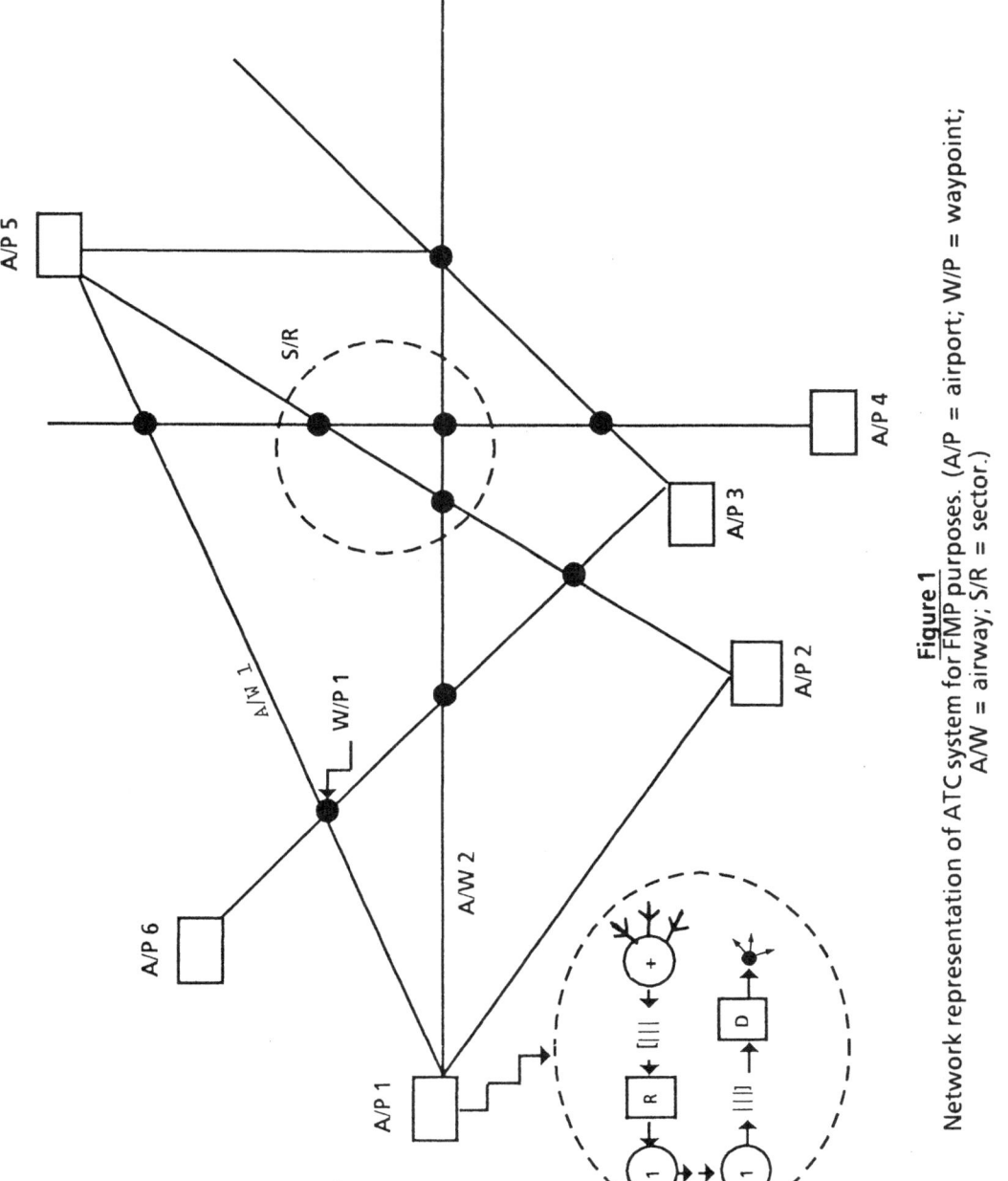

Figure 1

Network representation of ATC system for FMP purposes. (A/P = airport; W/P = waypoint;
A/W = airway; S/R = sector.)

In the instance of the FMP, airports are by far the most important of these elements since in the great majority of cases they constitute the principal "bottlenecks" of the network. Each airport has an arrivals capacity* and a departures capacity (see "servers" R and D associated with Airport 1 in Figure 1) defined as the maximum number of movements (arrivals or departures) that can be conducted at an airport per hour. These capacities are often highly variable and depend on numerous factors, the most important of which are: weather conditions (visibility, wind, precipitation); runway configuration in use (itself a function of weather); ATC separation requirements and procedures (also functions of weather and runway configuration in use); traffic mix (i.e., types of aircraft, such as wide-body jets, conventional jets, propeller and turbo-prop, private jets, etc., that compose the traffic) and fraction of each type in the traffic; runway geometry (location of runway exits, accessibility of neighboring taxiways); and human factors (performance of ATC controllers and pilots). For any given set of conditions regarding all of the above factors, existing airport models can estimate quite accurately the resulting capacity (see, for example, Simpson et al. (1986)). It is not unusual to encounter 2:1 and even 3:1 ratios between an airport's highest and lowest** capacities. For example, at Boston's Logan International Airport, the arrivals capacity under visual meteorological conditions (VMC) can be as high as 65 per hour and under instrument meteorological conditions (IMC) as low as 27, with many in-between values also possible-depending primarily on weather and runway configuration in use.

Due to the fact that the underlying factors, especially weather, change over time, so do airport capacities. Moreover, because of the high sensitivity of capacity to weather and in view of our limited ability to predict accurately weather conditions at a given specific location (the airport) even a few hours in advance, forecasts of capacity are subject to much uncertainty. One way to model this mathematically is by defining random variable

---

*Arrivals capacities are sometimes referred to as airport acceptance rates.

**We refer here to lowest capacities under "routine" poor weather conditions and not to cases of snowstorms, thunderstorms, heavy fog, etc. when capacity is sometimes reduced to zero.

$$\text{CAPR}(X, t_1|t_2) = \quad \text{the arrivals capacity at airport X at time } t_1 \text{ as forecasted at time } t_2 (t_2 \leqq t_1),$$

and similarly for the departures capacity, CAPD $(X, t_1|t_2)$. CAPR and CAPD could be viewed as discrete random variables (for example, for Boston CAPR would take integer values in the range of 27 to 65). The probability distributions associated with these variables are then summaries of our "state of information" at time t_2 regarding capacity at airport X at some future time t_1. (Presumably, as $t_1 - t_2$ decreases, these probability distributions should become "tighter" around the values of CAPR and CAPD which will actually materialize at the time of interest t_1, i.e., the central moments of the random variables should generally decrease with $t_1 - t_2$, although this relationship may not always be monotonic.)

Airways, waypoints and sectors all have their own capacities as well, measured by the maximum number of aircraft per unit of time that can traverse these elements or, in the case of sectors, by the maximum number of aircraft that can simultaneously occupy them. In practice, the capacities offered by these elements usually far exceed the traffic flows that can be generated by and absorbed at the network's sources and sinks - the airports. However, some strategically located sectors and, more rarely, some waypoints located close to a group of airports may occasionally be an exception to this rule.

If we now make the reasonable assumption that the capacity of airways in itself is almost never the cause of congestion, delays in the ATC system can occur as a result of an unfavorable demand-to-capacity relationship* at the following ATC elements:

i)   Airport of origin (due to congestion of the airfield on departure).

ii)   Airways en route due to congestion at a waypoint.

iii)   Airways en route due to congestion of an en route sector.

---

*There is little to be gained, for the purposes of this presentation, from drawing a distinction between: "stochastic delays" (i.e., delays due to the stochasticity of demand and capacity) in a queueing system which operates with a utilization ratio $\rho < 1$; and delays which are due to the fact that some ATC elements, such as busy airports, are sometimes operated for periods of time as long as a few hours with utilization ratios exceeding unity ($\rho > 1$).

iv)   Airport of destination* (due to congestion of the airfield on arrival.)

In a system without flow management any delays associated with (i) - (iv) would be suffered at the corresponding congested points: on the airfield, prior to take-off; on the airways, through airborne "holding" near a congested waypoint or prior to entering a congested sector; and in terminal areas, through airborne holding and "vectoring" prior to landing at the airport of destination. A flow management system, however, can use available information (subject to uncertainty) about the network and attempt to alleviate the impact and cost of delays by taking the following types of actions:

(1) Delaying the departure times of aircraft ("gate holds" or "ground holds") i.e., not allowing an aircraft to start its engines and leave its airport gate or parking area at the scheduled departure time even if the aircraft is otherwise ready to depart.

(2) "Metering" of traffic, i.e., regulating the rates of aircraft flow through specific points on the network such as departure runways or various en route waypoints.

(3) En route re-routing, i.e., modifying the en route flight plans of selected aircraft in order to by-pass congested en route areas.

(4) Imposing en route speed control restrictions, e.g., requesting en route aircraft to fly at less than cruising speed in order to properly time their arrival at a waypoint or the terminal area of an airport.

(5) High-altitude holding and path-stretching maneuvers in order to delay arrival at a congested terminal area and avoid more costly holding at low altitudes.

In addition, a flow management system may, naturally, rely on the usual queueing procedures (see (i) - (iv) above) for delay-taking.

The FMP, stated in most general terms, is then the problem of designing a flow management system which will minimize the cost of ATC delays (in the short-term

---

*To avoid drawing a distinction between en route sectors and terminal area sectors (i.e., airspace in the immediate vicinity of airports) we shall henceforth consider terminal area sectors as integral parts of airports. Thus, references to "sectors" imply en route sectors.

sense described in this section) subject to a set of operational and policy constraints.*

Clearly, the FMP can be addressed at several different levels of detail, ranging from macroscopic to microscopic, with the appropriate level depending on the types of flow management actions which are under consideration. When the concern is with strategic flow management, macroscopic models with a high level of aggregation are in order. Decisions of a more tactical nature require more detailed, microscopic models.

On considering flow management actions (1) - (5) described above, it is obvious that (4) and (5) are of a tactical ("fine-tuning") nature, (1) is of a strategic nature and (2) and (3) fall in-between**. Since it is useful for expository purposes to concentrate on a version of the FMP that contains all the essential concepts without being burdened by excessive detail, we shall describe next what we shall refer to as the generic FMP, a macroscopic model that examines only actions of type (1) as possible flow management tools. Thus the generic FMP can assist in resolving the most important, from the practical point of view, strategic choice in ATC flow management, namely the trade-off between ground-holding delays and airborne delays. The reason this trade-off is so important is that ground-holding delays (taken with the aircraft stationary and its engines off) imply much lower costs per unit of time than airborne delays - for reasons of fuel consumption, maintenance and depreciation costs, safety, etc. Hence, if all delays could somehow be taken on the ground prior to departure, enormous cost savings to the airlines and to the public in general might result.

---

*The term "cost" is used in a general sense here to imply a disutility function that includes such difficult-to-quantify aspects as safety and the cost of schedule disruptions in addition to direct delay costs. "Policy constraints" refer to such issues as the equitable sharing of delay costs among ATC users. These will be discussed further in Section 4.

**Note that (1) can also be viewed as a form of "metering" of traffic traveling to congested destinations, i.e., as action of type (2).

3. A GENERIC VERSION OF THE FMP

The generic FMP can be described with reference to the single-destination network shown in Figure 2, which is a much-simplified version of the network of Figure 1. We make the following set of assumptions:

(i) The only capacitated element of the ATC network is the arrival airport Z; all other elements of the network (departure airports/sources, airways, waypoints, sectors) have unlimited capacity.

(ii) Travel times of aircraft between each source and airport Z are deterministic and known in advance; delays can occur only as a result of congestion at Z, as indicated by the queueing system in front of airport Z on Figure 2.

(iii) At the beginning of the time-interval $[0, T]$ under consideration (T might typically be a period of 6-12 hours) we are given a complete list of scheduled flight departure times from all airports 1, 2, ..., N to airport Z for all $t \in [0, T]$.

(iv) At all times $t_2 \in [0, T]$, we have access to the probability distribution for the forecast capacity of airport Z, CAPR $[Z, t_1|t_2]$ for any $t_1 \in [t_2, T]$. (The range of values and the probability distribution of CAPR $(Z, t_1|t_2)$ can, of course, vary with t_1 and the probability distributions for the various values of t_1 can be revised over time, i.e., as t_2 changes.)

Consider now F_i, the i-th flight, according to some a priori indexing scheme. Assume F_i is originally scheduled to depart from airport j at time s_i and assume that its travel time to Z in the absence of any delays is $d_i(j, Z)$. At $t = s_i$, the flow management system must then decide whether to allow F_i to take off as scheduled or to postpone departure through ground holding until some later time s_i ($s_i > s_i$).

This decision depends on:

(a) The probability distribution of CAPR $(Z, t|s_i)$ for every $t \in [s_i, T]$.

(b) The times when each of the aircraft which are already airborne at s_i and are traveling toward Z will arrive at Z (or, more precisely, will be available to land at Z if there is no queue at Z).

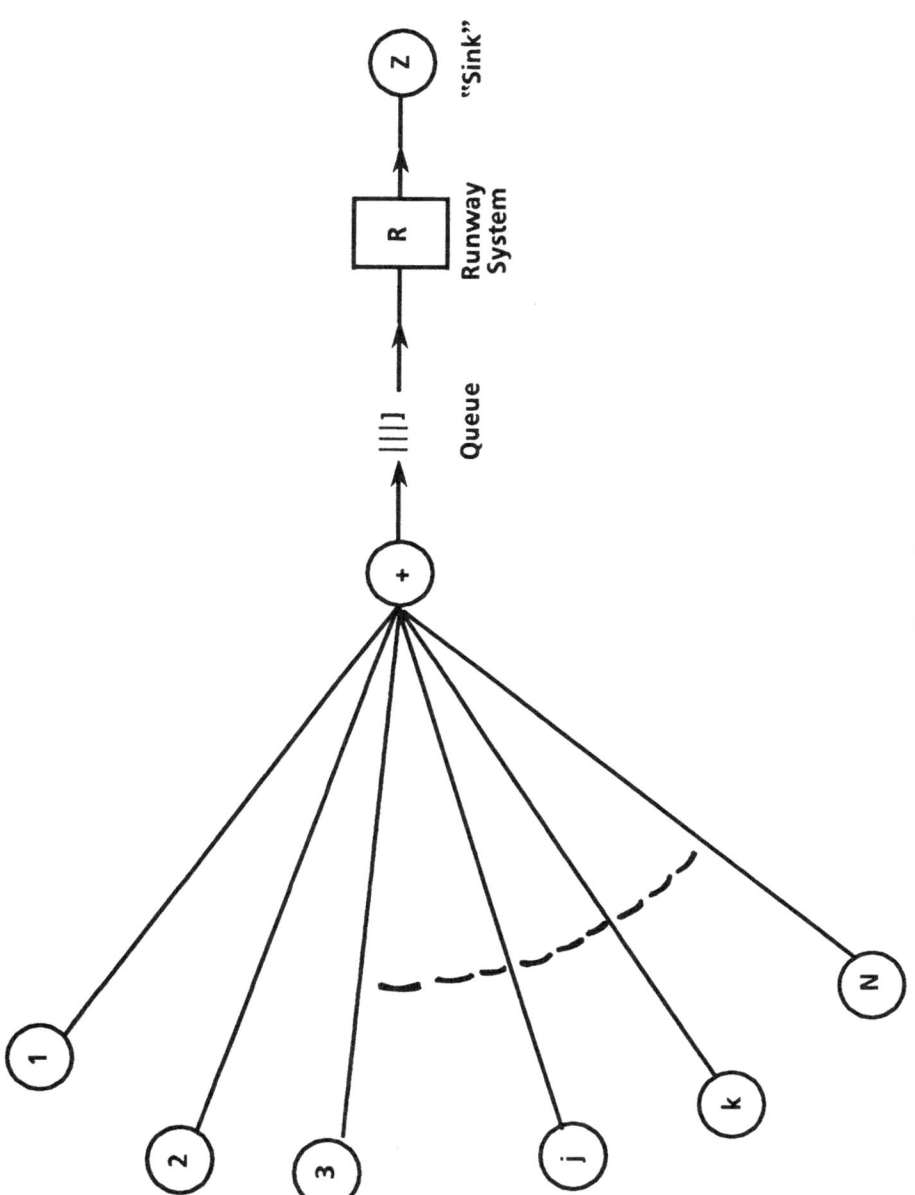

Figure 2

Schematic representation for the generic FMP. For each airport of origin 1, 2, ..j, ... N, the a priori schedule of flight departures to Z is given.

(c)   The current status (at $t = s_i$) of the queueing system at Z.

(d)   The scheduled departure times and scheduled arrival times at Z of all
        aircraft scheduled to depart Z at any time $t \in [s_i, T]$.

The key point to note here is this: If at s_i it were possible to compute exactly the
queueing time w_i of F_i at Z - assuming F_i sought to land at Z at t_i , as scheduled - then
it would be best for F_i to "take" all this delay as a ground-holding delay at the
airport of origin j. In other words F_i would then be assigned a new departure time
from j equal to $s'_i = s_i + w_i$. On departing at s'_i, the aircraft would then presumably
proceed to fly to and land at Z at $t_i + w_i$ ($= s'_i + d_i$ (j, Z)) with no airborne delay.

However, because of the uncertainty concerning CAPR for values of t in $[s_i, T]$, the
queueing time w_i cannot be predicted exactly at s_i and a stochastic decision problem
results. An overly "optimistic" strategy, i.e., allowing too many aircraft to depart on
time, may result in excessive airborne delays if CAPR turns out to be "low" at the
time of the arrival of the flights in the vicinity of Z. On the other hand, an unduly
"conservative" strategy, i.e., assignment of long ground-holding times to many
aircraft in order to avoid airborne delays, may result in unused airport capacity, if
CAPR turns out to be on the "high" side. In ATC language, airport Z will be
"starved", in this latter case, while flights are being delayed unnecessarily on the
ground at their airports of origin.

A second key observation is that at s_i (the time when a decision must be made on
how long, if at all, flight F_i should be held at the airport of origin) it will actually be
necessary to develop a tentative ground-holding strategy not only for F_i but also for
all flights that have not yet departed for Z. The reason is that the best strategy for F_i
can only be determined in the context of an overall strategy for all flights with
scheduled departure times in $[s_i, T]$ -see (d) above.

We can now define a strategy, R(t), at any time $t \in [0, T]$ as follows: R(t) is a vector
consisting of m elements, where m is the number of flights that are still left to
depart for Z from the airports 1, 2, ..., N during [t, T]. The element of R(t)
corresponding to flight F_i, $r_i(t)$, indicates the amount of time beyond its scheduled
departure time, s_i, that F_i will be required to hold on the ground ($r_i(t) = 0$ means
that F_i will be allowed to depart as scheduled at s_i).

The generic FMP is then concerned with developing a strategy $R_o(t)$ for $t \in [0, T]$ that
will minimize the expected total cost of ground-holding and airborne delays over
the time-period of interest T. Clearly the FMP is both a stochastic problem since

CAPR is a random variable and a <u>dynamic</u> one since the probability distribution of CAPR changes with time. Thus $R_o(t)$ will have to be revised over [0, T] as the probability distribution for CAPR changes and as aircraft depart from their airports of origin at their originally scheduled times, s_i, or at delayed departure times*s_i'.

The generic FMP if well defined for any given objective function related to airborne and ground-holding delays, as long as the queueing process at airport Z is well-defined. For example, the simplest model possible would be to assume that it takes a constant amount of time to "serve" each aircraft at Z and that the queue discipline is first-come, first-served according to the time of arrival of aircraft at the queue at Z. Thus, if for the time interval $[t_1, t_2]$, CAPR at Z turns out to be equal to 30 landings per hour, each aircraft would take exactly 2 minutes to serve. For any given schedule of arrivals at Z during the interval $[t_1, t_2]$ and initial conditions at t_1, we can then compute the airborne delays for each aircraft involved. The actual schedule of arrivals at Z during $[t_1, t_2]$ will in turn be determined by the flow management strategy $R_o(t)$ followed during the interval $[0, t_2]$.

4. DISTRIBUTIVE EFFECTS

A major difference between the FMP and other flow management and flow control problems in various settings is the great importance to the users of distributive effects resulting from FMP strategies. We shall illustrate this point through a simple example which will serve to elucidate further the generic FMP and will then discuss the issue in general.

Consider Figure 2 again and assume that at t = 5**, exactly two flights are scheduled a priori to arrive at Z, one from airport j, exactly 5 time units away, and the other from airport k, only 1 unit away. The flights will be conducted by two identical aircraft whose costs per unit of time of waiting on the ground ("ground-holding")

---

*There is nothing in the statement of the problem to prevent revising any delayed departure times s_i' several times. However, from the practical point of view, it would be highly undesirable to revise the departure time of any flight more than once or, at most, twice in a day.

**Throughout this example, we shall use a time axis which has been "discretized" into unit time periods t = 1, 2, ..., T.

and in the air are $c_g = \$1$ and $c_a = \$3$, respectively. For all $t \neq 5$, the capacity of airport Z will be assumed infinite, so that no delays can occur then. However, for $t = 5$, the capacity at Z will be either 1 (aircraft landing) or 2. If both flights seek to land at Z at $t = 5$ as scheduled and CAPR turns out to be equal to 1, one of the aircraft will have to be delayed while airborne for one time unit until $t = 6$.

Assume that at $t = 0$, the time when the flight from j to Z is scheduled to depart, the probability of the event "CAPR (Z, 5|0) = 1" is equal to p (and of "CAPR (Z, 5|0) = 2" equal to 1-p). Similarly let q and 1-q be the probabilities, respectively, of "CAPR (Z, 5|4) = 1" and "CAPR (Z, 5|4) = 2."

It is easy then to see that, for any $p, q \in [0, 1]$, an optimal strategy is to allow the j-to-Z flight to depart on time at $t = 0$ and then at $t = 4$ to either ground-hold the k-to-Z flight for one unit of time at k (if $3q > 1$) or to allow the k-to-Z flight to depart on time (if $3q < 1$). [The expected total delay cost of this strategy evaluated at $t = 4$ is min $(1, 3q)$.] Note that, under this strategy, all the ground delay costs, if any, will be borne by the short-range k-to-Z flight and, in fact, if we assume that the k-to-Z flight, if released at $t = 4$, will reach Z a small increment of time later than the j-to-Z flight, all delay costs will always be borne by the k-to-Z flight under a first-come, first-served discipline at Z.

This last observation can be generalized considerably to the more realistic case in which CAPR at Z at $t = t_0$ can take values in the range [m, h], where the integers m and h are, respectively, the minimum and maximum values of CAPR that have a non-zero probability under any a priori state of information, i.e., the event "CAPR (Z, $t_0|t_1$) > m-δ" has probability 1 and the event "CAPR (Z, $t_0|t_1$) > h + ϵ" has probability 0 for any positive values of δ and ϵ and for any $t_1 = 0, 1, 2, ..., t_0$. If n aircraft ($n > m$) are scheduled to arrive at Z at $t = t_0$ and under a scenario analogous to the one presented in our example (identical aircraft characteristics for all flights with $c_a > c_g$) there exists at least one optimal flow management strategy under which the m most "remote" (longest flight time) flights would never be subjected to any ground or airborne holding and thus all the delay costs are borne by the n-m "shortest-range" flights. This makes sense intuitively: as $t = t_0$ approaches, we allow the m earliest departing flights to leave their airports of origin as scheduled, since we are sure that we shall have enough capacity to accommodate these flights without any delays; then we exercise flow management on the n-m shortest-range flights, adjusting the flow of aircraft to best match the predictions of available capacity at

Z, also taking advantage of the fact that our "forecast" of the value of CAPR will presumably become more accurate as $t = t_0$ gets closer.*

In practice, such an approach would imply that long-distance flights are given preferential treatment. In fact, this is already partly true under today's Central Flow Control system in the United States: flights longer than 4 hours are almost never subjected to ground holding prior to departure. This means that non-stop flights from the West Coast or from Western Europe to the congested Eastern seaboard airports of the United States are exempt of many of the delays that short-and medium-distance flights within that part of the country often sustain. This can even result in "systematic" biases in favor of airlines that fly long-distance East-West routes and against those concentrating on short-distance North-South markets.

In more complicated scenarios where all aircraft do not have identical costs, the biases generated by strategies that seek to minimize total delay costs are not quite as straightforward as a simple short- vs. long-range distinction. However, as pointed out by the recent work of Andreatta and Romanin-Jacur (1987) - see also Section 6 below- a tendency to systematic biases (e.g., favoring large aircraft with large unit-time costs over small ones) is an almost inherent characteristic of strategies concerned with "optimizing" system performance with respect to a single aggregate criterion. While one should not be surprised to discover that a simple and "natural" optimization criterion often implies such systematic biases the resulting flow management strategies may be difficult or impossible to implement in an environment in which the principles of "equal access" to airports and of non-discrimination among users are universally accepted ones.**

---

*Note, however, that the result we have quoted does not require that the forecast improve and, in fact, holds for any evolution over time of the "forecast" i.e., the "state of information" regarding CAPR.

** On the other hand, one may note that, as airports have become more congested, they have also become more specialized, e.g., through virtual <u>de facto</u> exclusion of small general aviation aircraft from major commercial airports. Deviations from the long-established tradition of "first-come, first-served" sequencing are also becoming increasingly common in today's ATC systems especially in terminal areas and airports.

Thus, natural objective functions, such as minimizing the total cost of delays, would have to be modified in a practicable system by introducing criteria or constraints of a "distributive" (or "disaggregate") nature, such as ensuring, at least partly, "equitability" in sharing delay costs among individual aircraft and airlines.

5. OTHER COMPLICATIONS

Attempts to solve more realistic or more microscopic-level versions of the FMP on a network-wide basis must contend with a number of additional complications. For example - and most obviously - considering waypoints, sectors and departure runways (at the airports where flights originate) as capacitated elements increases greatly the complexity of FMPs. So does, to an even greater extent, the inclusion of alternative flow management tools such as metering, re-routing, speed control, etc. (Section 3) in addition to ground-holding with which the generic FMP is solely concerned.

When a truly high level of modeling accuracy is desirable, it should be recognized that, rather than being simply "sources" and "sinks" of aircraft flows, airports actually impose a set of rather unusual conservation-of-flow relationships over a period of time of any significant length (more than 2 or 3 hours). This is because individual flights are usually assigned to specific aircraft. Thus, a particular departure scheduled for mid-day at an airport X cannot take place unless the aircraft that will perform that flight (1) has already arrived at X (probably performing a flight with a different flight number) and (2) a time interval sufficient to "turn around" that aircraft has elapsed since its arrival at X. This means that arrivals and departures at any given airport are strongly interdependent - except possibly for the beginning of a day when a pool of "overnighting" aircraft is available at those airports where early-day flights originate.

An even more peculiar interaction between arrivals and departures often takes place with respect to determining airport capacity. At many airports arrivals and departures share the same runway (or runways); or, when arrivals and departures are conducted on different runways, these runways may intersect or be "close parallel" so that the rate of operations on the arrivals runway(s) may affect the rate of operations on the departures runway(s) and vice versa. This means that flow management strategies may themselves affect airport capacities. For example, if for some reason it is decided to impose a "ground-hold" on all departures from a particular airport during some time interval, this may enable that airport to increase its acceptance rate ("capacity") for arrivals during that interval. Similarly, a large flow of arrivals into an airport over some time interval, may restrict the ability of the flow management system to dispatch the preferred number of departures from that airport during that same interval. Thus, airport capacities, which we have treated so far in this paper as exogenous random variables may, in fact, sometimes constitute endogenous variables, i.e., may be determined in part by the flow management strategy adopted. Obviously, it is very difficult to handle analytically this complication.

A number of more narrow technical points also give rise to some additional difficulties. We give two examples: First, there are clearly bounds due to fuel capacity on how long an airplane can be made to wait while airborne. Second, at a more microscopic level of detail, the capacity of an airport is also a function of the sequencing of operations on the runways. For instance, if a wide-body jet is landing behind a small propeller aircraft in IMC weather, the minimum required separation in the United States between the two on final approach is 3 nautical miles, resulting in intervals of the order of 75-90 seconds between the successive touchdowns of the two aircraft on the runway. By contrast, in the reverse case when a small aircraft is landing behind a wide-body jet, the minimum required separation on final approach is 6 nautical miles at a point several miles from the runway and the corresponding interarrival interval can be 4 minutes or longer. In general, this means that capacity is affected by both the mix of aircraft that the flow management system directs into or out of any airport over (relatively short) periods of time and by the sequencing of these aircraft during such periods.

6. BRIEF SURVEY OF THE EXISTING LITERATURE

As noted at the outset of this paper, the existing literature on the ATC flow management problem is limited indeed.

In terms of background, it should be mentioned first that a Central Flow Control Facility (CFCF) has been operated for several years by the FAA in Washington, D.C. and is charged with carrying out strategically-oriented flow management functions (ground holds, some metering and some re-routing) for the ATC system in the United States.* CFCF is equipped with outstanding and ever-improving

---

*No analogous facility is currently in operation anywhere else in the world

information-gathering capabilities, including regional and local weather data and forecasts, up-to-the-minute data on the status of airborne traffic throughout the country and projections of traffic levels over a time-horizon of several hours. However, CFCF currently relies almost exclusively on the judgement of its expert air traffic controllers for decisions regarding flow management. Much of the ongoing research on the FMP in the United States is aimed toward developing algorithms and automation aids for these controllers so that they can manage more effectively the flow of air traffic.

Recognizing the complexity and practical importance of the FMP, a number of organizations and researchers have recently initiated the development of computer-based simulation "test beds" whose purpose would be to provide an environment for experimenting and evaluating alternative FMP strategies and algorithms. Simulations of this type are described by Butler (1987) and, in this volume, by Hormann (1987). The FAA is also in the process of preparing for flow management experimentation purposes, the Total Traffic Management (TTM) model, a highly detailed representation of the United States ATC system.

In the area of analytical studies, a paper by Attwooll (1977) represents one of the first efforts ever to examine ATC congestion from a network-wide perspective and offers several valuable insights. A number of papers from the Italian National Research Council(IASI-CNR) have also addressed strategic aspects of FMPs (see, e.g., Bianco et al. (1981)). More recently some progress has been made toward developing algorithmic approaches for the FMP. Sokkapia (1985) has proposed a very simple heuristic for dealing with the generic FMP. In the notation of Section 3, he suggests that at the scheduled departure time s_i of flight F_i, an estimate be made of w_i, the delay to be suffered by F_i upon reaching its destination at $t_i = s_i + d(j, Z)$, based on the probability distribution of CAPR $(Z, t_i|s_i)$ and a projection of traffic demand at Z. He then suggests "discounting" w_i by a certain factor, based on the identity of the airport of origin, j, and the flight time $d(j, Z)$. Call this discounted delay estimate $w_i'(< w_i)$. The "ground hold" time assigned to F_i should then be set equal to w_i' and F_i will leave its parking area at $s_i + w_i'$. A rule for computing the discount factor to be applied in each case has not been proposed and no indication is given that the algorithm has been implemented in any way, even in a simulation environment.

Andreatta and Romanin-Jacur (1986) are, to our knowledge, the only ones to have investigated in depth a simplified version of the generic FMP of Section 3. They have studied a one-time-unit problem with n flights to a single destination similar to our example of Section 4 (i.e., CAPR is finite and can possibly cause delay problems during only a single time period t_0 of unit length and CAPR is infinite for all other time periods before and after t_0). Another important assumption is that the probability distribution of CAPR $(Z, t_0 | t_1)$ does not change with t_1, i.e., remains the same for all $t_1 \leqq t_0$. For arbitrary ground holding and airborne unit costs, $c_g(i)$ and $c_a(i)$, for each flight, they then develop a $0(n^2)$ dynamic programming algorithm for deriving an optimal ground-holding strategy R (note that R is independent of t in this case and that the choice for each flight F_i is to either depart as scheduled or to delay departure for one time unit). This work seems to be the most advanced to date on the FMP.

Finally, in connection with the last complication mentioned in Section 5, the recent work of Trivizas (1987) on the sequencing of aircraft landings and take-offs in ATC terminal areas, extended to the case of multiple-runway airports the earlier work of Dear (1978) and Psaraftis (1980) and developed highly efficient sequencing algorithms which are particularly well-suited for parallel processing. Sequencing algorithms of this type will undoubtedly constitute an integral part of sophisticated flow management systems of the future, as they help maximize runway acceptance rates for any given traffic mix and combination of landing and takeoff demand levels.

7. CONCLUSIONS

We began by discussing the FMP in general qualitative terms and continued with a more specific description of the generic FMP, a version whose purpose is to explore the major "strategic" question faced by an ATC flow management system, namely the tradeoff between ground-holding and airborne delays. It is clear that the FMP should be solved on network models of the ATC system which, in order to capture essential aspects of the ATC environment, must have the following characteristics:

i)   be stochastic;

ii)   be dynamic;

iii)   represent each aircraft as a discrete entity (no continuous flows);

iv)   consider distributive effects of any flow management strategies on individual aircraft (in addition to aggregate effects);

(v)   account for a number of unusual conservation-of-flow relationships and demand/capacity interactions at airports.

Research on examining network-wide ATC problems such as the FMP is only now beginning, with much ongoing effort directed toward development of simulation test beds for evaluating FMP algorithms and only limited works so far, on "solving" the problem itself. Flow management problems in ATC are challenging, open and of major practical importance.

REFERENCES

Andreatta, G., and G. Romanin-Jacur (1987) The Flow Management Problem. Transportation Science 21, to appear.

Attwooll. V.W. (1977) Some mathematical aspects of air traffic systems. Journal of Institute of Navigation 30: 394-411.

Bianco, L. et al. (1981) New approaches to air traffic control management. Transport Research and Social and Economic Progress, London, S. Yerrel, Vol. 4: 2373-2384.

Butler, J.F. (1987) An air traffic control simulator for the evaluation of flow management strategies. Flight Transportation Laboratory, Report 87-5, Massachusetts Institute of Technology, Cambridge, MA.

Cohen, Dayl and A.R. Odoni (1985) A survey of approaches to the airport slot allocation problem. Flight Transportation Laboratory, Report 85-3, Massachusetts Institute of Technology, Cambridge, MA.

Dear, Roger G. (1976) The dynamic scheduling of aircraft in the near terminal area. Flight Transportation Laboratory, Report 76-9, Massachusetts Institute of Technology, Cambridge, MA.

Federal Aviation Administration (1986) The national aviation system plan. U.S. Dept. of Transportation, Washington, D.C.

Hormann, A. (1987) ATSAM (air traffic simulation analysis model). This Volume.

Psaraftis, H.N. (1980) A dynamic programming approach for sequencing groups of identical jobs. Operations Research 28: 1347-1359.

Simpson, R.W., A.R. Odoni and F. Salas-Roche (1986) Potential impacts of advanced technologies on the ATC capacity of high-density terminal areas. Flight Transportation Laboratory, Report 86-10, Massachusetts Institute of Technology, Cambridge, MA.

Sokkappa, B.G. (1985) Arrival flow management as a feedback control system, The MITRE Corporation, Washington, D.C.

Trivizas, D.A. (1987) Parallel parametric combinatorial search: its application to runway scheduling. Flight Transportation Laboratory, Report 87-4, Massachusetts Institute of Technology, Cambridge, MA.

ON-LINE MANAGEMENT AND CONTROL OF AIR TRAFFIC

by

André Benoît*

European Organisation for the Safety of Air Navigation
EUROCONTROL
Engineering Directorate
72, rue de la Loi, B-1040 Bruxelles

Foreword

This article provides a summary of the lecture delivered at the Advanced Workshop on "Flow Control of Congested Networks: The Case of Data Processing and Transportation", organised by the Consiglio Nazionale della Ricerche (Research Project on Transportation) in cooperation with the Scientific Affairs Division of the North Atlantic Treaty Organization (Capri, Italy, 12-17 October 1986).

Introduction

The continuous increase in air traffic density and complexity compels air traffic handling authorities to anticipate solutions which meet safety requirements and are efficient in terms of economy, capacity and the comfort of crews and passengers. In Western Europe, an appreciable proportion of flights have a duration of less than one hour. Accordingly, efficiency must be achieved through optimisation of the whole flight taken in relation to overall traffic, and not as a succession of quasi-optimum phases based on local conditions. The paper will discuss this idea further and outline some of the development work undertaken or needed towards this aim.

---

* Also Chargé de Cours,
    Faculté de Sciences Appliquées
    Université Catholique de Louvain
    B-1348 Louvain-la-Neuve

NATO ASI Series, Vol. F38
Flow Control of Congested Networks
Edited by A. R. Odoni et al.
© Springer-Verlag Berlin Heidelberg 1987

AIR TRAFFIC

Air traffic perceived as a collection of aircraft, whether carrying passengers, freight or mail, or accomplishing specific civil or military missions, exhibits a number of facets which might be divided into two main categories related to its <u>constitution</u> and its <u>handling</u>.

The two categories are not independent; for example the constitution of a given fleet clearly influences the means required to handle it.

The elements determining the traffic constitution include the distribution, nature and densities of connections and the fleet composition, namely types and quantities of aircraft, avionics included.

The national and international civil and military authorities will define and provide the means (facilities, techniques, procedures, man-power) to handle the actual traffic in a safe, smooth and efficient manner.

AIR TRAFFIC GROWTH (Overall ICAO traffic)

The overall growth in air traffic over the past 10 years is illustrated in Figure 1 on the basis of information provided by ICAO (Ref. 1).

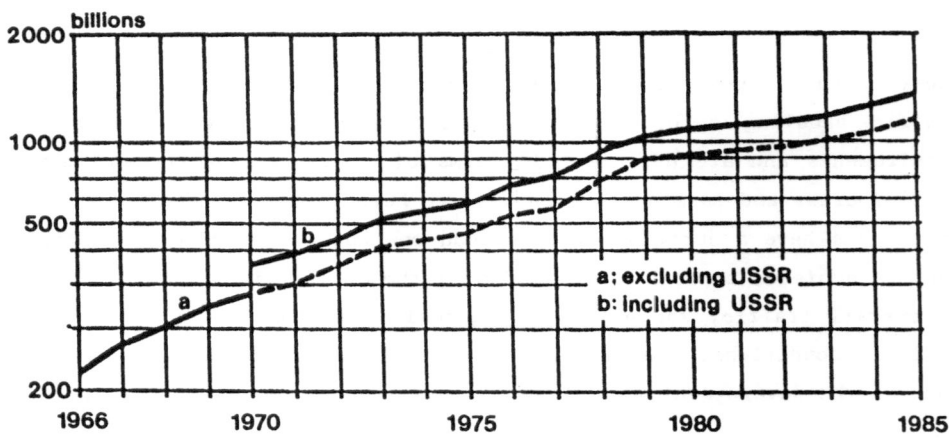

Passenger-kilometres for scheduled services

(Source: Ref. 1)

Figure 1

The average yearly rates of increase over this 10-year period are as follows:

(a) <u>passengers</u> :   - passengers carried   5%

                     - passenger-km   7%

(b) <u>freight</u>   :   - tonne-km   7%

(c) <u>mail</u>   :   - tonne-km   4%

In total, the yearly rate of air transport growth expressed in tonne-km and averaged over the past decade is of the order of 7%.

Although reductions in the 1985 yearly rates of increase have been observed for both freight and mail, passenger traffic (passenger-km) continues to grow at a yearly rate of the order of 7%, while the total rate of growth (tonne-km) remains slightly greater than 5%.

It has been argued that the increase in freight and particularly passenger traffic might be compensated in terms of air movements by the use of a larger number of wide-bodied aircraft. This certainly no longer applies today, and present trends even suggest the opposite. Indeed, the opposite is likely to result from the expected increased freedom in air transportation, also known in some circles as "deregulation", leading to a larger number of connections conducted with smaller aircraft (higher rate of connection on existing routes and additional connections on newly created routes).

Accordingly, it is to be expected that the density and complexity of air traffic will increase appreciably over the next 10 years: the handling of the overall movements will need to be performed with a high degree of efficiency if the cost and comfort of air transportation are to be maintained at reasonable levels. This becomes even more evident when viewed in terms of the impact of traffic growth on areas already suffering a high degree of congestion, as illustrated hereafter in the case of typical European medium-to-high traffic density airports.

TRAFFIC DENSITY AT EUROPEAN AIRPORTS

Figure 2 gives an illustration of the traffic density at three typical
European airports (Brussels, Belgium; Frankfurt, FRG; London, U.K.).

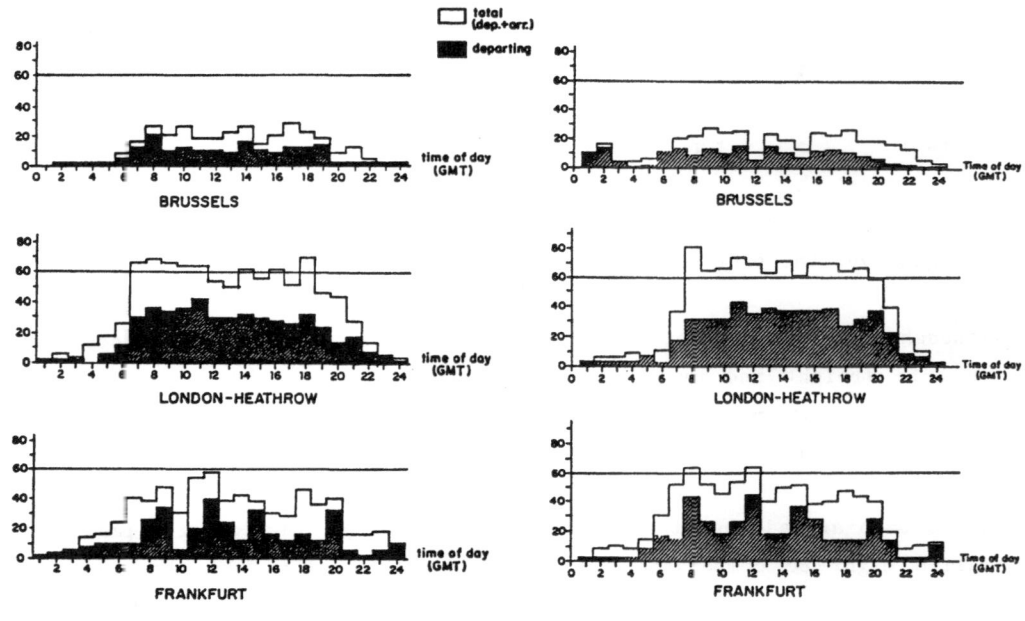

a) Summer 1974   b) September 1985

Traffic density as observed at European airports

Figure 2

(Source: EUROCONTROL Statistics Service)

At high traffic density airports, of which London, U.K., is the most
remarkable example in Europe, saturation is reached during practically the
whole period of operation - and not only during two or three main morning
and evening peaks -, which leads to appreciable delays currently being
absorbed at low altitude in the four arrival stacks and/or by low-level
path-stretching implying vectoring of the aircraft. The average delay per
aircraft amounts to more than 5 minutes (Ref. 2), the density of traffic
being of the order of 95% of capacity. For the present handling of air
traffic at that level of intensity, even a slight traffic increment leads
to appreciable increases in delays. It is accordingly essential to improve
the efficiency of traffic handling especially at these levels. It has been

shown that for a terminal area such as London, appropriate integration of en-route and approach control phases could lead, within the present departure/landing capacity, to a saving in excess of 25,000 tonnes of fuel per year for inbound traffic alone over an area extending some 150 nm around the airports (Heathrow and Gatwick) while at the same time reducing the discomfort experienced by crews and passengers during delays spent in orbit on arrival.

Clearly, for medium traffic density airports, of which Brussels is a typical example, the saving resulting from similar integration of control phases would be appreciably less; nevertheless, the potential benefit has been estimated and found to be of the order of 4 to 7% of the corresponding flight costs (Ref. 3). In addition, the traffic density at such airports is likely to increase appreciably in the near future either as a result of congestion suffered elsewhere or simply because, the environment being less critical, some of them will remain open during the night while heavy density airports close.

In this respect, over the past few years the traffic at Brussels National Airport has grown from some 300 up to over 500 daily movements. Although most of the increase takes place in the night period, this traffic growth nevertheless leads to peaks (from 9 to 10 a.m.; from 3 to 4 p.m.) of the order of 40 to 50 movements per hour, which already contrasts with Figure 2.

In conclusion, the increase in traffic density and complexity makes it necessary to anticipate and implement means to handle the traffic efficiently in terms of safety, capacity, cost and comfort. This paper places the emphasis on one particular element of the overall hierarchical management/control loop, namely the organisation of the traffic to be made on-line over an extended area surrounding and including a main terminal.

HANDLING OF AIR TRAFFIC

Management / control

The handling of air traffic includes two essential aspects currently called "management" and "control". In a previous lecture (Ref. 4), we have summarised the essential characteristic steps in the historical development of the "management" terminology. In short, management deals with collections of aircraft, while control applies to individual flights. Also, in terms of timescale, control constitutes the on-line component of air traffic handling while management operates "upstream", as far as 20 years ahead.

Clearly, whatever their subsequent impact on flight efficiency (safety, expedition, consumption, economy, capacity), measures undertaken several hours to several years before the flights concerned enter the area of control have no direct bearing on the on-line definition of the related trajectories. With respect to an area of control, this is the case for all measures undertaken by a management unit organising or limiting the traffic prior to its entry into the area concerned, the last of these units, chronologically, being the flow management cell upon which the area of control depends. By contrast, measures taken as a result of modifications of the overall traffic existing in a given area generate control directives to be implemented immediately and/or shortly thereafter. This management component, which operates on-line together with the actual control which results, will be discussed further later in this paper.

Timescale in air traffic handling

In order to situate and illustrate the relative positions of control and management components in the overall set of ATM/ATC actions, use is made of Ratcliffe's ATC scale (Ref. 5), duly adapted for the present purpose (Fig. 3). In this diagram, the main central line schematises the succession of steps taken for the handling of traffic and subsequent guidance of each individual flight.

Air traffic management

A number of management actions have been taken well before any aircraft enters the space/time area of control. These actions include the planning and implementation of new main facilities, the introduction of conceptual approaches reflecting advances in relevant technologies, the reorganisation of the airspace resulting from revisions of national and/or international

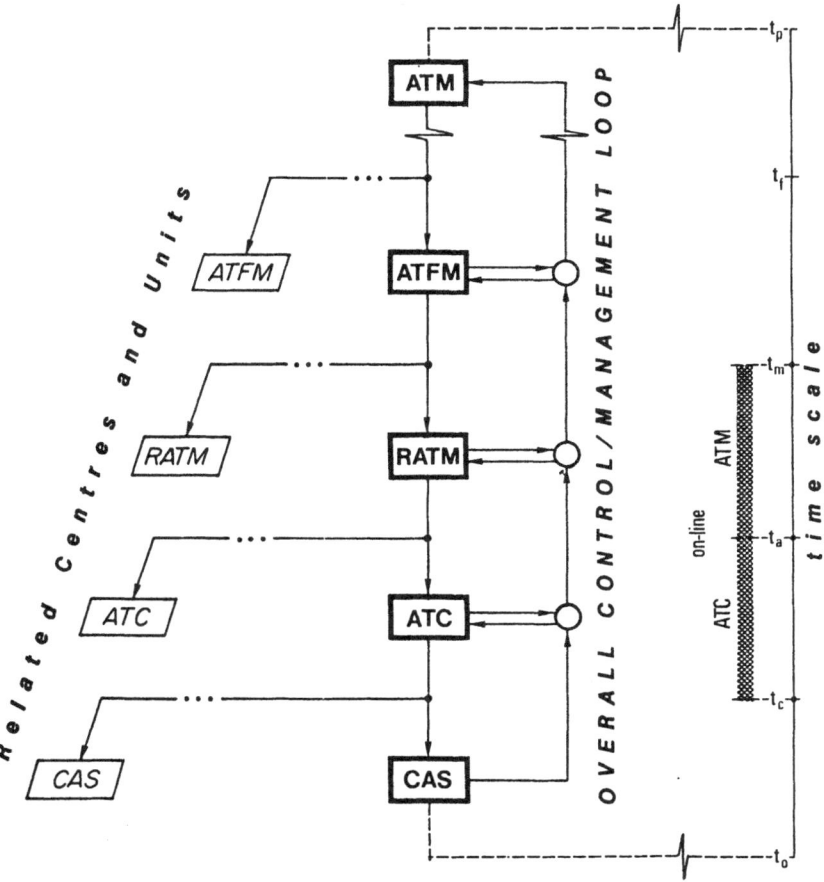

Legend

ATM : Air Traffic Management

ATFM : Air Traffic Flow Management

RATM : Regional Air Traffic Management (ex. ROSAS)

ATC : Air Traffic Control

CAS : Collision Avoidance

Schematic timescale in air traffic handling
(adapted from Ratcliffe's approach, Ref. 5)

Figure 3

policies and all measures taken well in advance in order to render demand compatible with potential or available services.

All these measures are included under Air Traffic Management (ATM) and the related look-ahead timescale extends from 20 years or so down to a few weeks. Examples of such actions include the planning and construction of additional airports, new routes and possibilities for area navigation, replacement or upgrading of navigation aids (such as the use of satellites (Refs. 6 and 7), introduction of a microwave landing system (Ref. 9), upgrading of an instrument landing system from one category to the next wherever applicable), considerations and development of advanced air traffic handling concepts and their progressive introduction into operation (Ref. 9), specification and design of new control centres (Ref. 10) and some of those long-term actions aimed at rendering demand compatible with available services.

Air traffic flow management

In the air traffic handling chronological and hierarchical control loop, the next series of measures includes the air traffic flow management (ATFM) actions. The relevant look-ahead time extends from a few months down to a few hours. Essentially, air traffic flow management aims at reconciling demand and capacity. It ranges from demand-scheduling management (3 to 8 months) to short-notice constraints (2 to 24 hours). The departure scheduling at main airports - with a look-ahead time of the order of 6 months - falls within this first category of ATFM actions. It was introduced in Europe, initially at London airport, U.K., as a result of V. Attwooll's work (Ref. 11).

On-line air traffic management

This particular component of air traffic handling deals with collections of aircraft in a way appreciably different from Air Traffic Flow Management. When aircraft enter the relevant handling area, an assessment of the overall situation is made (aircraft requests, traffic situation, capacity usage, overall and specific constraints) on the basis of which guidance directives are generated and issued directly or subsequently to these particular aircraft; guidance corrections for some of those aircraft already in the area may result.

The relevant timescale (t_m) depends on the extent of the handling area. In Western Europe, an adequate estimate would range from approximately 30 to 40 minutes for possible implementation in the not-too-distant future to approximately 1 to 2 hours for use in a truly integrated European Community. These considerations will be elaborated upon further in subsequent sections.

Air traffic control

The Air Traffic Control elements will implement the guidance directives generated by the RATM unit. Depending on the organisation of the airspace "jurisdiction", the related timescale (t_a) may be of the order of t_m or appreciably reduced as is the present practice.

Collision avoidance

There is no need to dwell on this particular safety component: it should come into action only if all the other separation measures have failed, that is to say when a pair of aircraft on a collision course have slipped through the meshes of a 2-to-5-minute safety net and, as a consequence, a possible disaster is imminent. Depending on the system and/or the geographical area, the avoidance directives could be generated either on the ground and automatically relayed to the aircraft, or directly in the cockpit. A typical timescale might vary from 60 to 20 seconds.

INTEGRATION OF CONTROL PHASES

Earlier developments

For over 15 years now, we have been advocating accurate prediction of aircraft trajectories over extended areas in order to provide optimum or quasi-optimum flights for the airlines and other operators (Refs. 12 to 17).

In 1975, Erwin (Ref. 18) proposed a detailed concept, in which "the Air Traffic Control systems defines four-dimensional tracks for all arrivals that will derandomize and space traffic for landing on the runway". Among other aspects, the author rightly places emphasis on the controllability of flight time in terms of total path length and speed versus altitude aero-performance limits. It is worth noting that even at that time Erwin esti-mated the range of time-of-transit controllability on the basis of total flight distance to touch-down to be of the order of 200 nm. This clearly suggests a real integration of the control phases, covering landing, approach, descent and cruise, or part thereof.

Most of the efforts initiated in the same period, or slightly later, aimed at reducing the stacking penalties make either explicitly or merely impli-citly the same basic assumption: integration of control phases over an extended area corresponding to flight lengths, from touch-down, of the order of 150 to 300 nm. Previously, an approach such as Erwin's was often qualified as "strategic" (Refs. 20 to 22). Today, for the same look-ahead time, the term on-line regional management is preferred by the author, implying without ambiguity that both prediction and subsequent trajectory control are covered in a tight overall control loop, as described below.

European scene

In Europe, Brussels (Belgium) is less than one hour's flying time from the main airports in the adjacent countries. For the purpose of investigating the impact of excess route lengths on fuel consumption in Western Europe, an international network was considered made up of those routes connecting the capitals (or cities where the main airport is located) of the eleven States that participate in the EUROCONTROL Route Charges System. In such a European network, the length of a flight averaged over the yearly traffic is found to be of the order of 300 nm (Ref. 23).

This average length would be further reduced if the complete network, i.e. including domestic traffic, were taken. Accordingly, "given the geogra-phical and time scales involved, it should be possible to organise flights in Europe, including the departure and arrival sequences, in such a way that any aircraft cleared to depart would land at its destination after a flight conducted in accordance with airline policy, without alternations resulting from short-term planning and subsequent directives that are disruptive to the air traffic" (Ref. 24). This prompts recommendation of an approach consisting in central on-line management and regional control of air traffic, as described hereafter.

On-line handling of air traffic

The future handling of air traffic in an extended continental area such as Western Europe (C in Figure 4) can be schematised as follows. The airspace external to this extended area is labelled E.

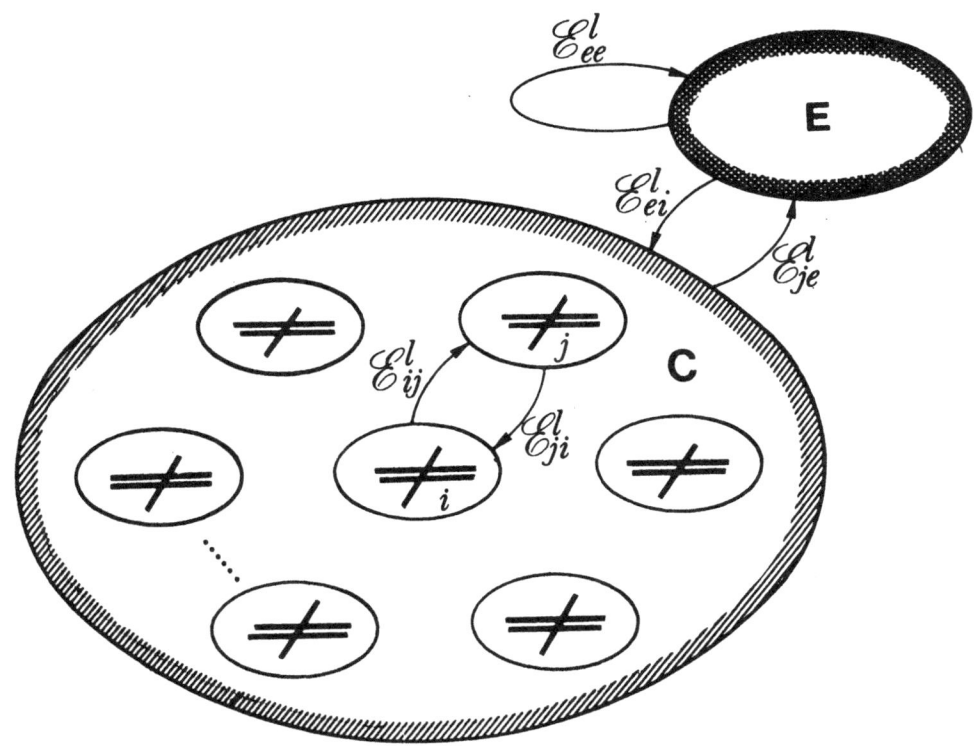

Flows of traffic in an extended management/control area

Figure 4

The overall area (C) includes a set of regional sub-areas (such as sub-areas i or j in Figure 4). Each sub-area includes a main terminal and possibly additional secondary airports. Four main categories of traffic can be identified according to origin and destination, viz:

- origin and destination inside one and the same sub-area (regional traffic);
- origin and destination located in two different sub-areas (Western European traffic);
- origin only or destination only inside overall area C (traffic originating or terminating outside the area);
- origin and destination both located outside overall area C (overflying traffic).

In order to handle the traffic efficiently, a centralised on-line overall management unit will require to be set up. For each aircraft entering the system, i.e. entering the extended area C either from inside (departure from one of the sub-areas) or from outside (either with a destination inside C or not), this unit generates the relevant 4-D trajectory and possibly amends the trajectories of some of the aircraft already in the area, so as to minimise the global traffic cost as incurred by the operators.

The main control variables include the take-off and landing sequences over the area, with the associated transit times, these two sets of variables being subject to quite different sets of constraints. The related trajectory control directives are dispatched to the regional control units for implementation.

From a scientific viewpoint, methods are available to optimise that complexity and, computerwise, these can be conducted on-line - compatibility with real-time operation appearing feasible. Preliminary investigations have indeed been made in this direction.

An analysis of the overall control problem is presented in Reference 25 which also contains an extensive - although not exhaustive - bibliography on the subject.

Technically, the global optimisation approach is certainly feasible, but a number of areas will need basic research and fundamental development. This applies in particular to the ground movements, the 4-D update of meteorological information, and the generation of fully reliable directives for accurate guidance of aircraft of all categories.

The advent of Mode-S in the field of surveillance will very likely be sufficient to complete the technological support required, data link capacity included.

The essential man/system interface problem will probably be much easier to solve than is expected today. The new generation of traffic handlers, either supervising a central unit or implementing the guidance directives regionally, will dialogue with the overall traffic/aircraft guidance optimiser as he would with a friendly adviser, in simple terms and in full confidence. As a consequence, the operational conduct of traffic should take place smoothly.

Greater difficulties would certainly appear at the organisational level in a number of areas; these include the definition, collection, transfer and on-line processing of information, the standardisation of protocols for automatic exchange (ground/ground; ground/air) of information to be used on-line, structuring the jurisdiction of the corresponding reorganised airspace, etc.

At some stage of European development, it will be possible to effect such an integration of air traffic handling - with on-line management and control, departure and arrival sequences as essential parts of the control variables - and it is now opportune to initiate the basic developments required to meet such a stimulating challenge.

THE ZONE OF CONVERGENCE (ZOC)

Geographical meaning

The Zone of Convergence (ZOC) concept originates from the same basic idea as that which guided Erwin in his original work (Ref. 18). It was proposed as an efficient short-term measure to bring about a significant improvement in the handling of air traffic in Western Europe.

The geographical area covered by ZOC includes and surrounds a main terminal and, possibly, secondary airports, extending some 100 to 300 nm from the runways. The term Zone of Convergence was coined in the initial development (A. Benoît and A. Fossard) to avoid any ambiguity with current practice, and in particular with the structure of the en-route/approach control in and around a terminal. Examples of actual Zones of Convergence have been given for both Belgium (main terminal: Brussels) and the United Kingdom (main terminal: London) in previous publications on ZOC, inter alia, in References 26 and 27. These two areas were selected to reflect both medium and high traffic densities in Europe.

Conceptual outline

This approach can be considered as a first step towards the integration of all control phases in a large region of Western Europe. Over the ZOC-area, i.e. the extended area surrounding and including the main terminal concerned, the on-line handling of traffic will exhibit two fundamental, closely interrelated components, namely a traffic management and an aircraft guidance module. These components are outlined in the following paragraphs.

(a) A traffic management module comes into operation every time an aircraft enters the ZOC airspace from either outside or inside. This optimiser determines the best 4-D trajectory for the new aircraft and possibly modifies the trajectories of some of the aircraft already in the ZOC airspace so as to minimise the overall deviation from the operator's trajectory requirements. This component has been described to some extent in previous papers, in particular in Reference 28.

(b) An accurate guidance technique (operational procedure and reliable algorithm) is used to conduct each flight through the area, whatever the aircraft navigation equipment. This component will be described further in the next section.

Expected benefits

Although the initial optimiser was essentially designed to minimise the overall flight cost (consumption and flight time combined), it was soon confirmed that appreciable improvements ensued in terms of expeditiousness, capacity (maximum use of available landing capacity), and safety (References 22, 28 and 29).

ZOC control variables

For each flight, the ZOC approach integrates the control over all phases of the entire part of the trajectory flown in the area. The control variables include the landing sequence with adjustments of the departure sequence, and the set of corresponding times of transit through the Zone. Maximum use is made of the aeroperformance speed control range, completed whenever necessary by modifications to the geographical track, in particular for the definition of base and final turn prior to ILS interception. The 4-D guidance procedure developed and assessed in a realistic environment will now be summarised.

GROUND-BASED 4-D GUIDANCE OF AIRCRAFT

4-D guidance of individual aircraft: a prerequisite

The quality of the ZOC optimiser will depend greatly on the possibility of guiding aircraft accurately through the area (in the case of an inbound flight for instance, from entry to touch-down) in a practical and reliable manner, so as to maintain the actual landing time within seconds of the relevant value determined at entry, i.e. 15 to 60 minutes beforehand.

Basic requirements

The calculation, prediction, guidance and control of flights should be made on-line. The techniques selected should use aircraft performance information readily available from manufacturers and/or operators. The guidance accuracy should be such as to deliver the aircraft at critical points within seconds of the initial prediction. Further, in operational use, horizontal and vertical profile alterations must be allowed for. The guidance directives should be clear, unambiguous and readily relayed by the "air traffic controller" to the aircraft. In addition, they should be compatible with present day R/T communications but adaptable to future data link operation.

Development, assessment and results

In accordance with these basic requirements, techniques and procedures for the guidance and control of aircraft have been developed and assessed in real-time operation (Ref. 24). The accuracy relies on the availability of suitably located DME stations.

The computer/controller/pilot dialogue is described in Reference 30. The guidance directives are generated automatically by the computer and displayed on the radar screen for the controller's use.

The remarks concluding the validation experiments (Ref. 24) have been confirmed by recent additional tests: it is expected that the prediction/guidance/control tool under development will provide ATC with the means to control aircraft time-of-arrival to within 10 seconds of the schedule established on entry into the zone, in a way consistent with present-day operation and valid for all aircraft equipped with distance measuring equipment.

Those aircraft equipped with Flight Management Computer System configured for 4-D navigation could possibly receive the messages in a form duly adapted for onboard computer input.

Provision is made for adaptation to future air/ground data link operation: preliminary tests in this direction are projected in 1988.

CONCLUSIONS

Notwithstanding the benefits which would result from the integration of the air traffic handling phases on a Western European scale, a number of reasons still prevent the on-line optimisation of the traffic, including departure and arrival sequence adjustment, on a global basis.

Nevertheless, at this stage, it is appropriate to develop, assess and implement concepts intended as intermediate steps towards this ultimate aim. Optimal control in a Zone of Convergence is one example of this: it integrates the control of all phases of flight over an extended area including and surrounding a main terminal, the "radius" of the relevant geographical area being of the order of 100 to 300 nm or more, that is to say adaptable to the scale of a European country. The results include substantial benefits in terms of economy (fuel consumption and flight time combined), expedition, and the comfort of crews and passengers, elimination or reduction of delays on arrival, enhancement of safety (reduction of potential conflicts) and maximum use of available handling capacity.

This component of traffic handling, which operates on-line and integrates traffic management and individual flight control over an extended area, could certainly become one of the most valuable contributory factors to the quality of air traffic services in the future. Futhermore, the tools developed to this end, in particular the 4-D control of trajectories program, are directly suited to subsequent developments.

Disclaimer

The views expressed in this paper are those of the author; they do not necessarily reflect the policy of the Agency.

REFERENCES

1. ICAO (1986) "Scheduled airline passenger-traffic growth increase in 1985", ICAO Bulletin, June 1986.
2. Attwooll, V. and Benoît, A. (1984) "Fuel economies effected by the use of FMS in an Advanced TMA". The Journal of Navigation, Vol. 38, No 1, January 1985; Also EUROCONTROL Report 842003, February 1984.
3. Benoît, A. and Swierstra, S. (1983) "Potential fuel consumption savings in medium to high density extended terminal areas". EUROCONTROL Report 832004, March 1983.
4. Benoît, A. (1986) "Air Traffic Management: Development of techniques and procedures for the control of individual flights and air traffic in an advanced TMA". Lecture delivered at the seminar on "Air Navigation Aids", Associazione Elettrotecnica e Elettronica Italiana, Milan, Italy, 27 February 1986.
5. Ratcliffe, S. and Gent, H. (1974) "Quantitative description of a Traffic Control Process". The Journal of Navigation, Vol. 27, No 3, 1974.
6. Carel, O. (1986) "Le choix des futurs systèmes de la navigation aérienne civile". AGARD CP-410, pp. 11-1 to 11-7, December 1986.
7. Poritzky, S.B. (1986) "FANS - A U.S. perspective". AGARD CP-410, pp. 12-1 to 12-7, December 1986.
8. Seifert, R. (1986) "MLS: Its technical features and operational capabilities". AGARD CP-410, pp. 42-1 to 42-10, December 1986.
9. Vachiery, V. (1986) "La navigation aérienne et l'avion à l'horizon 2000". AGARD CP-410, pp. 13F-1 to 13F-8, December 1986.
10. Vandenbroucke, A. (1986) "Belgium moves ahead in air traffic control". AGARD CP-410, pp. K-1 to K-10, December 1986.
11. Attwooll, V. (1975) "The optimization of traffic flow around a network". AGARD CP-188, pp. 15-1 to 15-6, February 1976.
12. Benoît, A. et al. (1970) "Aircraft Trajectories. An approach to the Calculation of Aircraft Trajectories for Possible Application in Air Traffic Control". Half-yearly Information Review of the European Organisation for the Safety of Air Navigation, EUROCONTROL Vol. II, No. 2, pp. 11-17, June 1970.
13. Benoît, A. (1971) "Applicability of the EROCOA Trajectory Prediction Module to Actual Scheduled Flights". EUROCONTROL Report 722016, September 1971.
14. Benoît, A. and Martin, R.H.G. (1972) "On the Generation of Accurate Trajectory Predictions for Air Traffic Control Purposes". Paper presented at the IATA 19th Technical Conference "Handling the Air Traffic of the Long Term Future", Dublin, October 1972. Also EUROCONTROL Report 722032, August 1972.
15. Benoît, A., Swierstra, S. and Storey, J. (1975) "The Introduction of Accurate Aircraft Trajectory Predictions in Air Traffic Control". Paper presented at the 20th International Symposium of the Guidance and Control Panel of AGARD, Cambridge, Mass., U.S.A., May 1975. AGARD CP-188, pp. 16-1 to 16-28, February 1976. Also EUROCONTROL Report 752011, April 1975.
16. Benoît, A. et al. (1977) "An evolutionary application of advanced flight path prediction capability to the control of air traffic". Paper presented at the International Conference on Electronic Systems and Navigation Aids, Paris, November 14-18 1977. Also EUROCONTROL Report 772016, August 1977.

17. Benoît, A. and Swierstra, S. (1980) "Optimum use of cruise/descent control for the scheduling of inbound traffic". International Conference sponsored by Flight International on "Fuel Economy in the Airlines". Royal Aeronautical Society, London, U.K., April 1980. Also EUROCONTROL Report 802013, February 1980.

18. Erwin, R.L., Jr. (1975) "Strategic Control of Terminal Area Traffic" Paper presented at the 20th International Symposium of the Guidance and Control Panel of AGARD, Cambridge, Mass., U.S.A., May 1975. AGARD CP-188, pp. 3-1 to 3-13, February 1976.

19. Benoît, A. (1973) "A Concept of Air Traffic Control Based upon Accurate Aircraft Trajectory Prediction". Paper presented at the XIth General Assembly of EUROCAE, December 7, 1973. Also EUROCONTROL Report 732041.

20. Andreussi, A., Bianco, L. and Ricciardelli, S. (1981) "A simulation model for aircraft sequencing in the near terminal area". European Journal of Operational Research, Vol. 8, pp. 345-354, 1981.

21. Dear, R.G. "The dynamic scheduling of aircraft in the near terminal area". FTL Report T. 76, 9 M.I.T., Cambridge, Mass. USA, Sept. 1986.

22. Imbert, N., Fossard, A.J. and Comes, M. (1979) "Gestion à moyen terme du trafic aérien en zone de convergence". IFAC/IFOR symposium, Toulouse, France, March 6-8, 1979.

23. Benoît, A. and Devry, H. (1981) "Impact of excess route length on fuel consumption in a European air-route network". EUROCONTROL Report 802021, December 1981. Abstract: EUROCONTROL 802021-A, December 1981.

24. Benoît, A. (1985) "4-D control of current air carriers in the present environment". Paper presented at the Seminar on "Informatics in Air Traffic Control", Consiglio Nazionale delle Ricerche, Progetto Finalizzato "Trasporti", Capri, Italy, October 1985. Also EUROCONTROL Report 862013, March 1986.

25. Khan Mohammadi, Sohrab (1983) "Gestion automatisée du trafic aérien sur un ensemble d'aéroports". Thèse n° 81/1983, Ecole Nationale Supérieure de l'Aéronautique et de l'Espace, Toulouse, France, October 1983.

26. Benoît A., Sauer, P. and Swierstra, S. (1983) "Estimates of nugatory fuel consumption in an extended terminal area (Traffic inbound to Brussels National)". EUROCONTROL Report 8120038, March 1983.

27. Benoît, A., Sauer, P. and Swierstra, S. "Estimates of nugatory fuel consumption in an extended terminal area (Traffic inbound to London Heathrow)" EUROCONTROL Report 812021, March 1983.

28. Benoît, A. and Swierstra, S. "Air Traffic Control in a Zone of Convergence: Assessment within Belgian airspace". Paper presented at the International Council of the Aeronautical Sciences (ICAS), Toulouse, France, September 10-14, 1984. Also, EUROCONTROL Report 842009, April 1984.

29. Attwooll, V.W. and Rickman, D. (1984) "A Study of conflicts within a Zone of Convergence". Proceedings of the EUROCONTROL SPACDAR 11th Meeting, Brussels, September 25-27, 1984.

30. Benoît A., Swierstra, S. and De Wispelaere, R. "Next Generation of Control Techniques in Advanced TMA. Automatic Assistance for the Controller/Pilot dialogue". Paper presented at the AGARD Guidance and Control Panel 42nd Symposium on "Efficient Conduct of Individual Flights and Air Traffic", Brussels, Belgium, 10-13 June, 1986. Also EUROCONTROL Report 862016, June 1986.

An Integrated View of Air Traffic Management Problems

Robert W. Simpson
Flight Transportation Laboratory, Room 33-412
Department of Aeronautics and Astronautics
Massachusetts Institute of Technology
Cambridge, Massachusetts U.S.A. 02139

Abstract

The problems of ATM (Air Traffic Management) are described in a systematic way in order to provide a coherent basis for future research. The capacitated elements of the ATC system are identified as the Runway Systems at major airports, and Airspace Sectors within the airways network. Control over congestion can be exercised through ATM "directives" identified as Speed Reduction, Aircraft Reroute, Flight Initiation Delay, and Restructure Airspace Geometry. Sub-processes in ATM are identified as Arrival Flow Management, Terminal Area Departure Flow Management, Airport Ground Flow Management, and Sector Flow Management. The primary role of uncertainty in predicting traffic flow rates and air traffic capacities is explained in the operation of a real-time, dynamic ATM process.

An Integrated View of Air Traffic Management Problems

This paper is designed to provide a common basis for understanding the problems in air traffic flow management and thereby allow the identification of key areas for research activities aimed at introducing advanced automated methods.

1.   Definitions

Air Traffic Control - ATC - deals with controlling the separations between all pairs of aircraft in order to ensure safe operations. It is tactical and real time.

NATO ASI Series, Vol. F38
Flow Control of Congested Networks
Edited by A. R. Odoni et al.
© Springer-Verlag Berlin Heidelberg 1987

Air Traffic Management - ATM - deals with the control of traffic flow rates for various streams of aircraft to minimize congestion, and thereby minimize the need to exercise ATC. It is strategic, and real time.

Congestion - describes a high density of traffic in terms of aircraft/cubic volume. Since ATC requires a minimal density, through its separation standards, aircraft must be delayed to avoid congestion.

Delay - for each aircraft, delay is the difference in time between arrival at some point under congestion conditions and the normal time of arrival which would have been realized if there were no traffic.

Capacity Flow Rate - given desired ATC separations and traffic speeds, traffic mix, etc., there is a long-term average value for the maximum traffic flow rate which can be achieved for each of a number of elements of the ATC system. These values are the capacity flow rates associated with each of these elements.

2.   Capacity, Delay, and Congestion - Queueing Processes

Queueing theory is a field of applied mathematics that deals with queueing processes. While it cannot be applied rigorously to ATC processes (for various reasons), there are a number of simple observations that are pertinent to air traffic flow management. Figure 1 shows a simple queueing process consisting of an arrival flow of traffic, a queue, and a server which requires a variable processing time to service each arrival, before releasing it into an output flow.

The arrival flow has a known average flow rate and statistical distribution for interarrival times. The average flow rate is averaged over some specified duration, and may be time varying in a longer term. In ATC, we normally talk in terms of hourly flow rates, and there are "hour of the day" and "day of the week" cycles in traffic flow rates. While queueing theory often deals with random uncontrolled arrival flows, ATC flows may be well spaced, i.e., controlled such that a minimum interarrival interval is maintained.

Queues are difficult to define in ATC. They are visible at takeoff runways, and within airborne holding stacks, but when ATM issues directives, it avoids congestion by forming a distributed queue of aircraft at various physical locations. When the server is an airport, the queue awaiting service consists of aircraft located in the terminal area, in holding stacks, enroute to the terminal area if under speed reduction directives, and at multiple airports where aircraft are awaiting departure times. Queueing theory normally assumes that the queue is located at the server and that there is zero time to travel from the queue to the server.

Servers are capacitated processes with a known "saturation" capacity flow rate that is the maximum flow rate achievable as a longer-term average. This rate can often be exceeded in the shorter term. In the ATC system, capacities are not constant, changing with weather, traffic mix, available facilities, human operator skills, etc. Capacity values may change by factors of 1/2 or 1/3 quite unexpectedly.

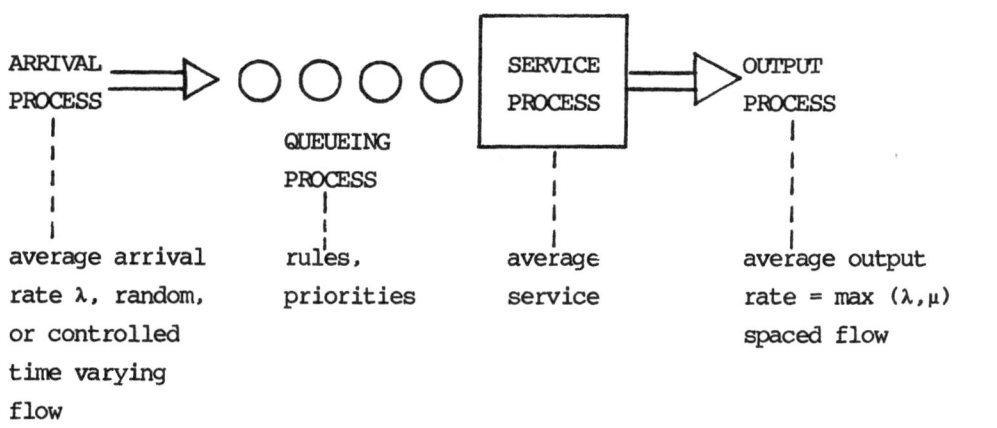

FIGURE 1 - A SIMPLE QUEUEING PROCESS

It should be clear that the only cause of delay is the capacity of the server. While ATM can avoid congestion by managing the queue, it cannot reduce or eliminate delay caused by a capacitated element of the ATC system. ATM only transfers the delay to occur in another location. If ATM is efficient, it will not create further delay because of this queue management. If not efficiently performed, it may cause further delay. Arrivals must be continuously delivered by ATM to the capacitated element so as to prevent it from becoming idle. Otherwise, there is delay due to traffic management.

Delay occurs even if the average arrival rate is less than the capacity flow rate. Due to the variability of interarrival times at a constant, longer term average arrival rate, it is possible for two or three aircraft to arrive simultaneously, or during a "busy period" when the server is processing prior arrivals. Since the average arrival flow rate is less than the average capacity rate, the delays incurred in this mode of operation will smooth the short-term peaking of the arrival flow and cause a regular output flow during busy periods. In ATC, this short-term mode of operation is called "metering" where the delays incurred are small, and many arrivals arrive with no delay.

Whenever the average arrival rate exceeds the capacity flow rate, the average queue size will be growing and average delays will be increasing. Later, when the arrival rate is less than the capacity rate, the server is still busy working to deplete the queue. In ATC it is possible to control the arrival flow rate by means such as slowing aircraft down, path stretching, or keeping them on the ground. This mode of operation is called "flow control." Traffic Flow Management consists of both "metering" and "flow control" modes of operation. There is a continuous transition from metering to flow control as the average traffic rate approaches and then exceeds the average capacity flow rate. We can arbitrarily separate metering from flow control by defining metering to occur within a span of control of 15 minutes or less before arrival at the capacitated element.

3. Capacitated Elements of the ATC System

There are only two types of elements that currently reach saturation in today's ATC System: runway systems at major airports; and sectors of terminal area or enroute airspace. There are no saturated airway segments of the airways network as occur in railroad or highway networks, since the collection of segments, crossing points, etc., that make up a "sector" in ATC will saturate first.

a) Runway System Capacity

At major airports, there are multiple runways in operation simultaneously with some assignment of aircraft mix and takeoff/landing operations to each runway. Each such runway operating configuration has an hourly capacity based on ATC separation criteria for total operations, as well as sub-capacities for landings and takeoffs separately, and perhaps other subcapacity values expressed by class of aircraft such as small piston aircraft. The values of these capacities may vary by hour or day as configurations, weather, and traffic mix are varied. The landing capacity at major airports is the primary cause of delay and congestion in the US ATC system.

b) ATC Sector Capacity

For any sector in the ATC system, the sector controller has a routine workload directly proportional to traffic flows along the airways and through the intersections of that sector, and also a monitoring and intervention workload proportional to the square of the density of traffic in that sector. Each individual controller may have a higher or lower capacity to handle sector traffic, which may be expressed either as an overall traffic flow rate, or as a maximum "occupancy." Both measures need to be expressed as an average over some duration such as 5 minutes, or 15 minutes, or one hour. There is a need for better definition of ATC sector workload capacities. Sector capacities may vary strongly when radar surveillance or communication equipment fails, and when severe weather phenomena occur.

4.   Congestion Controls Exercised by Air Traffic Management Directives

Unlike normal processes in queueing theory, the arrival flow of air
traffic can be controlled. There are four basic methods of exercising
control. (We shall call these ATM control messages "directives" to be sent
by a flow manager to an ATC controller.)

a) Speed Reduction

To prevent a queue from forming, ATM may impose a speed
reduction on aircraft so as to control its arrival time at the server
or capacitated facility. Speed increases are not normally possible
since aircraft cruise near their maximum speed.

b) Reroute Aircraft

It may be possible to specify another path from present
location of the aircraft to the capacitated element, or to bypass it
completely. The path changes may include an altitude change which
avoids entering a busy sector of airspace (sectors are defined both
in geographic and vertical dimensions). This directive includes
placing aircraft in a "holding pattern".

c) Delay Initiation of Flight

While it may create ground congestion at a departure airport,
ATM may defer entry of an aircraft into the airspace, thereby causing
the delay to be taken at that airport rather than in the airspace.
In the U.S.A., this directive is called a CDT (Controlled Departure
Time) and is used when the delay at arrival is expected to exceed one
hour for a period of a few hours.

d) Restructure Airspace Geometry

When the capacitated element is a sector of the ATC system, ATC
managers may elect to institute short-term changes in airspace
structures and procedures that redistribute ATC workload and flows.

5.   Sub-Processes within Air Traffic Management

There are a small number of basic flow management processes that have
been identified in ATM. Today, they operate as independent processes, and
any aircraft can be affected by them simultaneously or sequentially.
Successive aircraft flowing along an airway may (or may not) be under
directives from one or more of these subprocesses.

a)   Airport Arrival Flow Management - AFM

Since landing capacity is a prime source of delay, this is a
common subprocess for all aircraft destined into a given airport. It
can be divided into two parts - AFC (Arrival Flow Control) and AFS
(Arrival Flow Smoothing). As defined previously, AFC attempts to
control the arrival flow rate over periods of 15 minutes or longer,
and AFS smooths or meters the actual arrival flow in the shorter
term.

b)   Terminal Area Departure Flow Management - DFM

At major terminal areas, where there may be multiple airports,
the workload in departure sectors may be a capacity limiting element.
Again we may distinguish between DFC (Departure Flow Control),
issuing longer term delays, and DFS (Departure Flow Smoothing), which
issues shorter term, small departure delays to smooth the departure
flow into any sector, or through any fix in that sector.

c)   Airport Ground Flow Management - GFM

At a major airport, where there may be aircraft delayed
awaiting departure clearance times, different separation criteria for
aircraft departing in different directions, and DFM directives to
control or smooth departure sector workloads, there is a "pushback"
or "gate departure" management problem in getting aircraft to arrive
at the takeoff runways at the correct time and in the correct
sequences. This subprocess is concerned with integrating the
directives issued by AFM and DFM, and managing the outbound ground
traffic flow to execute those directives.

d) Sector Flow Management - SFM

This subprocess is concerned with managing the airborne flows such as to control workload levels in any sector. It issues delays and/or reroutings to control (SFC) or smooth (SFS) sector traffic flows. Since the aircraft are airborne, this subprocess must concern itself with workloads in several adjacent sectors to ensure that avoiding high workloads in one sector does not transfer the problem to adjacent sectors. Reroutings in the airways network must spread workload levels by finding a new path along airways segments that minimizes delays to aircraft (or makes use of aircraft due to be delayed at destination anyway) and which controls the workload levels in all sectors so as to be under their capacity levels. For example, an overloaded arrival sector at a major terminal area may cause reroutings well back into cruise which direct aircraft along different airways into another arrival sector. These reroutings must be chosen in time and location such as not to overload any other enroute sector, or the second arrival sector, and to avoid causing any further delay to the arrival aircraft.

It is believed that all ATM processes can be classified into one of the above subprocesses. The AFM, DFM, SFM processes create flow management "directives" that use the various congestion control methods mentioned above. These directives are passed to the ATC process for execution. A particular ATC sector may receive directives from any of these processes for certain of the aircraft currently controlled by that sector. While the subprocesses can operate more or less independently, there can be strong interactions between them, which argues for an integrated, systemwide ATM process. For example, in the current U.S. ATM system, the Enroute Metering process for arrival flow smoothing ignores any prior delay imposed through CDT's issued by the current version of AFC. If two simultaneous arrivals occur, it is equally likely that the aircraft with larger CDT penalties will be sequenced second and incur further delay. In general, the integration of ATM subprocesses will require that all aircraft are tagged with two delay measures, Current Trip Delay (CTD) and Expected Further Delay (EFD). These would be updated regularly as the various ATM subprocesses issue directives affecting the aircraft.

6. Uncertainty in the Prediction of Future Traffic Flow Rates

ATM processes need to predict traffic flow rates for an airport or sector for several hours into the future. There is always some degree of uncertainty associated with these predictions, which decreases with decreasing prediction time. ATM decision making must account for these uncertainties. Three phases of information uncertainty can be identified:

a) Phase 1 - Prior Knowledge

In air traffic, the traffic flows exhibit a daily and weekly, and a seasonal cycle. Past experience gives some guidance as to hourly traffic expectations for today. As well, there is a strong component of scheduled airline traffic with published schedules for departure and arrival times. While the departure schedule usually provides good information, the arrival schedule necessarily depends on winds, so that there will be daily variations from the stated arrival schedules, particularly for longer-haul flights. The daily routings flown by scheduled aircraft will vary also so that sector workloads are not known from this information.

b) Phase 2 - Flight Plan Information

Presently, each aircraft is required to file a flight plan at least one-half hour before planned departure. This confirms a schedule, or a prior historical expectation for a nonscheduled flight. It also provides an updated routing and arrival time from an ETE (estimated time enroute) based on current weather predictions which allows a forecast of sector loadings. It is also desirable to gather information on cancellations or planned delays in scheduled operations to improve Phase I information for ATM decisions. Unfortunately, there is no way to know whether the expected nonscheduled traffic will not materialize. As Phase 2 information arrives, it reduces the uncertainty of traffic information, thereby allowing better decision-making by ATM processes.

c) Phase 3 - Flight Initiation

The next update of information is the request to activate the flight plan by the pilot, and the subsequent initiation of the flight. Due to normal operational reasons, flights may not start exactly as filed, or may suffer departure traffic delays in getting airborne. Once airborne, and particularly once established in cruise, the prediction of future ETA's at sector boundaries significantly improves and can be predicted within a few minutes, even on long-haul flights. The flight plan ETE can be updated based on actual ground speeds experienced once at cruising altitude. Note that filed flight plans may not be initiated for various reasons.

In predicting future traffic flow rates for an airport or ATC sector, the information for a future hour is always a mix of these three phases. As prediction time decreases, more flight plans will have been filed, and more aircraft will have activated and reached cruise conditions. Cancellations and nonappearance of nonscheduled traffic reveal themselves as time progresses. Since traffic may be departing from airports as close as fifteen minutes from destination, the arrival flow smoothing (AFS) process may still have to contend with Phase 2 information.

Notice that the imposition of ATM directives changes the expected traffic flow rates. Subsequent decision making should be aware of the modifications to expected traffic. It is difficult to predict the effect of delay directives on subsequent scheduled flights by the same airline aircraft. If an airline aircraft is one hour late in arriving at a station, the subsequent flight for that aircraft may or may not be delayed. The airline may be able to substitute another aircraft and crew at a home station, or the aircraft might be scheduled for extra time on the ground, which allows it to turn around and depart more or less on time. Rapid resubmittal of revised flight plans and cancellations is necessary if the ATM processes are to continue to have up-to-date data for dynamic decision making later in the day.

7.   Uncertainty in the Prediction of Capacities

The two types of capacities in the ATC system are Runway System
Capacities and Sector Workload Capacities. Both are strongly dependent on
weather and equipment status, and uncertainties in future capacities are
strongly related to the accuracies of weather forecasting. Changes in
equipment status are unexpected events and are not predictable.

a) Runway System Capacity

The average maximum traffic service rates for a specific
operating configuration of runways at an airport are a function of
ATC separation criteria, ATC procedures, ATC controller skills, etc.
It probably can be estimated in the longer term (several hours) with
an accuracy of better than 10%, although in the shorter term (15
minutes) there may be wider variations. The uncertainty problem in
predicting runway capacity is predicting the availability of runway
configurations, and moving the airport to its highest capacity
available configuration as weather changes.   Wind speed and
direction determines crosswinds and tailwinds for runways. When
these exceed certain values the runway becomes unavailable and all
configurations that use that runway are then unavailable. Ceiling
and visibility determines the need for landing approach systems and
different levels of ATC procedures to maintain safe approach and
departure operations. Approach navaids and lighting must be
serviceable to establish the ceiling and visibility limits. At major
airports, there may be dozens of runway operating configurations in
assigning landings and takeoffs by class of aircraft to multiple
simultaneous runways.

Weather is the prime source of uncertainty in predicting runway
system capacity changes. Changes in windspeed and direction, ceiling
and visibility are often associated with passages of frontal weather
at the airport, and predicting the time of such passage is critical.
Runway surface conditions affect available configurations for "hold-
short" operations, so that predicting rainfall and snowfall rates is
also important. Wet surfaces eliminate "hold-short" landing
operations, and runways may be out of service for short periods to

allow snow removal. Since visual approach conditions affect landing capacities, the prediction of afternoon haze and smog on otherwise-good-weather summer days is important. Afternoon thunderstorms passing through approach areas or over the airport are difficult to predict in the longer term.

The second source of uncertainty is the unexpected failure of ATC facilities. This is not predictable, and requires that ATM processes be able to react quickly to prevent local congestion in the short term.

b) Sector Workload Capacity

While it is possible to define runway system capacity rather precisely in terms of ATC separation criteria and traffic mix, it is not possible to state Sector Workload capacities rigorously. The problem is our general state of knowledge on mental workloads of human operators. There are, however, a number of methods based on experimental tests and operator opinion ratings that relate ATC workloads classified as "high", "medium", "low", etc., to various measures of traffic activity in the sector. It is not possible to pass judgment on the accuracy of such methods, and they are likely to depend strongly on the skill and attitudes of the test operator. Despite these problems, it may be possible to get local ATC controllers and supervisors to establish "safe workload" levels for every sector (and perhaps even for the currently assigned operator) expressed in terms of traffic flow rates per hour, or simultaneous occupancy averaged over some short duration. Given such locally-determined capacity limits on traffic flow, ATM processes would then have a well-defined value to use in decision making, even if they cannot be validated.

There still would be uncertainties arising from weather and equipment failures. It is likely that lower capacity limits (including zero capacity) will exist when frontal weather, icing, thunderstorms, etc., are present in any sector. A failure of radar surveillance increases controller workloads and separation criteria and would indicate lower capacity values for such periods of time. Communication failures would mean zero capacity until the failure is overcome, hopefully within a short time.

8.　Summary – ATM Processes and Problem Areas

The problems in air traffic management are stochastic, real-time, dynamic control problems where ATM directives are generated based on the current and predicted future states of the system such as to control traffic flow rates through the multiple capacitated elements of the ATC system. They are not deterministic problems (although current ATM methods assume this to be true) since ATM decision making must be based on uncertain estimates of future traffic and capacity flow rates. Introduction of methods to deal with uncertainty is badly needed to improve the average performance of ATM processes. These methods will involve the timing and distribution of ATM directives throughout the systemwide network, and a "discounting" of average expected delay towards a smaller, more certain value.

There are some interactions between local and global flow control processes, although in many cases they can operate autonomously. It is not clear how an integration of ATM processes should be pursued. It may be possible to have flow management directives issued by local traffic flow managers and to make the effects of these directives available to all other local flow managers as needed on a very timely basis. It is desirable to make use of the real-time data processing and data communication capabilities available at the present time to move ATM processes towards real-time operation by distributed flow managers close to local problems and limitations.

Recent advances in various technologies have much to offer to the creation of improved ATM processes. Computer workstations linked to a global communications network provide considerable power in maintaining an up-to-date database on traffic flows of interest to local flow managers. They are able to create graphics presentations easily digested by humans, allow data to be easily modified, to create exploration of "what if" scenarios, etc., which allow flow managers to participate in an interactive mode with management information processes. Beyond that, it is possible to extend the human decision-making process by developing various automated decision-making processes. These can be of the form of DSS (Decision

Support Systems), which assist the human by providing automated elements of the complete decision process, or they can be of the form of "expert" systems, which try to consolidate the expertise of other, perhaps more experienced, air traffic managers in an automated system which acts as an advisor or consultant in solving current problems.

A COMBINATORIAL OPTIMIZATION APPROACH TO AIRCRAFT SEQUENCING PROBLEM

L. Bianco, G. Rinaldi, A. Sassano
Istituto di Analisi dei Sistemi ed Informatica
Consiglio Nazionale delle Ricerche
Viale Manzoni, 30
00185 - Roma
Italy

ABSTRACT

In this paper a combinatorial optimization approach to aircraft sequencing problem is proposed.
In particular the single runway case, with the hypotesis that airplanes wait to land at different times, is considered.

It is shown that the problem of maximizing the runway utilization can be modeled as a n job-one machine scheduling problem with non zero ready times, sequence dependent processing times, and with the objective of minimizing the maximum completion time.

A solution algorithm is outlined and tested by various examples and the computational results are discussed.

Implementation issues are also considered and suggestions on how improve the algorithm performances are made.

NATO ASI Series, Vol. F38
Flow Control of Congested Networks
Edited by A. R. Odoni et al.
© Springer-Verlag Berlin Heidelberg 1987

1. Introduction

One of the main problems airlines must face nowadays is flight delay caused by air traffic congestion. This phenomenon is due to the particolar traffic distribution which, owing to users requirements, is highly concentrated on the main airports. The effects of this situation are felt mostly in the arrival and departure phases. In fact, in the last years it has been assessed that only 10% of the delays occurs along the aircraft routes, while the corresponding percentages for arrivals and departures are, respectively, about 60% and 30%. For this reason, recently, specific attention has been devoted to the management of Terminal Area (TMA) where the overall system efficiency becomes nearly as important as safety. Moreover, due to the continuous growth of air traffic, a reduction of human intervention in the operations management is needed and, consequertly, automated systems have been proposed. From this point of view the TMA problem is *the automation of the aircraft flow control and sequencing in the proximity of the airport so as to satisfy an optimality criterion.*

As a consequence, the need arises to define a model which can be dealt with mathematically and to analyze all those elements which constitute a semi or fully automated control system.

To contribute in solving the TMA control problems, in this paper we consider in particular the Aircraft Sequencing Problem (ASP).

In section 2, after a brief TMA description, the ASP formulation is given and the most significative solution proposed in the literature are reviewed. To overcome some limits of these approaches, in section 3 the ASP is reformulated in terms of job - machine scheduling model. Sections 4 and 5 are devoted to illustrate, respectively, the optimization algorithm developed and the computational results.

2. The Aircraft Sequencing Problem

During traffic peak periods, the control of aircraft arrivals and departures in a TMA becomes a very complex task.

Air traffic controllers, among other things, must garantee that every aircraft, either waiting to land or preparing to take off in such a congested area, maintains the required degree of safety. They have to decide also what aircraft should use a particular runway, at what time this should be done and what manoeuvres should be executed to achieve this. The viable accomplishment of such a task becomes more difficult in view of the fact that aircraft are continuously entering and leaving the system and that, at peak periods, the demand for runway occupancy may reach, or even exceed the capabilities of the system.

It is at such periods that excessive delays are often observed, resulting in passenger discomfort, fuel waste and disruption of the airlines' schedules.

Under such "bottleneck" conditions, an increase in collision risk can logically be expected as well . As a consequence, because of safety considerations, the structure of TMA is rigidly defined and all aircraft must fly satisfying prefixed procedural constraints.

To simplify the understanding of the problem we refer to a TMA as shown in fig. 1 and fig. 2 and we consider only landings, even if the approach proposed in the following allows to simultaneously take into account takeoffs.

Then the following aspects must be underlined:

a) every aircraft must approach the runway for landing, flying along one of the prestructured paths of TMA;

b) the runway can be occupied by only one aircraft at a time;

c) every aircraft must fly along the common approach path following a standard descent profile;

d) during all the approach phases a separation standard between every couple of subsequent aircraft must be maintained;

e) the sequencing strategy used by almost all major airports of the world is still today the *First-Come - First-Served* (FCFS) discipline.

As it is well known, FCFS strategy is very simple to implement, but it is likely to produce excessive delays. Therefore, any effort must be made to minimize the delay or optimize some other measure of performance related to the passenger discomfort, without violating safety constraints.

Consequently the TMA problem can be stated as follows: *Given a set of aircraft entering the TMA and given, for each aircraft, the Preferred Landing Time (PLT), the runway occupancy time, the cost per unit time of flight, the geometry of the approach path and glide path and the corresponding aircraft speeds, assign to each aircraft the starting time from the fix and the approach path in such a way that the procedural constraints are satisfied and a system performance index is optimized.*

With the TMA operating in the aforementioned way, the TMA problem can be decomposed into the two following sub-problems:

1) given the constraints on the aircraft performance, the initial and final states (position and speed) and the pre-established flight time, determine the optimum trajectories which connect these states with the specified flight time;

2) given a set of PLT and the maximum admissible delay, determine the Actual Landing Times (ALT) sequence which satisfies the procedural constraints on the runway and the glide path and optimizes a system performance index.

To a large extent these two problems are independent. In fact, as the required controls to follow the approach paths can be calculated in advance, it is possible to predetermine the optimal flight path. Therefore, the need of "real-time" calculations is limited only to sub-problem 2, called Aircraft Sequencing Problem, which is the topic discussed here.

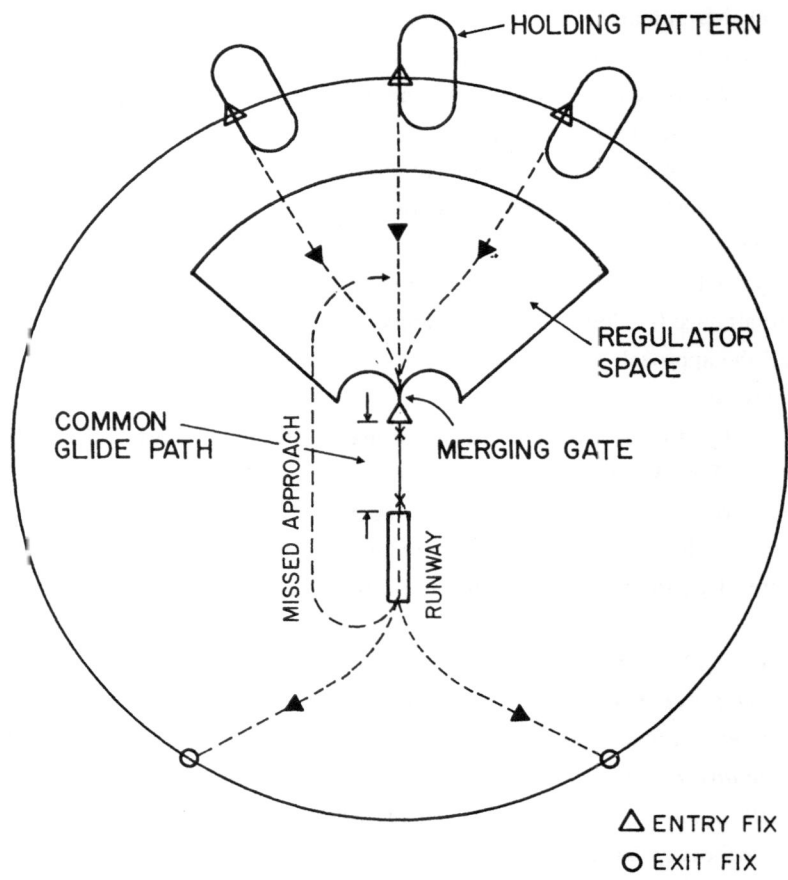

Fig. 1-STRUCTURE OF TMA

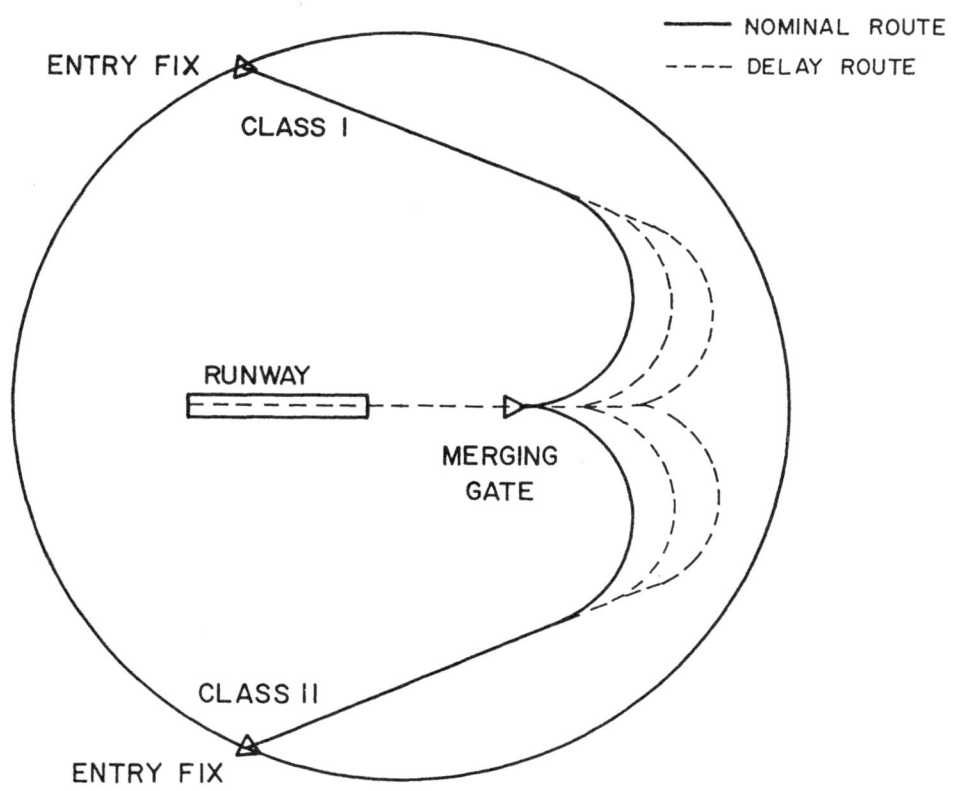

ENTRY FIX

CLASS I

NOMINAL ROUTE

---- DELAY ROUTE

RUNWAY

MERGING
GATE

CLASS II

ENTRY FIX

Fig. 2-DELAY ROUTES

The goal of solving the ASP is, at least theoretically, achievable for two reasons.

First, safety regulations state that any two coaltitudinal aircraft must maintanin a "minimum horizontal separation", which is a function of the types and relative positions of these two aircraft.

Second, the "landing speed" of an aircraft is generally different from the landing speed of another aircraft.

A consequence of the variability of the above parameters (minimum horizontal separation and landing speed) is that the *minimum permissible time interval* between two successive landings is a variable quantity.

Thus, it may be possible, by rearranging the initial position of the aircraft, to take advantage of the above variability and obtain a landing sequence that results in a more efficient use of the runway as compared with that obtainable by using the FCFS discipline. In fact, an optimal sequence does exist; it is theoretically possible to find it by examining all sequences and selecting the most favorable one.

The method suggested above to determine the optimal sequence is safe, but extremely inefficient, because the computational effort associated with it is a factorial function of the number of aircraft and it is not possible to evaluate all combinations in a short time interval (as the nature of ASP requires) even on the fastest computer. To give an idea of the difficulty, it is sufficient to consider that, with only 10 aircraft, we would have to make 3,628,000 comparisons and with 15 aircraft 1,307,674,368,000 comparisons.

It should be also pointed out that while the main factor that suggests the existence of an optimal landing sequence is the variability of the minimum permissible time interval between two successive landings, it is the same factor that makes the determination of this optimal sequence a nontrivial task.

Moreover, the real world problem involves many other considerations, especially as far as the implementation of sequencing strategies is concerned.

For these reasons, the relevant literature on the subject has been till now considerable and growing [see for example Odoni (1969), Tobias (1972), Park and others (1972, 1973), Pardee, Bianco and others (1978, 1979), Trivizas (1985).].

Two papers, in particular, seem to offer an adequate solution to the operational needs.

The first is an excellent investigastion of the ASP made by Dear (1976). He pointed out that, in order to determine the landing order, we need to consider all aircraft currently in the system. As this number can be very large (20 or even more simultaneously) he put forward serious doubts on the possibility of reaching an optimal solution in real-time even with pseudo-enumerative techniques. Therefore, he resorts to a simulation model where identical arrival streams, under various sequencing strategies, are compared. In particular he proposes a *Constrained Position Shifting* (CPS) strategy instead of FCFS strategy. That is, no aircraft may be sequenced forward or rearward more than a prespecified number of positions (Maximum Position Shifting) from its FCFS position.

In the second paper Psaraftis (1978, 1980) takes into account Dear's proposal about the CPS management concept, but he develops an exact optimization algorithm based on the dynamic programming approach and referring to the "static" case when all aircraft are supposed to wait to land at a given time. In reality, every aircraft entering TMA has an earliest landing time (which is the PLT) depending on the characteristics of TMA, the aircraft speed, pilot preferences and so on. Therefore, the aircraft to be sequenced wait to land at different times.

Fig. 3 shows an example of a real world situation. In this case the Psaraftis approach cannot be easily utilized.

3. A Job-Scheduling Formulation of the ASP

Suppose that the air traffic controller is confronted with the following problem: A number n of aircraft are waiting to land at different PLT_i at a single runway airport. His task, then, is to find a landing sequence for these aircraft, so that a certain measure of performance is optimized, while all problem constraints are satisfied.

We now make the problem statement more specific:

1) It is assumed that the pilots of all aircraft are capable and willing to execute the instructions of the controller given enough prior notice.

2) The measure of performance selected is the *Last Landing Time* (LLT). The corrresponding objective is then to find a landing sequence such that the aircraft that lands last does this as quickly as possible.

3) Concerning the problem constraints, only the satisfaction of the *minimum interarrival time* constraints is required. This means that the time interval between the landing of an aircraft i, followed by the landing of an aircraft j, must not be less than a known time inteval t_{ij}.

4) The composition of the set of aircraft waiting to land is, of course, assumed to be known. For each ordered pair (i, j) of aircraft, the minimum time interval t_{ij} is also known.

5) At any stage of sequencing procedure, the air controller is free to assign the next landing slot to any of the remaining aircraft. This means that we ignore the initial positions which the aircraft had when they arrived at TMA.

At this point it is not difficult to see that the problem described above can be represented by means of a particular *job-machine scheduling* model.

In fact, with the aforementioned assumptions, the following analogy may be established:

a) to each landing operation is associated a job;

b) the runway corresponds to a machine with capacity one;

c) the PLT_i of aircraft i corresponds to the ready time r_i of job i;

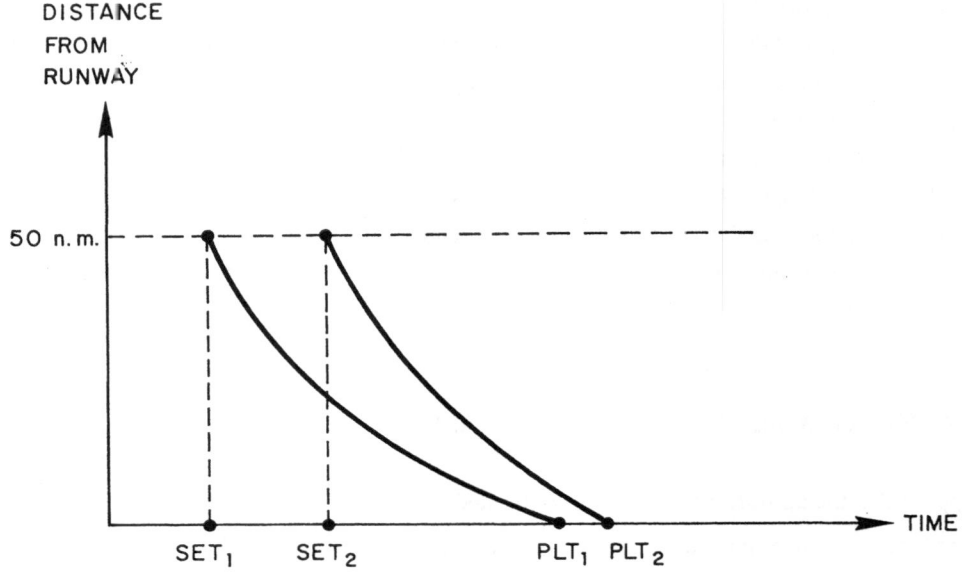

Fig. 3 - EXAMPLE OF AIRCRAFT CONFLICTING ON THE RUNWAY

d) the ALT$_i$ of aircraft i corresponds to the start time t$_i$ of job i;

e) the LLT corresponds to the maximum completion time C$_{max}$ of the job schedule;

f) the minimum time interval t_{ij} between the landing of aircraft i, followed by the landing of aircraft j, corresponds to the processing time p_{ij} of job i when it depends on job j following in the sequence.

Therefore, the ASP, as defined here, can be mathematically reformulated as the $n|1|r_i|$seq-dep$|$C$_{max}$ scheduling problem. For this problem the approach proposed by Bianco et others (1985) can be utilized.

4. The n| 1| r$_i$| seq-dep| C$_{max}$ Model

Let $J = (1,2,...,n)$ be a set of n jobs to be processed on a single machine, and denote by r_i the *ready time* of the job i. A matrix (p_{ij}), with $i, j \in J0 = J \cup \{0\}$, is given where p_{ij} $(i \neq 0)$ is the *processing time* of the job i if job j is the successor of i in the sequence $(j=0$, if i is the last job in the sequence) and p_{oi} is the *setup time* of the machine when the sequence starts with job i.

The $n|1|r_i|$ seq-dep$|$C$_{max}$ problem can be formulated using the following integer programming model:

$$\min (s + \sum_{i \in J0} \sum_{j \in J0} p_{ij} \, x_{ij}) \tag{1}$$

subject to

$$t_i + \sum_{j \in J0} p_{ij} \, x_{ij} - \sum_{k \in J0} \sum_{j \in J0} p_{kj} \, x_{kj} - s \leq 0, \qquad i \in J0 \tag{2}$$

$$r_i - t_i \leq 0, \qquad i \in J \tag{3}$$

$$(p_{ij} + T_{ij}) \, x_{ij} + t_i - t_j - T_{ij} \leq 0, \qquad i \in J0, j \in J, j \neq i \tag{4}$$

$$\sum_{j \in J0} x_{ij} = 1, \qquad i \in J0 \tag{5}$$

$$\sum_{i \in J0} x_{ij} = 1, \qquad j \in J0 \tag{6}$$

$$s \geq 0, \tag{7}$$

$$t_0 = 0, \tag{8}$$

$$x_{ij} \in \{0,1\} \qquad\qquad i \in J0, j \in J0, j \neq i \tag{9}$$

$$t_i \geq 0, \qquad\qquad i \in J0 \tag{10}$$

where:

$x_{ij} = 1$ if job i directly precedes job j

$x_{ij} = 0$ otherwise

t_i is the start time of job i

s is the machine idle time

T_{ij} are chosen to make constraints (4) redundant whenever $x_{ij} = 0$

Remark 1. This problem is NP-hard and in the case of zero ready times it reduces to the asymmetric travelling salesman problem (ATSP). Moreover, constraints (4) both prevent subtours in the ATSP solution and avoid two jobs to be simultaneously processed.

Remark 2. The above formulation can be considered as an ATSP with unlimited time windows $(r_i \leq t_i < + \infty)$.

5. Outline of the Solution Algorithm

The solution of the above problem can be obtained by a pseudo-enumerative procedure exploiting both some peculiar properties of the problem and efficient lower and upper bound.

Branching phase

Denote by $C(\sigma)$ the completion time of a subsequence σ, the following two dominance criteria, proved by Bianco and others (1985), are utilized for tree pruning:

Theorem 1: A subsequence $\sigma \pi k$ is dominated if there exists a permutation π' of π such that

$$C(\sigma \pi k) > C(\sigma \pi' k)$$

Theorem 2: A subsequence $\sigma j \sigma' i$ is dominated if

$$\text{a) } r_j > max \{r_i , C(\sigma)\} + p_{ij}$$

$$\text{b) } p_{ki} + p_{ih} \geq p_{hk}, \quad \forall h, k \in J - \{i\}$$

Then the following branching strategy can be stated.

At each node k of the enumeration tree is associated a partial sequence σ in which the first k jobs have been fixed. As first step theorem 2 is used to define, among the unscheduled jobs, a set of candidates to the $(k+1)$-th position in the optimal sequence. Subsequently, for each candidate i, the procedure checks whether or not it is dominated according to theorem 1. If it is dominated, a new candidate is examined; otherwise the lower bound is computed and, if it is less than an upper bound of the optimal solution, a new node, representing the subsequence σi, is generated. At this point the ready times of the unscheduled jobs which are smaller than $C(\sigma i)$ are set to $C(\sigma i)$.

Backtracking takes place whenever a complete sequence has been produced or all the candidates at a given level have been examined.

Bounding phase

At each node of the enumerative tree we have a new problem with job-set $J-\{\sigma\}$, new ready times and new setup times. To this problem two lower bounds and an upper bound are applied.

1) Lagrangean Lower Bound (LLB)

It is obtained as the solution of the lagrangean problem obtained by dualizing constraints (2) and (3), and dropping constraints (4) in the original model.

2) Alternative Lower Bound (ALB)

As the LLB tends to be weak in case of small variations in the matrix $\{p_{ij}\}$, the following heuristic rule can be utilized:

(i) to each job $i \in J$ associate the processing time $p_i = \underset{j}{min} \{p_{ij}\}$;

(ii) schedule the jobs using the FCFS rule and take the completion time as the ALB of the original problem.

Remark. Observe that ALB is the optimal solution of our problem when processing times do not depend on the sequence, and hence, in the general case, ALB is a lower bound on the optimal solution.

3) Upper Bound (UB)

As observed in the previous remark, EST (Earliest Start Time) rule is optimal if the processing times are sequence-independent. Consequently, a reasonably good feasible solution can be obtained by associating to each job an *average* processing time $p_h = (\sum_{j \in JO} p_{ij})/n$ and sequencing the jobs according to the EST rule. The upper bound UB is then obtained by computing the completion time of the EST sequence using the original processing times. Next we improve the UB value by applying two procedures based on Theorem 1.

The first one performes exchanges among pairs of adjacent jobs till a 2-exchange local optimum is reached.

To the resulting sequence, a second procedure is applied. It works as follows: Given a parameter m, for i=0,...,$n-m$, the current sequence is replaced by the best among the sequences obtained by permuting the jobs in the positions from $i+1$ to $i+m$. Notice that the resulting sequence is not in general an m-exchange local optimum.

6. Computational Results and Conclusions

The algorithm has been coded in Pascal and the tests have been carried out on a Vax-11/780.

Randomly generated test problems

Different series of test problems, each one containing 50 problems, have been performed by considering the number of jobs n=10 and n=15. For each n, processing times uniformly distributed on the interval [1,10] and ready times uniformly distributed on the three intervals [0,25], [0,40], and [0,50] are considered. The results are reported in table 1 and table 2, where in particular, the percentages of the node eliminated by the bound procedure and the dominance rules are shown.

Table 1. (n=10)

Ready times	0÷25	0÷40	0÷50
NG = Nodes generated	114	457	964
Computation time (msec.)	4,265	10,609	20,666
(LB/C$_{max}$) x 100	85.2	90.9	93.4
(UB/C$_{max}$) x 100	109.3	104.3	102.5
(nodes eliminated by LB) x 100/NG	68.5	48.8	36.2
(nodes eliminated by Th. 1&2) x 100/NG	31.5	51.2	63.8

Table 2. (n=15)

Ready times	0÷25	0÷40	0÷50
NG = Nodes generated	935	2,845	17,111
Computation time (msec.)	26,186	73,258	409,562
(LB/C_{max}) x 100	81.3	90.0	93.2
(UB/C_{max}) x 100	130.2	119.5	108.1
(nodes eliminated by LB) x 100/NG	67.0	58.0	49.5
(nodes eliminated by Th. 1&2) x 100/NG	23.0	42.0	50.5

Computational experience showed that the ratio β between the minimum and the maximum processing time turns out to be a critical parameter for the performances of the algorithm. For example if β is more realistically chosen to be 2 or 3, for a 10-job problem the number NG and the computation time become negligible. To give an idea of the behaviour of the algorithm for problems of larger size, we report a test on a 20-job problem with processing times uniformly distributed in the interval [5,10] (β=2) and ready times in the interval [0,100]. The same performance indices of Tables 1 and 2 are reported in Table 3.

Table 3. (n=20)

Ready times	0÷100
NG = Nodes generated	3,463
Computation time (msec.)	115,339
(LB/C_{max}) x 100	92.8
(UB/C_{max}) x 100	109.2
(nodes eliminated by LB) x 100/NG	2.0
(nodes eliminated by Th. 1&2) x 100/NG	98.0

Real-world ASP problems

To test the efficiency of the algorithm when applied to real-world problems, we must consider that aircraft, waiting to land, can be classified into a relatively small number m of distinct *categories* according to speed, capacity, weight and other technical characteristics.

As a consequence, the minimum interarrival times between two successive aircraft is a function only of the categories they belong to.

We have exploited this clustering of aircraft into categories to drastically reduce the size of the enumeration tree. In fact, it can be easily seen that, at the k-th iteration, only the earliest aircraft of each category, is eligible to be scheduled at position $(k+1)$-th.

The following table 4 represents the minimum interarrival times relative to the main categories of commercial aircraft, while tables 5 and 6 illustrate the results of two realistic large scale problems with 30 and 44 aircraft respectively.

The CPU times required to find the optimal solution have been respectively 373 sec. and 1,956 sec. Even though those times might seem not compatible with real time requirements, we want to point out that our code is experimental and very little has been spent to improve its efficiency. Neverthless we believe that the algorithm could be implemented to fit into real-time environment by using a faster machine, more sophisticated data structures and implementation techniques.

On the other hand, we want to stress that, as shown in tables 5 and 6, the optimal solution allows the saving up to about 20% on the runway utilization.

Table 4 - t_{ij} (sec); m=4

$1 \equiv B\ 743\ ;\ 2 \equiv B\ 727\ ;\ 3 \equiv B\ 707\ ;\ 4 \equiv DC\ 9$

i \ j	1	2	3	4
1	96	200	181	228
2	72	80	70	110
3	72	100	70	130
4	72	80	70	90

Table 5 (n=30)

Aircraft Number	Category	Nominal Landing Time (sec.)	FCFS Landing Time (sec.)	Optimal Sequence	Optimal Landing Time (sec.)
1	1	0	0	1	0
2	1	79	96	2	96
3	1	144	192	3	192
4	2	204	392	5	288
5	1	264	464	6	384
6	1	320	560	4	584
7	2	528	760	8	656
8	1	635	832	11	752
9	2	730	1032	7	952
10	2	766	1112	9	1032
11	1	790	1184	12	1104
12	1	920	1280	14	1332
13	3	1046	1461	10	1412
14	4	1106	1591	15	1492
15	2	1136	1671	16	1572
16	2	1166	1751	13	1642
17	2	1233	1831	18	1714
18	1	1642	1903	19	1810
19	1	1715	1999	20	1991
20	3	1770	2180	17	2091
21	1	2074	2252	23	2201
22	1	2168	2348	21	2273
23	4	2259	2576	22	2369
24	2	2427	2656	25	2465
25	1	2481	2728	24	2665
26	2	2679	2928	26	2745
27	3	2883	2998	27	2815
28	2	2982	3098	28	2915
29	1	3046	3170	29	2987
30	1	3091	3266	30	3083

Table 6 (n=44)

Aircraft Number	Category	Nominal Landing Time (sec.)	FCFS Landing Time (sec.)	Optimal Sequence	Optimal Landing Time (sec.)
1	1	0	0	1	0
2	1	79	96	2	96
3	2	144	296	3	296
4	2	204	376	4	376
5	2	264	456	6	448
6	1	320	528	7	544
7	1	528	624	8	640
8	1	635	720	5	840
9	2	730	920	9	920
10	1	766	992	11	1000
11	2	790	1192	13	1080
12	1	920	1264	14	1160
13	2	1046	1464	16	1240
14	2	1106	1544	17	1320
15	1	1136	1616	19	1400
16	2	1166	1816	10	1472
17	2	1226	1896	12	1568
18	1	1233	1968	15	1664
19	2	1286	2168	18	1760
20	2	1418	2248	21	1856
21	1	1642	2320	22	1952
22	1	1715	2416	24	2048
23	2	1749	2616	29	2144
24	1	1770	2688	30	2240
25	2	1809	2888	32	2236
26	2	1869	2968	34	2432
27	2	1929	3048	35	2528
28	2	1989	3128	20	2728
29	1	2074	3200	23	2808
30	1	2168	3296	25	2888
31	2	2229	3496	26	2968
32	1	2259	3568	27	3048
33	2	2326	3768	28	3128
34	1	2427	3840	31	3208
35	1	2481	3936	33	3288
36	2	2488	4136	36	3368
37	2	2565	4216	37	3448
38	2	2657	4296	38	3528
39	1	2679	4368	44	3608
40	1	2883	4464	39	3680
41	1	2982	4560	40	3776
42	1	3046	4656	41	3872
43	1	3091	4752	42	3968
44	2	3153	4952	43	4064

References

Bianco L., Nicoletti B., Ricciardelli S. (1987). An algorithm for optimal sequencing of aircraft in the near terminal area. Optimization techniques (Stoer J. ed.). Lecture Notes in Control and Information Science: 443-453, Springer-Verlag.

Bianco L., Ricciardelli S., Rinaldi G., Sassano A. (1979). The aircraft optimal sequencing as a N job-one machine scheduling problem R 79-36. Istituto di Automatica - Università di Roma and CSSCCA - CNR.

Bianco L., Ricciardelli S., Rinaldi G., Sassano A. (1985). Sequencing tasks with sequence dependent processing times. TIMS XXVI International Meeting, Copenhagen, R. 120 IASI-CNR Rome.

Dear R.G. (1976). The dynamic scheduling of aircraft in the near terminal area. FTL R 76.9. Flight Transportation Laboratory, MIT, Cambridge, Massachussets.

Odoni A.R. (1969). An analytical investigation of air traffic in the vicinity of terminal areas. Technical Report No. 46, Operations Research Center, MIT, Cambridge, Massachussets.

Pardee R.S. An application of dynamic programming to optimal scheduling in a terminal area air traffic control system. TRW Computers Company.

Park S.K., Straeter T.A., Hogge J.E. (1972). An analytic study of near terminal area optimal sequencing and flow control techniques. 14th Meeting of the Agard Guidance and Control Panel, Edinburgh, Agard CP 105: 12-1 - 12-18.

Park S.K., Straeter T.A. (1973). Near terminal area optimal sequencing and flow control as a mathematical programming problem. Mathematical Programming Society, Sumposium on Non-linear Programming, George Washington University.

Psaraftis H.N. (1978). A dynamic programming approach to the aircraft sequencing problem. FTL R 78-4, Flight Transportation Laboratory, MIT, Cambridge, Massachussets.

Psaraftis H.N. (1980). A dynamic programming approach for sequencing group of identical jobs. Ops. Res. 6: 1347 - 1359.

Tobias L. (1972). Automated aircraft scheduling methods in the near terminal area. AIAA Paper No. 72-120.

Trivizas D. (1985). Optimization of runway operations. Working paper. Flight Transportation Laboratory, MIT, Cambridge, Massachussets.

ATSAM (Air Traffic Simulation Analysis Model)
A Simulation-Tool to Analyze, Develop and Optimize Automated Air Traffic Flow Management Procedures

Andreas Hörmann

Technical University of Berlin
Institute of Aeronautics and Astronautics
D-1000 Berlin 10, Marchstrasse 14, Germany

ABSTRACT

ATSAM (Air Traffic Simulation Analysis Model) is a general-purpose simulation-tool for Air Traffic Control-related research regarding en-route traffic operations. ATSAM is designed as fast-time simulation with a detailed continuous model to allow the analysis of Air Traffic System scenarios with respect to safety and economy for almost any boundary condition. The program system comprises a data base system to model the airspace structure and organisation, the simulation of the air traffic system process (traffic input/generation, traffic planning/coordination, air traffic simulation) and evaluation software with numerical and graphical output. Currently ATSAM is used to analyze, develop and optimize air traffic flow control and management procedures for the German airspace.

1. INTRODUCTION

Due to safety reasons, future concepts for Air Traffic Management (ATM) have to be analyzed very thoroughly before being introduced into Air Traffic Control (ATC) operations. Thus, ATC-related research depends on the availability of flexible simulation-tools to model Air Traffic System (ATS) scenarios with the required level of precision for evaluating the performance of designed control procedures.

The software package ATSAM (Air Traffic Simulation Analysis Model) is such a simulation-tool with respect to the analysis of en-route air traffic. It offers a variety of software features to support the modeling of ATS scenarios for almost any boundary condition. ATSAM was developed at the Technical University of Berlin and sponsored by the German national research council (Deutsche Forschungsgemeinschaft) as part of a research project on air traffic flow control and management procedures.

Assessing ATC procedures embedded into a realistic ATS scenario is a problem of safety (e.g. collision avoidance) and economy (e.g. flight profiles, traffic flows). ATSAM, therefore, was designed as fast-time simulation with a continuous model to simulate and monitor the dynamic

NATO ASI Series, Vol. F38
Flow Control of Congested Networks
Edited by A. R. Odoni et al.
© Springer-Verlag Berlin Heidelberg 1987

behavior of the different aircraft along their flight paths. Descrete events are used to model ATC commands, system inconsistencies or error influences.

The program system comprises a data base system to model the airspace structure and organisation, the simulation of the Air Traffic System process (traffic input/generation, traffic planning/coordination, air traffic simulation) and evaluation software with numerical and graphical output. Simulation runs and sample scenarios may be completely defined and controlled by data input. Furthermore the modular layout allows an easy adaption of the ATSAM model to different ATC research tasks.

The simulation analysis currently performed is based on the scenario of the Air Traffic System of the Federal Republic of Germany including the adjacent air corridors to Berlin. Simulations are run on a CRAY-1S computer, requiring 3.2 MB of core memory.

2. ASSUMPTIONS ON THE ATC OF THE FUTURE

Today, the concentration of the air traffic at certain critical points of the traffic net regularly leads to situations requiring an increase in capacity. To handle a future (tremendously) increased air traffic volume in the most safe and efficient way the more or less computerized Air Traffic Control system of today has to be completed by some kind of automated tactical and strategical planning function.

Figure 1 shows a (simplified) organisational concept of the Air Traffic Control process within a future (semi-)automated system distinguishing between several steps of system development (full and dashed lines). The control concepts developed by means of ATSAM are designed for this future environment.

The essential functions of such an ATC-system of the future comprise a module to determine flight trajectories, a module to predict conflicts (planning phase) and a module to develop optimal solutions including a control action test function. Already existant in todays system are the monitoring function, the tactical control functions, the flight plan coordination function and a basic traffic flow control function (capacity management).

After intensive tests of the different algorithms as planning tools to support the controller and an intermediate period with verification of decisions by the controller, such a highly automated system seems to be realizable at last, changing the tasks of the controller from the ´manual´ control of the actual air traffic to systems management. According to experience, such structural changes of complex systems requiring international standardization will take at least 10 to 20 years to be finally established.

343

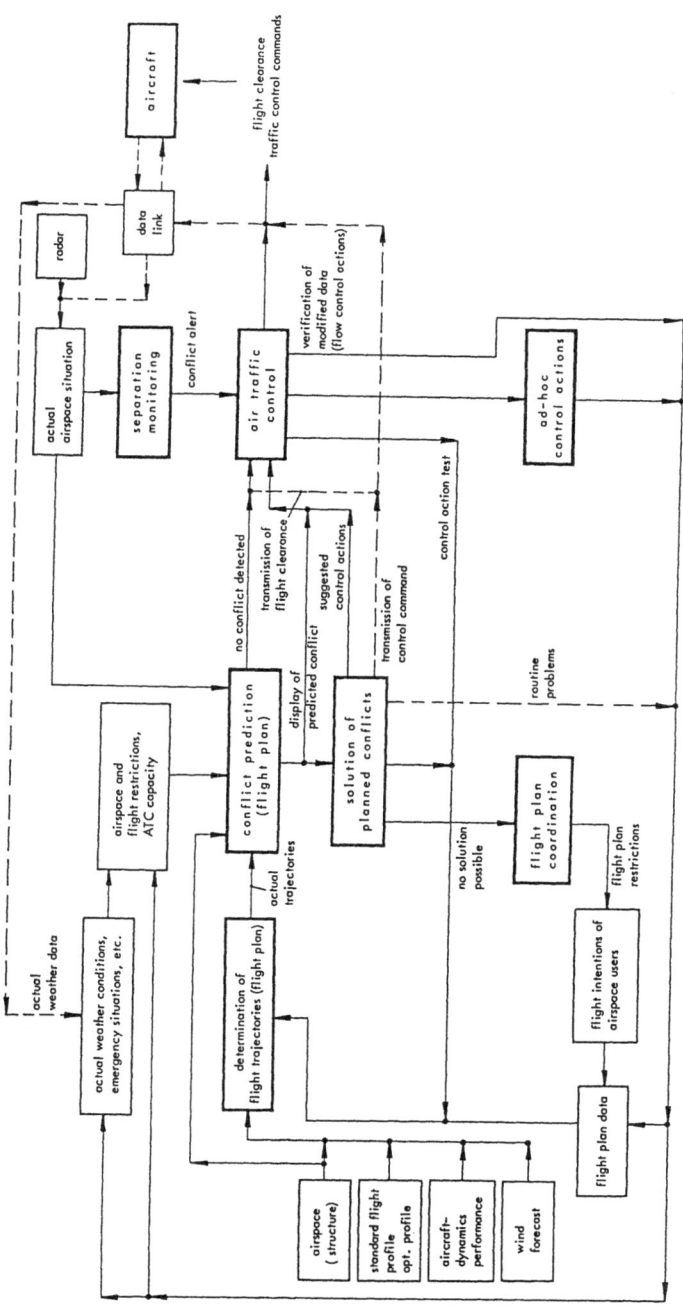

Figure 1: organisational concept of the computer-aided ATC process

3.   **ATSAM - A SOFTWARE-TOOL TO SIMULATE EN-ROUTE ATC SCENARIOS**

3.1 Software Requirements

Though the development of ATSAM started with the goal to have a program
available for the analysis, development and optimization of computerized
air traffic management concepts the software-package was designed as a
flexible, general purpose simulation-tool for ATC-related research. The
basic software requirements considered in the developing process were:

- precise modeling of reality
- generalized, problem-independent program design
- modular layout of the model system
- file-oriented scenario definition and program control
- optimized program code (CPU-time, memory)
- software portability

With respect to modeling en-route ATC scenarios the developed model
simulates all relevant system elements regarding flight operations (e.g.
aircraft dynamics) and considering Air Traffic Control (e.g. surveillance, control procedures) under conditions close to reality (i.e. error
influences modeled in Monte Carlo technique).

The analysis of large, complex traffic scenarios (currently the simulation analysis performed is based on the example scenario of the Air
Traffic System of the Federal Republic of Germany including the adjacent
air corridors to Berlin) during several "simulated" hours of flight
operations using such a "meso"scopic model system to determine the
required statistical data for assessing the results with sufficient
precision, has to be performed on a high-speed computer which usually
does not allow real time applications due to the applied timesharing.
Therefore, ATSAM was designed for fast-time simulations. However, on-
line programs are available for data analysis, e.g. a dynamic radar-like
display for the air traffic situation data recorded during the simulation.

As far as possible the program code has been generalized using a highly
modular structure, independent of special Air Traffic Control related
problems. The module concept allows an easy "exchange" of different Air
Traffic Control philosophies or any other modification to adapt the
simulation model to new developments in aeronautics (e.g. 4D-FMS, MLS).

Simulation runs and sample scenarios may be completely defined and
controlled by data input. ATSAM only offers the basic elements required
to model stucture, organisation, traffic demand and environmental influences of the scenario used as testbed for the analysis of ATC procedures.

ATSAM is written in standard FORTRAN 5 and thus implementable on (al-most) any other computer fulfilling the requirements of core memory and computing power. Currently a reduced model is being developed allowing the simulation of air traffic within a limited traffic area on a 16-Bit micro-computer.

3.2. Architecture of the ATSAM Program

Figure 2 shows the basic architectural concept of ATSAM in a very sim-plified way (missing are for example the numerous auxiliary programs to process raw data as input to the simulation data base and to analyze the result data). Due to the given task of simulating air traffic operations using a realistic, "meso"scopic model in any airspace structure, ATSAM can be subdevided into four major program sections:

● the data base module

Within Figure 2 the program pre-run to establish the data base required for the simulation (aeronautical data regarding structure and organisation of the scenario, performance data for flight planning and profile optimization purposes) is shown as a seperate part.

● the traffic input and generation module

This module models the traffic demand as it exists in reality, i.e. the Air Traffic System load.

● the (off-line) ATC planning module

This module represents all elements of the Air Traffic System re-garding the traffic planning process including strategic control. The program section contains the long-term planning functions of a flight plan or traffic coordination and the functions of a strate-gic Air Traffic Control (Air Traffic Flow Management).

● the simulation of the actual traffic

Besides the execution of each planned or non-planned flight this section models the Air Traffic Control including all monitoring and tactical control functions. Medium- and long-term traffic flow planning functions are implemented into and/or executed in coopera-tion with the planning module (strategic air traffic control).

3.3 System Functions of ATSAM

Subsequently the different functional units of the ATSAM program are de-scribed.

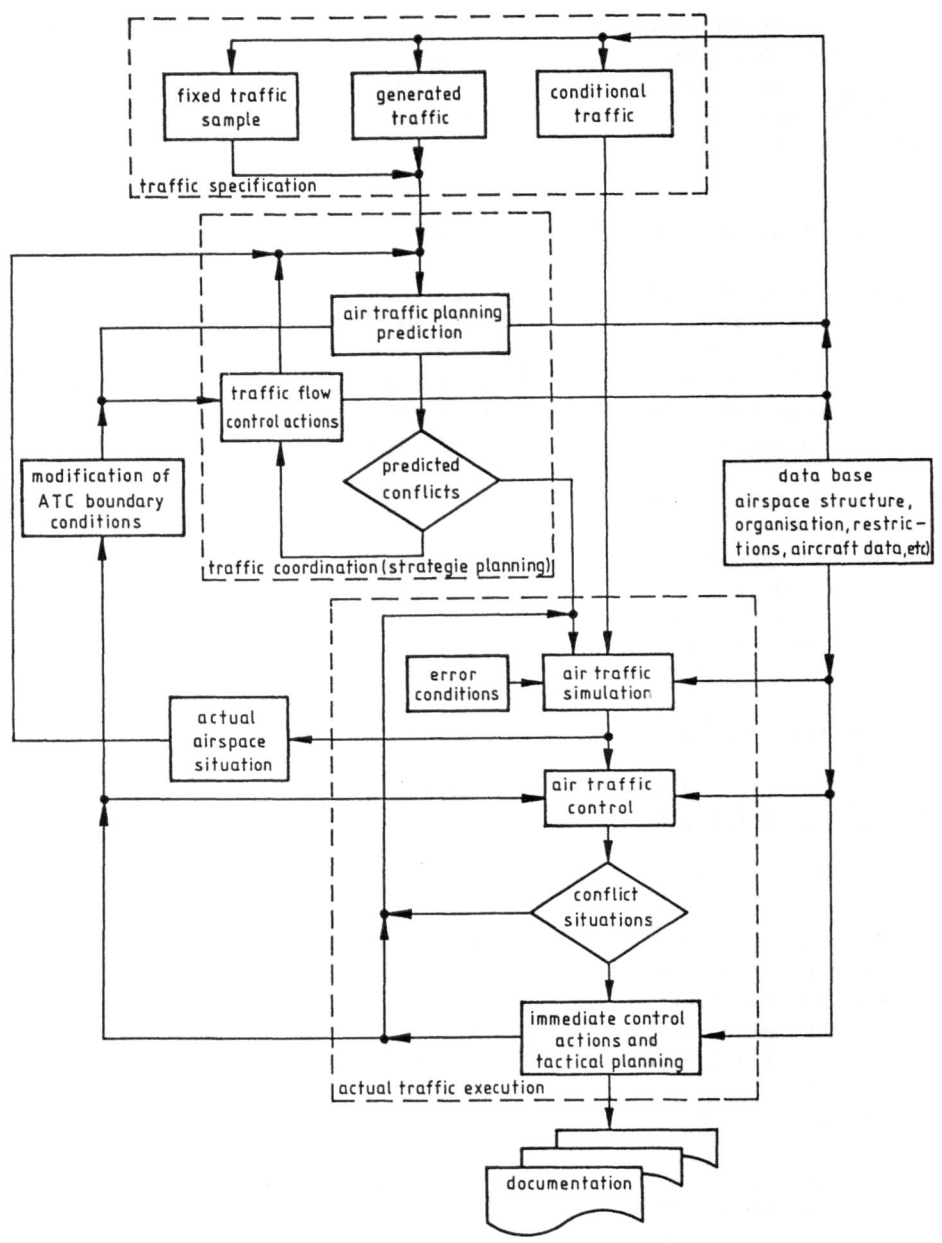

Figure 2: basic architectural concept of ATSAM

a)   The Aeronautical Data Base System

Each ATSAM program run starts by processing the ATSAM data base required
to generate traffic demand, to plan profiles/procedures and to simulate
flights. The data base module converts the sequential input data files
describing the airspace structure, the airspace organisation, the aircraft performance and the aircraft dynamics into the corresponding
direct access data base files. If automated control concepts should be
applied in the real Air Traffic Control process, this program section
could be used for the routine update of the required data base. However,
ATSAM is designed as a program system for research purposes requiring
frequent changes of the analyzed scenario.

Flight operations are simulated in a cartesian coordinate system. Therefore, all geographic coordinates of specified points (waypoints, navigation system locations, airport reference points, runway coordinates,
etc.) are transformed by means of a stereographic projection into a
rectangular coordinate system (x: eastern direction in nautical miles,
y: northern direction in nautical miles, z: altitude in feet).

The aeronautical data are processed in a hierarchical order to allow
internal logic consistency checks (the numbers listed in brackets refer
to the current example scenario).

● navigation systems and locations (339 nav-aids)
● waypoint coordinates and waypoint definition by nav-aids
  (686 waypoints)
● entry and exit points of considered scenario
  (108 airports/airfields, 169 entry/exit points)
● airways (134 airways)
● standard instrument arrival and departure routes
  (585 SIDs/STARs)
● standard and company routes (approx. 1350 routes)

First step in modeling the airspace structure is the definition of the
radio navigation stations. Navigation systems and techniques are modeled
by ATSAM as position errors resulting from inaccuracies of the respective position determination. If current standard navigation systems are
used, default accuracy values are assigned to simplify the data input.

The waypoints required for modeling the air traffic routes are defined
geographically in the horizontal plane (longitude, latitude). If sta-
tion-related radio location techniques are to be used for navigation,
the waypoints additionally have to be defined in relation to one (overhead, combined station) or two (seldom more) navigation system(s). Depending on the available stations, there may be several possibilities
for defining waypoints resulting in different position accuracies. Pre-
suming area navigation, the location of the waypoints not necessarily
coincides with the position of the corresponding navigation systems even
though, due to historical reasons, most of the published waypoints are
defined as overhead waypoints. Furthermore, the ATSAM model comprises

ground station independent navigation systems (e.g. inertial navigation, satellite navigation). An implemented generalized definition of position accuracy allows a system-independent determination of accuracy requirements for navigation systems.

Next step in defining the air traffic net structure is the specification of the spatial limitation of the considered example scenario. Neglecting the non-predicted traffic inserted to produce defined bottle-neck situations, the system entry and exit points represent at the same time the sources and sinks of the traffic load. The model differentiates between airports (begin or end of the arrival and departure routes) and entry/exit points as begin or end of the airways, to define the geographical limits of the simulated airspace.

Then the airways are defined to connect the waypoints. To connect the airports with the airway system standard departure and arrival routes (SIDs, STARs) are defined in the Aeronautical Information Publication (AIP). Due to the emphasis on the en-route ATC, a module to model the so-called radar vectoring technique usually applied by approach controllers has not been implemented yet.

The aeronautical data base is finished by processing the data of the standard routes for the different traffic relations. Generally for each traffic relation there exist several route definitions. Furthermore, the so-called company routes (routes preferred by air carriers) may be included into the data base.

b)   Modeling of Traffic Demand

The ATSAM model offers three different ways to specify traffic demand:

- the specification of fixed traffic samples (flight plan list),
- the generation of traffic according to specified distribution parameters and
- the specification of conditional traffic input.

Through combining these three modules any required traffic sample may be composed. The input of the data required for the traffic generation is integrated into the program pre-run and stored in the direct access data base.

Fixed samples are used to work with real data (e.g. flight strip planning data recordings) or to analyze special traffic situations. A flight plan list, however, contains only part of the data required for simulation. Missing parameters, as e.g. aircraft weight or level of avionic equipment, are determined or generated in combination with a data logic check of the input and the additional information.

The generation of the traffic demand usually takes place before the simulation starts, to combine traffic samples consisting of fixed flight plans and generated traffic. The comparability of the generated traffic with a real sample depends on the quality of the used data base for generation and on the computer effort regarding the applied data logic checking. However, there are a lot of problems in ATC-related research which require the analysis of non-typical traffic samples (e.g. to simulate critical situations).

The traffic input section passes flight intentions to a module coordinating at the system entry points by assigning entry times according to a user-specified coordination strategy. In the case of a combined sample the generated traffic is integrated into the fixed flight plan list. Flights of the conditional input device are passed directly to the simulation process without coordination or planning control to generate defined overload situations.

c)   Route Selection and Flight Planning

In a first approach a flight route (and the corresponding waypoints) are selected out of the aeronautical data base (standard routes of the AIP). If, which is the rule, several route definitions exist, the selection of a specific route considers the given boundary conditions of the route. In addition, different routes as preferred by the airspace user (company routes) may be used.

The economic (or optimal) altitude and speed profile is selected according to the performance of the respective aircraft type with fuel consumption being the parameter to be minimized. The algorithm for profile determination is based on the integrated range table concept used for flight planning in real flight operations. As for the route selection, airline specific flight operations may be considered.

The ´maximization of system capacity/throughput´ is an optimization goal of the Air Traffic Management concept. Thus, these functions are integrated into the off-line planning of the Air Traffic Control.

d)   Off-line Planning and Coordination

The planning and coordination module predicts the airspace situation, starting with the current traffic situation, the recorded flight plans for the respective time interval and the knowledge of the actual airspace and/or traffic restrictions and the predicted weather.

The program control of this functional unit comprises the functions: take-over of the flights from the simulation and the traffic demand list, pre-simulation, traffic monitoring, strategic Air Traffic Control, data recording, etc.. Due to the desired model accuracy the first ATSAM release simulates the flights for planning purposes. The development of reduced planning models is part of the research to be performed with ATSAM.

If no conflict situation has been detected, the respective flight plan will be passed unchanged to the simulation section. Otherwise the recorded flight plan will be changed according to a control action table until a conflict free traffic flow has been achieved. Due to short-term modified Air Traffic Control capacity values, sometimes even confirmed flight plans have to be changed.

All functions of the Air Traffic Control for air traffic planning purposes are summarized in one module to allow the program user an easy modification of different control procedures/concepts. Situations requiring control actions are differentiated into predicted conflicts and into detected bottle-neck situations. The available different types of control action by the ATC have to be defined to specify the respective control command sequence and criteria for their application considering the traffic situation. Complex control commands are defined using the realized basic elements of control (altitude command, heading command, speed command).

e)   Simulating Flight Operations

ATSAM includes a general, abstract aircraft class model with a maximum of ten main performance classes and a further differentiation into a maximum of ten subclasses to refine the considered aircraft data. Figure 3 shows the general concept of the aircraft model integrated into the overall simulation model.

The different hierarchically ordered levels begin with the traffic planning and coordination phase, using the flight intentions of the airspace users as input data and leading to coordinated flight plans, which may be modified by interventions of the Air Traffic Control.

The flight management system determines the profile parameters corresponding to the given flight plan considering the weather forecast. To integrate errors into the model, it determines at the same time deviations from the ideal profile parameters due to the error conditions to be considered. The model then commands the perturbed values. A flight path control algorithm leads the aircraft along the desired trajectories.

The aircraft dynamics are represented by a simplified dynamic model with three control parameters (air speed, path angle, bank angle), modeling the dynamic aircraft motion through the air space with sufficient accuracy compared with results basing on the complete differential equations of motion. The simulation/integration uses a 4th-order Runge-Kutta method. During the simulation, actual weather data are considered which may deviate from the forecast. Output of the aircraft model is the actual aircraft state and its position.

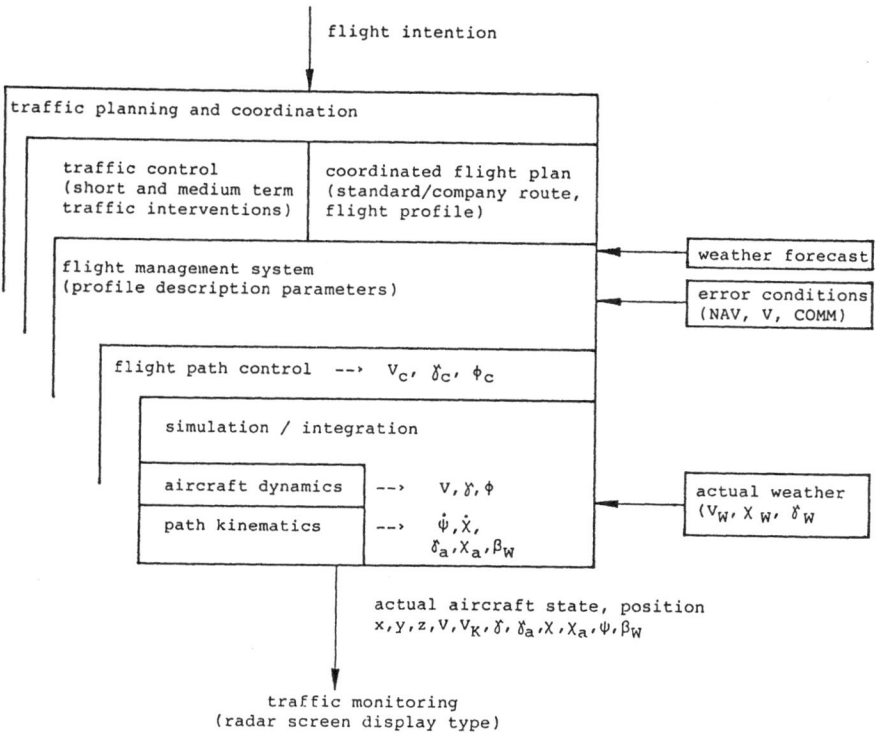

<u>Figure 3</u>: concept of the ATSAM aircraft model

f)   On-line Air Traffic Control

Main difference between simulation and prediction/planning is the consideration of error influences on the traffic flow and the flight path
accuracy (wind, navigation, operational influences). Furthermore,
through using the conditional traffic input, flights with time or situation related entry conditions and fixed flight plans may be set at any
point into the traffic flow, e.g. to define overload situations.

During the simulation run the Air Traffic Control monitors the traffic
and intervenes into the traffic flow if a conflict has been detected.
The on-line Air Traffic Control routine differentiates between the
situations: short-term actual danger, potential danger and bottle-neck
situation. As for the planning process the user has to specify the
respective control actions using the basic control action elements in
combination with a logic table. In addition, interdependencies between
the on-line Air Traffic Control and the off-line planning phase have to
be considered.

Depending on the pre-selected time interval, ATSAM switches between planning phase and simulation until the time reaches the defined simulation end.

g) Result Data Assessment

Besides a realistic representation of the Air Traffic System within the simulation model, a careful evaluation of the achieved result data is required to develop, analyze and optimize future concepts for Air Traffic Management.

Due to the complexity of the considered systems, usually graphical evaluation techniques are advantageous. Furthermore, graphics play an important role in controlling the extensive input data to define the airspace structure and organisation. Analytical methods in general are reduced to statistics or to the determination of quality function parameters.

With respect to the air traffic characteristics (safety, economy, regularity) and with respect to basic traffic statistics, ATSAM offers the feature of an on-line (statistical) analysis. Figure 4 shows an example of an on-line graphic result presentation feature: the radar-like display output ("snapshots" of the air space situation at constant time intervals) which is used besides the numerical output to achieve a better impression of what was happening in the air space during the simulated flight operations. At the same time all relevant data may be recorded on magnetic tape allowing a further off-line assessment by specific software to analyze the simulated traffic under any aspects.

4. EXAMPLES OF RESEARCH APPLICATIONS

As indicated, ATSAM was designed for the analysis, the development and the optimization of future air traffic control and management concepts. Subsequently the major applications of ATSAM within this research project are listed:

• analysis and evaluation of todays Air Traffic Scenario

• sensitivity analysis of different control actions for Air Traffic Management

• development and optimization of ATC control actions for Air Traffic Flow Management and Air Sspace Management

• development of different concepts and comparative assessment (traffic flow coordination / control)

• optimization of flight operations with respect to safety, economy and regularity

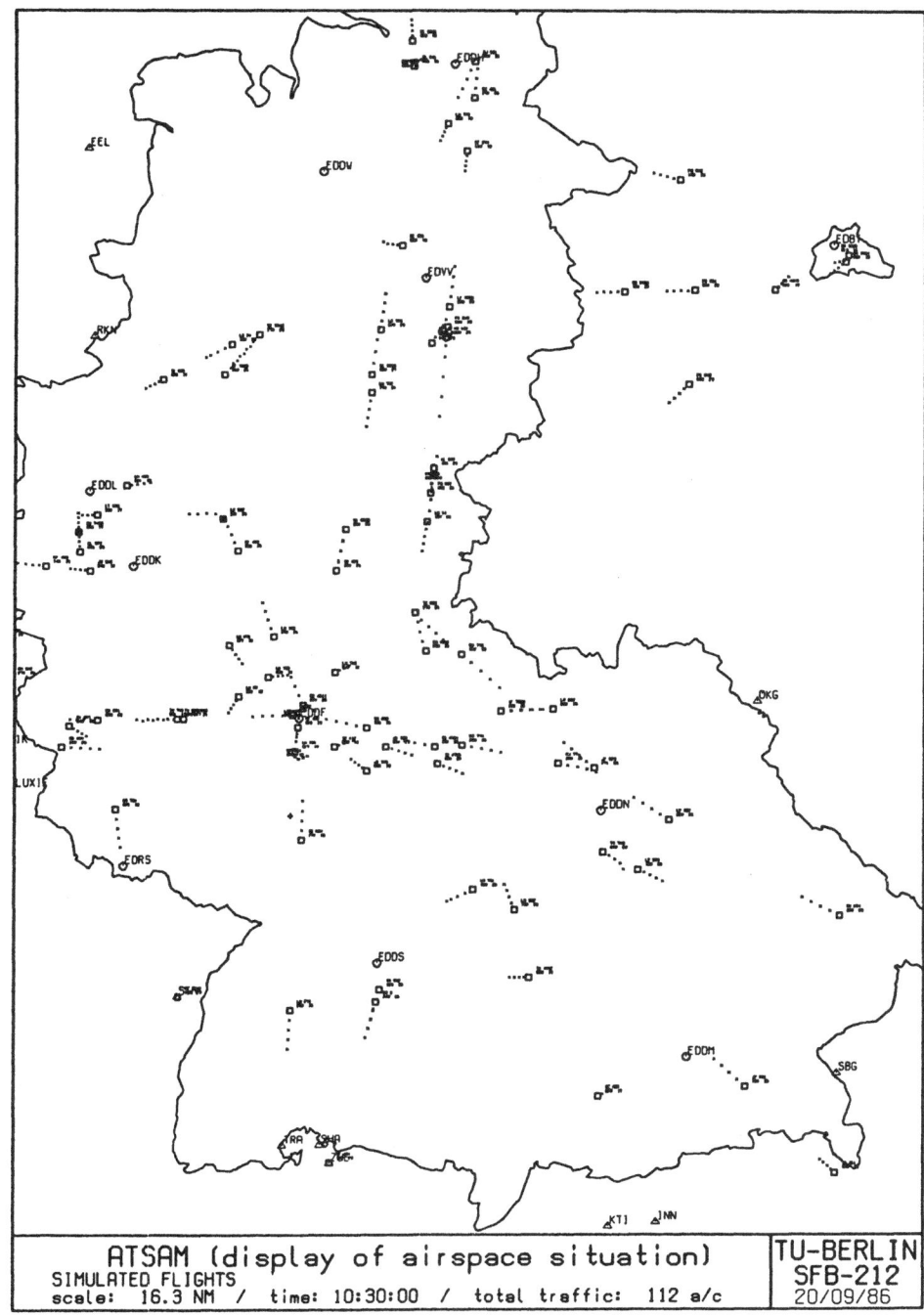

Figure 4: ATSAM on-line radar display (example)

To develop operationally applicable control algorithms which could be integrated into the existing Air Traffic Control planning software the concepts developed by means of ATSAM have to be refined by some kind of real-time simulation with special emphasis on human factors problems, e.g. the design of the man-machine interface.

5. SUMMARY

ATSAM (Air Traffic Simulation Analysis Model) was designed as general-purpose (fast-time) simulation-tool for Air Traffic Control-related research regarding en-route traffic operations.

Due to the developed "meso"scopic modeling technique with a continuous model to simulate and monitor the dynamic behavior of the different aircraft along their flight paths and discrete events used to model ATC commands, system inconsistencies or error influences, ATSAM allows the analysis of Air Traffic System scenarios regarding safety-related and economy-related aspects.

Basic features of the program package are a data base system to model the airspace structure and organisation, the simulation of the Air Traffic System process (traffic input/generation, traffic planning/coordination, air traffic simulation) and evaluation software with numerical and graphical output. As indicated while describing the different functional units, ATSAM is a highly variable and adaptable simulation-tool.

Due to numerous input data-controlled features to define the airspace structure and organisation, the aircraft model and the scenario boundary conditions, ATSAM allows the performing of simulations for almost any traffic area and sample (scenario) under various aspects of analysis without modification of the program code. Only the controlling logic of new Air Traffic Control concepts has to be integrated at defined interface points into the respective subroutines. ATSAM realizes the basic elements for the handling of air traffic. Thus, the user-specified program code may be reduced to an adequate combination of these generalized elements and to the specification of a more or less complex decision table to define the application of possible control actions.

Currently ATSAM is used to analyze, develop and optimize air traffic flow control and management procedures on the basis of a scenario covering the Air Traffic System of the Federal Republic of Germany, including the adjacent air corridors to Berlin.

355

6. REFERENCES

/1/ D.L. Adams : Preliminary functional Description of Inte-
S.M. Alvania grated Flow Management
R. Brubaker FAA-EM-82-7, October 1981

/2/ M. Fricke : Simulation of Automated Approach Procedures
A. Hörmann Considering Dynamic Flight Operations
to be published, AgardoGraph AG 301

/3/ L. Goldmuntz : Automated En-Route Air Traffic Control
et. al. (AERA) Concept
FAA-EM-81-3, March 1981

/4/ A. Hörmann : Konzeption und Entwicklung des Programmsystems ATSA
J. Stritzke (Air Traffic Simulation Analysis Model)
TU Berlin, ILR-Mitt. 157/85, September 1985

/5/ A. Hörmann : The Role of Simulation in the Evaluation of Future
Air Navigation Systems
ICAO-FANS Scenario Group Meeting,
Braunschweig, April 1986

NATO ASI Series F

NATO ASI Series F